T0140108

Studies in Computational Intelligence

Volume 869

Series Editor

Janusz Kacprzyk, Polish Academy of Sciences, Warsaw, Poland

The series "Studies in Computational Intelligence" (SCI) publishes new developments and advances in the various areas of computational intelligence—quickly and with a high quality. The intent is to cover the theory, applications, and design methods of computational intelligence, as embedded in the fields of engineering, computer science, physics and life sciences, as well as the methodologies behind them. The series contains monographs, lecture notes and edited volumes in computational intelligence spanning the areas of neural networks, connectionist systems, genetic algorithms, evolutionary computation, artificial intelligence, cellular automata, self-organizing systems, soft computing, fuzzy systems, and hybrid intelligent systems. Of particular value to both the contributors and the readership are the short publication timeframe and the world-wide distribution, which enable both wide and rapid dissemination of research output.

The books of this series are submitted to indexing to Web of Science, EI-Compendex, DBLP, SCOPUS, Google Scholar and Springerlink.

More information about this series at http://www.springer.com/series/7092

Gintautas Dzemyda · Jolita Bernatavičienė ·
Janusz Kacprzyk
Editors

Data Science: New Issues, Challenges and Applications

 Springer

Editors
Gintautas Dzemyda
Institute of Data Science
and Digital Technologies
Vilnius University
Vilnius, Lithuania

Jolita Bernatavičienė
Institute of Data Science
and Digital Technologies
Vilnius University
Vilnius, Lithuania

Janusz Kacprzyk
Systems Research Institute
Polish Academy of Sciences
Warsaw, Poland

ISSN 1860-949X   ISSN 1860-9503   (electronic)
Studies in Computational Intelligence
ISBN 978-3-030-39252-9   ISBN 978-3-030-39250-5   (eBook)
https://doi.org/10.1007/978-3-030-39250-5

This Springer imprint is published by the registered company Springer Nature Switzerland AG
The registered company address is: Gewerbestrasse 11, 6330 Cham, Switzerland

Preface

Modern technologies allow us to store and transfer large amounts of data quickly. They can be very diverse—images, numbers, streaming, related to human behavior and physiological parameters, etc. Whether it is just raw numbers, crummy at first sight images, or will help solve the current problems, predict the future, depends on whether we can process them and analyze them. Data science is a rapidly developing discipline. However, it is still very young and immature.

In particular, data science is related to visualizations, statistics, pattern recognition, neurocomputing, image analysis, machine learning, artificial intelligence, databases and data processing, data mining, big data analytics, knowledge discovery in databases. There are also many interfaces with optimization, block chaining, cyber-social and cyber-physical systems, Internet of things (IoT), social computing, high-performance computing, in-memory key-value stores, cloud computing, social computing, data feeds, overlay networks, cognitive computing, crowdsource analysis, log analysis, container-based virtualization, and lifetime value modeling. Again, all the directions listed are highly interrelated. Data science finds new fields of applications: chemical engineering, biotechnology, building energy management, materials microscopy, geographic research, learning analytics, radiology, metal design, ecosystem homeostasis investigation, and many others.

The most important challenges in data science currently may be highlighted:

- seamless integration of technologies supporting data sciences into complex cyber-physical–social systems;
- further development of data-driven intelligence methods and methodologies;
- the mingling of the "physical and digital worlds," within the container of data science, artificial intelligence, and machine learning.

With the growing demand for data science specialists and data analysts, new study programs are being introduced in most universities, and research groups are being established.

This book contains 16 chapters by researchers working in different fields of data science. They focus on theory and applications in language technologies, optimization, computational thinking, intelligent decision support systems, decomposition of signals, model-driven development methodology, interoperability of enterprise applications, anomaly detection in financial markets, 3D virtual reality, monitoring of environment data, convolutional neural networks, knowledge storage, data stream classification, security of social networking. In all the cases, the papers highlight different issues or applications of data science, and their contents will now be briefly summarized.

Egor S. Ivanov, Aleksandr V. Smirnov, Igor P. Tishchenko, and Andrei N. Vinogradov ("Object Detection in Aerial Photos Using Neural Networks") describe a new method for object detection on aerial photographs by neural networks using sliding windows of different sizes. The user can set a possibility threshold for the classification. The new approach proposed can be used with any type of a neural network. For numerical experiments, a convolutional neural network is used for aerial photographs. Moreover, as an extension, an algorithm is proposed that makes it possible to post-process data obtained from the use of the neural network as mentioned above. For illustration, a search of an aircraft in images is considered with image processing using a distributed data processing system proposed.

Vytautas Kaminskas and Edgaras Ščiglinskas ("Modelling and Control of Human Response to a Dynamic Virtual 3D Face") propose an application of identification and predictor-based control techniques for the modeling and design of control schemes of the human response as a reaction to a dynamic virtual 3D face. Two experiment plans are employed, the first one in which the 3D face is observed without a virtual reality headset, and the second one in which the 3D face is observed with a virtual reality headset. A human response to the stimulus (the virtual 3D face with a changing distance between the eyes) is observed using EEG-based emotion signals (excitement and frustration). Experimental data is obtained by observing the stimulus and response in real time. A cross-correlation analysis of data shows that there exists a dynamic linear dependency between the stimuli and response signals. The dynamic system identification is applied which ensures the stability and a possible higher gain for one-step prediction models. Predictor-based control schemes with a minimum variance or a generalized minimum variance controller with a constrained control signal magnitude and change speed are developed. The numerical selection of the weight coefficient in the generalized minimum variance control criterion is based on a closed-loop stability condition and an admissible value of the systematic control error. The results show a high prediction accuracy and control quality of the model proposed.

Ilona Veitaite and Audrius Lopata ("Knowledge-Based Transformation Algorithms of UML Dynamic Models Generation from Enterprise Model") are concerned with the fact that in the present-day organizations, there exists a big gap between business and information technologies. The information technology strategy planning is a multistage process, and the information system development relies on the implementation of an information system life cycle at each stage. A knowledge-based information system engineering methodology is proposed using system modeling and decision-making tools and methods. Participants of the information system project such as a developer or a programmer are allowed to use not only knowledge on the project, which is collected in a traditional CASE tool storage, but also the knowledge storage in which the subject area knowledge is collected according to formal criteria. Tools and techniques of the Unified Modeling Language (UML) model generation of diverse knowledge-based models combining frameworks, workflow patterns, modeling languages, and natural language specifications are used. A knowledge-based subsystem, as a CASE tool component with an enterprise metamodel (EMM) and enterprise model (EM), is shown to be of a significant help for the UML dynamic models generation using transformation algorithms which gives a possibility to operate additional models validation methods. The main purpose of the paper is to present how the EM can be used in the UML models' generation process, with an account for original issues related to knowledge-based IS engineering in business and an IT alignment process. Some elements of variations of the UML models after their generation from EM are shown. The main result is a proposal of an UML sequence model generation from the EM process. Some examples are shown to illustrate the approach.

Dalė Dzemydienė and Vytautas Radzevičius ("An Approach for Networking of Wireless Sensors and Embedded Systems Applied for Monitoring of Environment Data") provide a new approach for an extension of the capacities of a wireless sensor network (WSN) by enabling possibilities of monitoring sea environment parameters which is related to the control of a recognition process of hydrometeorological situations by providing a proper structure of the decision support system (DSS) which can work under specific real-time conditions. In the multilayered buoy, the infrastructure based on the WSN mechanisms of interconnected devices has to work under extreme weather conditions. Moreover, the modern embedded systems to be used in the problem considered must have possibilities of an adequate evaluation of risky situations. This implies a need for the use of the WSN based on the technology of the Internet of things (IoT). The main results of this work are a proposal of a new architectural solution with extended capacities of integrated embedded systems and other gathering sources with an assessment of various water pollution situations. The solution of issues related to network allocation, system resource allocation, and system architecture-related solutions, which make it possible for the wireless channels to accommodate big streams of data, is a considerable contribution. The experimental results show that the proposed prototype of a multilayer system can well work under real conditions of sea with an interconnected network of sensors and controllers. The method proposed is used for the evaluation of functionality of the embedded systems and the WSN. The results of experiments show a functional effectiveness of collection and gathering of data under restricted and limited capacities of the embedded systems for operative control.

Kamil Ząbkiewicz ("Non-standard Distances in High Dimensional Raw Data Stream Classification") presents a new approach for classifying high-dimensional raw (or close to raw) data streams. It is based on the k-nearest neighbor (kNN) classifier. The novelty of the proposed solution is based on the use of non-standard distances which are computed by using compression and hashing methods. Basically, as opposed to standard distances, exemplified by the Euclidean, Manhattan, Mahalanobis, etc., ones which are calculated from numerical features of the data, the non-standard distance is not necessarily based on extracted features and one can well use raw (not preprocessed) data. The proposed method does not need to select or extract features. Experiments on datasets with the dimensionality of larger than 1000 features show that the proposed method in most cases performs better than or similar to other standard stream classification algorithms. All experiments and comparisons are performed in the Massive Online Analysis (MOA) environment.

Maria Visan, Angela Ionita, and Florin Gheorghe Filip ("Data Analysis in Setting Action Plans of Telecom Operators") are concerned with problems related to a fierce battle of the telecom operators to win the communication service market by defining suitable value proposals, structuring the right technologies, and "go-to-market" partnerships. The telecoms possess a huge volume of data that can be analyzed and used for preparing better decisions. The authors present a detailed analysis of the "battlefield" between the telecoms emphasizing the context, main issues of telecom operators that support data for these services, examples of services and potential users of them, possible solutions to the architectural and methodological implementation, etc. An analysis of these aspects in practical cases is presented and analyzed.

Saulius Gudas and Andrius Valatavičius ("Extending Model-Driven Development Process with Causal Modeling Approach") are concerned with the model-driven development which is considered to be the most promising methodology for the cyber-social systems (CSS), cyber-enterprise systems (CES), cyber-physical systems (CPS), and many other types of complex systems. Causality is an important concept in modeling as it helps to reveal the properties of the domain hidden from the outside observer. The subject domain of the CES as well as of the CSS is a complex system type named an "enterprise." The aim of this article is to enhance the model-based development (MDD) process with a causal modeling approach which aims at revealing the causality inherent to the specific domain type and to represent this deep knowledge on the CIM layer. This implies a need to add a new layer of the MDA, namely the layer of domain knowledge discovery. The traditional MDA/MDD process uses the external observation-based domain modeling on the CIM layer. From the causal modeling viewpoint, an enterprise is considered to be a self-managed system driven by internal needs. A specific need creates a particular causal dependence of activities—a management functional dependence (MFD). The concept of the MFD denotes some meaningful collaboration of activities, namely the causal interactions required by the definite internal need. The first step is the conceptualization of the perceived domain causality on the CIM layer. A top-level conceptual causal model of the MFD is defined as a management transaction (MT). The next step is a detailed MT modeling when an elementary control cycle (EMC) is created for each MT. The EMC reveals the internal structure of MT and goal-driven interactions between internal elements of the MT, that is, a workflow of data/ knowledge transformations. The results of the authors' study help better understand that the content of the CIM layer should be closely aligned with the domain causality. The main contribution is an extended MDA scheme with a new layer of the domain knowledge discovery and the causal knowledge discovery (CKD) technique tailored for the enterprise domain. The method applies uses a twofold decomposition of management transaction: a control view based and a self-managing view based. The outcome of the method is a hierarchy of management transactions and their internal components: lower-level management functions and processes, goals, knowledge, and information flows. A causal knowledge discovery technique is illustrated using the study program renewal domain.

Algirdas Lančinskas, Pascual Fernández, Blas Pelegrín, and Julius Žilinskas ("Discrete Competitive Facility Location by Ranking Candidate Locations") deal with a competitive facility location which is a strategic decision for firms providing goods or services and competing for the market share in a geographical area. In the literature, there are many different facility location models and solution procedures which differ in ways various aspects and parameters, for instance the location space, customer behavior, objective function(s), etc., are reflected. The work focuses on two discrete competitive facility location problems: a single-objective discrete facility location problem for a firm entering the market, and a bi-objective discrete facility location problem for a firm that plans expansion. Two random search algorithms for the discrete facility location based on the ranking of candidate locations are proposed, and numerical tests are performer to show their performance. It is shown that the ranking of candidate locations is a suitable strategy for the discrete facility location problems since the algorithms can yield optimal solutions.

Povilas Treigys, Gražina Korvel, Gintautas Tamulevičius, Jolita Bernatavičienė, and Bożena Kostek ("Investigating Feature Spaces for Isolated Word Recognition") are concerned with issues related to the appropriateness of a two-dimensional representation of a speech signal for speech recognition based on deep learning. The approach combines the convolutional neural networks (CNNs) and a time-frequency signal representation converted to the feature spaces investigated. In particular, waveforms and fractal dimension features of the speech signal are chosen for the time domain, and three feature spaces are considered for the frequency domain, namely the linear prediction coefficient (LPC) spectrum, Hartley spectrum, and cochleagram. Since deep learning requires an adequate training set size of the corpus and its content may significantly influence the outcome, the dataset produced is extended with mixes of the speech signal with noise with various signal-to-noise ratios (SNRs) to augment the data set. For the evaluation of the applicability of the implemented feature spaces for the isolated word recognition task, three experiments are conducted, i.e., for 10-word, 70-word, and 111-word cases.

Anita Juškevičienė ("Developing Algorithmic Thinking Through Computational Making") is concerned with issues related to algorithmic thinking, which is the main component of computational thinking, and is necessary for computer programming. If we go further, we can notice that computational thinking has much in common with many digital age skills, but it is still a challenge for educators to teach computational thinking in an attractive way. The author provides first a comprehensive and critical literature review on computational thinking in education and its implementation. The results show that modern technologies are widely used for enhancing the effectiveness and efficiency of learning, and improving algorithmic thinking. This confirms that modern technologies can facilitate effective and efficient learning, attaining computational thinking skills and learning motivation.

Krzysztof Kąkol, Grażina Korvel, and Bożena Kostek ("Improving Objective Speech Quality Indicators in Noise Conditions") are focused on the modification of speech signal samples and on testing them with objective speech quality indicators after mixing the original signals with noise or with an interfering signal. Modifications that are applied to the signal are related to the Lombard speech characteristics, i.e., pitch shifting, utterance duration changes, vocal tract scaling, and manipulation of formants. A set of words and sentences in Polish, recorded in silence, as well as in the presence of interfering signals, i.e., pink noise and the so-called babble speech, also referred to as the "cocktail-party" effect, is utilized. Speech samples are then processed and measured using objective indicators to check whether modifications applied to the signal in the presence of noise increase the values of the speech quality index, i.e., Perceptual Evaluation of Speech Quality (PESQ) standard.

Dalius Mažeika and Jevgenij Mikejan ("Investigation of User Vulnerability in Social Networking Site") deal with an important problem of the vulnerability of social network users which is a serious social networking problem. A vulnerable user might place all friends at risk so that it is important to know how the security of the social network users can be improved. The authors address issues related to user vulnerability to a phishing attack. First, short text messages of social network site users are gathered, cleaned, and analyzed. Moreover, the phishing messages are built using social engineering methods and sent to the users. The k-means and mini-batch k-means clustering algorithms are evaluated for the user clustering based on their text messages. A special tool is developed to automate the user clustering process in the presence of a phishing attack. An analysis of the users' responses to phishing messages using different datasets and social engineering methods is performed, and some conclusions related to the user vulnerability are derived.

Tatjana Sidekerskienė, Robertas Damaševičius, and Marcin Woźniak ("Zerocross Density Decomposition: A Novel Signal Decomposition Method) develop a new zerocross density decomposition (ZCD) method for the decomposition of nonstationary signals into subcomponents (intrinsic modes). The method is based on the histogram of zerocrosses of a signal across different scales. The main properties of the ZCD and parameters of the ZCD modes (statistical characteristics, principal frequencies, and energy distribution) are analyzed and compared with those of the well-known empirical mode decomposition (EMD). For the analysis of the efficiency of decomposition, the partial orthogonality index, total index of orthogonality, percentage error in energy, variance ratio, smoothness, ruggedness, and variability metrics are employed, and a novel metric distance from a perfect correlation matrix is proposed. An example of a modal analysis of a nonstationary signal and a comparison of decomposition of randomly generated signals using the stability and noise robustness analysis is provided. The results show that the proposed method can provide more stable results than the EMD.

Florin Gheorghe Filip ("DSS—A Class of Evolving Information Systems") discusses the evolution of a particular class of information systems, called the decision support systems (DSSs), under the influence of several technologies. First, a description of several trends in automation is provided. Decision-making concepts, including consensus building and crowdsourcing-based approaches, are then presented. Next, basic aspects of the DSSs, which are meant to help the decision-maker to solve complex decision-making problems, are critically reviewed. Various DSS classifications and taxonomies are described from the point of view of specific criteria, such as the type of support, the number of users, and the type of a decision-maker. Several modern information and communication technology (ICT) tools, techniques, and systems used in the DSS design are addressed. Special attention is paid to the use and role of artificial intelligence, including cognitive systems, big data analytics, and cloud and mobile computing. Several open problems and challenges, concerns, and cautious views of scientists are revealed as well.

Andrius Valatavičius and Saulius Gudas ("A Deep Knowledge-Based Evaluation of Enterprise Applications Interoperability") basically address the enterprise which is a dynamic and self-managed complex system, notably from the point of view of the integration and interoperability of enterprise software which is a core problem that determines system efficiency. The authors deal with the interoperability evaluation methods for the sole purpose of evaluating multiple enterprise applications interoperability capabilities in the model-driven software development environment. The characteristic feature of the method is that it links the causality modeling of the real world (domain) with the traditional model-driven architecture (MDA). The discovered domain causal knowledge transferring to the CIM layer of MDA forms the basis for designing application software that is integrated and interoperable. The causal (deep) knowledge of the subject domain is used to evaluate the capability of interoperability between software components. The concept of a management transaction reveals causal dependencies and the goal-driven information transformations of the enterprise management activities (in-depth knowledge). An assumption is that autonomic interoperability is attained by gathering knowledge from different sources in an organization, and that an enterprise architecture and software architecture analysis through Web services can help gather required knowledge for automated solutions. In this interoperability capability evaluation research, 13 different enterprise applications are surveyed, initially using four known edit distance calculations: Levenshtein, Jaro-Winkler, the longest common subsequence, and Jaccard. Combining these results with a bag-of-words library gathered from "Schema.org" and included as an addition to the evaluation system. The method has been improved by moving more closely to semantic similarity analysis.

Marius Liutvinavicius, Virgilijus Sakalauskas, and Dalia Kriksciuniene ("Sentiment-Based Decision Making Model for Financial Markets") are concerned with the effects of sentiment information for triggering unexpected decisions of investors and incurring anomalies of financial market behavior. The proposed model postulates that the sentiment information is not only influential to investment decisions but also has a varying impact on different financial securities and time frames. The algorithm and simulation tool are developed for including the composite indicator and designing adaptive investment strategies. The results of simulations by applying different ratios of financial versus sentiment indicators and investment parameters make it possible to select efficient investment strategies which outperform the well-known ones based on the standard and poor (S&P) index.

We hope that a broad coverage of the volume, the inspiring and interesting contributions, included in this volume, which present both the state-of-the-art and original contributions, will be of much interest and use for a wide research community.

We wish to express our deep gratitude to the contributors for their great works. Special thanks are due to anonymous peer referees whose deep and constructive remarks and suggestions have greatly helped improve the quality and clarity of contributions.

And last but not least, we wish to thank Dr. Tom Ditzinger, Dr. Leontina di Cecco, and Mr. Holger Schaepe for their dedication and help to implement and finish this important publication project on time, while maintaining the highest publication standards.

Vilnius, Lithuania   Gintautas Dzemyda
Vilnius, Lithuania   Jolita Bernatavičienė
Warsaw, Poland   Janusz Kacprzyk

Contents

Object Detection in Aerial Photos Using Neural Networks

Egor S. Ivanov, Aleksandr V. Smirnov, Igor P. Tishchenko
and Andrei N. Vinogradov

Abstract The paper describes the method of objects detection on aerial photographs using neural networks. The aim of this paper is to present an object detection algorithm by neural networks using sliding window. Main idea and main benefit of this concept is that image processing by sliding window with different sizes and that user can set possibility threshold for neural network classifying. This approach can be used with any neural network types because the goal of neural network in the method is to classify current part of image. For our experiments we took convolutional neural network and aerial photos. Also in this paper described the extension of this method. It's an algorithm that allows post processing of data obtained as a result of the operation of neural networks. The problem of searching for aircraft in images is considered as an example. Some results of aircraft detection are presented in this paper. Image processing took place in distributed data processing system that also described in this paper.

Keywords Remote sensing of the earth · Recognition · Image analysis · Neural networks

1 Introduction

Data from remote sensing images can be used to solve a variety of tasks: detection of forest fires; monitoring and evaluation of the effectiveness of reforestation activities; environmental management (deforestation, construction of quarries, illegal landfills, assessment of rationality in the extraction of natural resources, etc.); creation of geo maps; inventory of objects in protected areas; monitoring of construction sites, etc.

E. S. Ivanov · A. V. Smirnov · I. P. Tishchenko
Ailamazyan Program Systems Institute of RAS (PSI RAS), Petra-I st. 4a, s. Veskovo, Pereslavl
District, Yaroslavl Region 152021, Russia

A. N. Vinogradov (✉)
Department of Information Technologies, Peoples' Friendship University of Russia (RUDN
University), Miklukho-Maklaya str. 6, Moscow 117198, Russia
e-mail: vinogradov-an@rudn.ru

© Springer Nature Switzerland AG 2020
G. Dzemyda et al. (eds.), *Data Science: New Issues, Challenges
and Applications*, Studies in Computational Intelligence 869,
https://doi.org/10.1007/978-3-030-39250-5_1

1

The number of processed images coming from Earth remote sensing satellites is now growing rapidly. A large amount of aerial images brought up a need for their fast analysis and accurate classification in order to efficiently facilitate their usage in practical civil and military domains.

The necessity of automatic extraction of valuable information from aerial images encouraged the development and further improvement of various processing methods with a specific purpose (Ševo and Avramovic 2016). Objects detection is a popular task in image processing, especially in remote sensing images or aerial photographs. The term 'object' used in this survey refers to its generalized form, including man-made objects (e.g. vehicles, ships, buildings, etc.) that have sharp boundaries and are independent of background environment, as well as landscape objects, such as land-use/land-cover (LULC) parcels that have vague boundaries and are parts of background environment (Cheng and Han 2016).

Objects detection and recognition in images is a difficult task. In connection with the fact that the number of processed images, coming from Earth remote sensing satellites, is now growing rapidly, it is necessary to ensure the automatic ability to detect different objects of interest in images.

To solve objects detection problem, various algorithms are used, from threshold binarization (Chen et al. 2008; Pitknen 2001) to the application of directional gradient histograms (Rosebrock 2015) and Viola-Jones methods (http://docs.opencv.org/trunk/d7/d8b/tutorial_py_face_detection.html). Local feature extraction methods for object detection are effective in scale-invariant feature transform, histogram of oriented gradients. Local features are often used together with bag-of-visual-words paradigm to calculate the final descriptors, which represent the histogram of occurrence of different local features detected in one specific image (Ševo and Avramovic 2016).

For object detection and classifying also using neural networks (Hinton et al. 2012b; Cheng et al. 2016) which have appeared in the second half of the 20th century, have gained widespread acceptance. Various architectures of neural networks are now available, which are used not only to find objects of interest, but also to classify and recognize objects. The use of neural networks also allows you to find several types of objects of interest.

In this paper we will consider an example of using a neural network to find various types of aircraft located on the airport. The resulting information after post processing, can be used in monitoring tasks runways, for example, to track the movement of aircraft or to count their number.

Information processing took place in the distributed data processing system (Kondratyev and Tishchenko 2016). The system has a modular architecture. The system automatically determines which modules to run st the current moment and provides data exchange between modules.

2 Distributed Data Processing System

The described system is very useful software tools for users who don't have knowledge about parallel programming. The main goal of the system is to provide data processing in a pipelined parallel mode. This approach speeds up data processing using a supercomputer. At the input, the system receives a task description, which contains a list of processing modules and launch parameters. Such tasks presented by XML-file.

The distributed data processing system model contains of two parts: the Worker and the Scheduler. The Worker concept is closely linked with the concept of a computing node. The Worker is the same thing as computing node where it was launched. It describes computing resources, data processing modules from those nodes.

The Worker could execute most of data processing modules. It depends on the Worker capabilities. As example could be named GPU support. Also the Worker could execute commands from command list which describes data processing computation. Worker state consists of information about usage of CPU, memory, hard drive and data processing modules activity.

The distributed data processing system Scheduler decides commands for process by using information about resources from all known Workers. It generates commands for Workers from incoming data processing task description. Scheduler does task decomposition, load balancing and provides a mechanism for module communication in computing environment. It has information about computing environment—computing node list with information: CPU load, memory usage, network usage, connection and module state, threads status.

Graphical user interface of the distributed data processing system is presented at Fig. 1.

Data processing application is described by special task (scheme of data processing). Scheme consists of data processing modules and transfer channels between them. Scheme describes data processing conveyor. Usually task consists of several modules where could be operations branching operations and data combining. But task always has initial and finishing modules. Scheduler creates internal representation for task and verifies it.

The processing division into stages allow to process data independently with several threads. It may be perfect choice for quickly making data processing application for distributed system.

The system is based on a modular structure. By module is meant a software component that processes data arriving on the input channel and transmits the processed information to the output channel. Each component has its own role in the system but all of them are intended for one goal. The system is the environment for running program modules in the pipeline parallelism mode.

Each module can be presents as a minimal program with some processing function and may have several input and output channels. The system considers each module as one function with input parameters. It provides an ability to organize calculation process with pipeline parallelism for each single module.

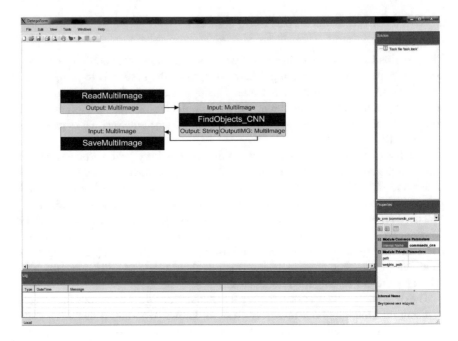

Fig. 1 Graphical user interface

Prepared scheduling algorithms allow obtaining acceleration in various problems. The tools have several drawbacks. The best results can be achieved with coarsely granular concurrency. Scheduler has numerous leaks of time when choosing the next action. The other disadvantages are (Kondratyev and Tishchenko 2016):

1. Not all algorithms are suitable.
2. Maximum load of computing resources is not always the best choice.

The concept of the distributed data processing system has worked well as for single computer and for the computer network.

3 Convolutional Neural Networks

At all the use of neural networks is due to their similarity with the successful operation of biological systems, which, in comparison with other systems, consist of simple and numerous neurons that work in parallel and have the opportunity to learn (Kriesel 2007). For our experiments, a convolutional neural network was chosen—a special architecture of artificial neural networks proposed by Jan Lekun and aimed at effective image recognition.

Convolutional networks use relatively little pre-processing compared to other image classification algorithms. This means that the network learns the filters that in traditional algorithms were hand-engineered. This independence from prior knowledge and human effort in feature design is a major advantage (Romanov 2018).

They have applications in image and video recognition, recommender systems, image classification, medical image analysis, and natural language processing.

A convolutional neural network consists of several layers. These layers can be of three types: convolutional, subsampling layers, and fully connected. By combining the dimensions of the input images, the sequence and the number of layers, you can get different architectures of neural networks.

The number of layers in the used neural network was 13. Kernel size for convolutions operations was 3. As activation function was used Leaky ReLU. Scheme of neural network used for experiments is presented in Fig. 2.

To train a neural network, it is necessary to prepare a training dataset samples. Creating a training dataset is an important step, because it determines how the network learns and, as a result, it affects the final recognition result.

The training took place on a sample consisting of four classes: roads, vegetation, urban buildings and airplanes. Quantity of samples varies from 1000 to 1500 in each class. The total number of sample images for the training sample was about 5000.

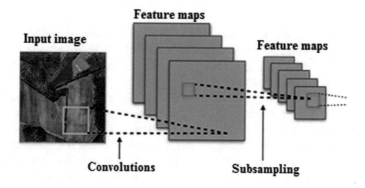

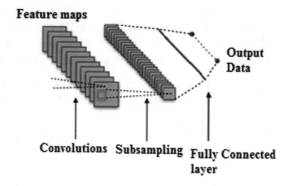

Fig. 2 Scheme of used neural network

Fig. 3 Samples of training set

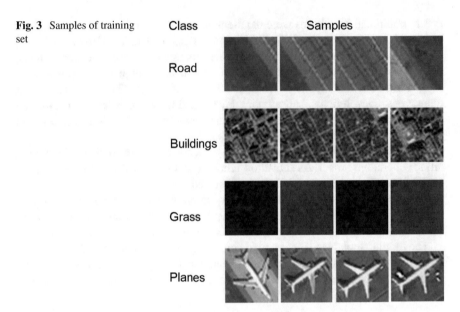

Class | Samples
Road
Buildings
Grass
Planes

The size of each image from training dataset (if necessary) was converted to 32 × 32, because images in created can be of any size, not only 32 × 32, but neural network requires special image size, so we need to convert size before learning process.

Some examples of the training sample are showed in Fig. 3.

The number of learning epochs was 5000. During each iteration (epoch) neural network tries to calculate neuron weights that more correctly describe current object. And after changing weights neural network runs test, it takes some random images of the training dataset and classifies them. At this step neural network knows classes of each image because all images in dataset are labeled. So neural network can calculate quantity of right- and wrong-classified images in each test. Percent of wrong classified images from total images quantity in the current test calls Learning error. Usually learning error decreases with learning iterations. But it almost impossible to have 0 error value. During 5000 iterations error was decreased and finally it reached value that wasn't changed during about 50 epochs. The learning error was 0.2.

A scheme describing the solution to the task of learning a neural network in the distributed data processing system consists of the sequential execution of modules:

ReadMultiImage_array. A module that reads many standards from the specified directories and remembers their belonging to one or another class, according to the read data.

FindObjects_CNN. A module that contains the neural network architecture with the ability to change network hyper-parameters, such as the number of convolution layers, the number of classes, etc.

FindObjects_CNN module also has a parameter, the value of which determines whether the network will be trained (to learn) or will go into the mode of searching for objects of interest. Additional training allows you to train a neural network on a small amount of data while maintaining the state of the network at the completion of previous training. In this task, the parameter is set for training.

4 Object Detection Methods by Convolutional Neural Network Review

A lot of approaches related to finding an object of a given class in images using neural networks based on local features extracting methods and consist of the following steps: selecting objects (most often using SIFT, HOG histograms), after which the selected object is fed to the neural network for recognition (Dorogii 2012; Sarangi and Sekhar 2015; Ren et al. 2017).

The benefit of this approach is that this approach is scale-invariant. The disadvantage of this approach is that not always objects can be distinguished using such histograms, as a result—not all objects can be found in the image (an error is possible both in the selection of objects and in recognition). Also color images are usually transformed into grayscale images so it is impossible to use color information for object detection such way.

Network-based approach incorporates a deep learning technique, widely used in machine learning, which hierarchically extracts local features of high order. cNNs usually contain a number of convolutional layers designed to detect features as well as a classification layer, all together capable of learning deep features based on the training data. This kind of approach has two major advantages compared to the feature based approach. It is able to learn higher order local features by combining low-level features and inherently exploiting spatial dependence among them (Ševo and Avramovic 2016).

To avoid an error with selecting objects using histograms, the following object search algorithm usually implementing. The image is bypassed with sliding window of specified sizes N $\times$ N in N/2 (or some different) increments. Each of the image fragments under a sliding window is served by a neural network for recognition like it presented in Ševo and Avramovic (2016), https://ru.coursera.org/lecture/convolutional-neural-networks/convolutional-implementation-of-sliding-windows-6UnU4, https://medium.com/machine-learning-world/convolutional-neural-networks-for-all -part-ii-b4cb41d424fd. But image processing with constant window size has a drawback: objects can be too big or too small for the selected window size, and so objects can't be detected in images. So such algorithms are non-scale-invariant.

5 Object Detection

We proposed modified algorithm of sliding window to turn it in scale-invariant way.

Image are processing by sliding window and part of image are classifying by neural network. If the probability of assigning the considered fragment to a certain class K is greater than the specified threshold value p0, then it is considered that a class K object is depicted in a given area. After processing full image such way all the fragments are assigned to a class. This approach like a method presented in (Ševo and Avramovic 2016) but the difference is that we use sliding window with different size: we use a set of sliding window sizes and process image all of them. So our method is no-sensitive to object size.

The search pattern using a neural network consists of the sequential execution of modules:

ReadMultiImage. Image reading module.
FindObjects_CNN. Module of convolutional neural network. In this case it runs with parameters launched for object detection. It's possible to set object classes that will be detect in image.
SaveMultiImage. The module for storing results found using a neural network. It saves image and write information about possibilities of each found object (object's class and coordinates).

The task of objects detection of different scales is quite difficult at present (Girshick et al. 2014; Hinton et al. 2012a). In Figs. 4 and 5 presented the results of the "airplane" class objects detection with the different sizes of sliding windows: 150×150, 100×100, 80×80, 64×64, 32×32.

We use different sizes of sliding window to provide guaranteed object detection in current window: object can be too much for 32×32 window and in this small window can contained only part of whole object, so this image fragment can be wrong classified and labeled. For example, if we have a big gray aircraft (about 150 $\times$ 150) and we use only fragments 32×32, almost each fragment will be contain

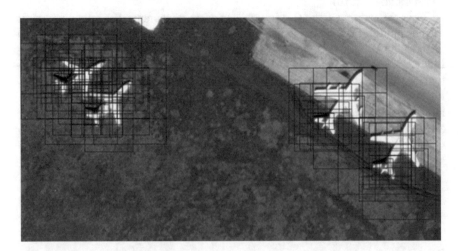

Fig. 4 Results of objects "airplanes" detection

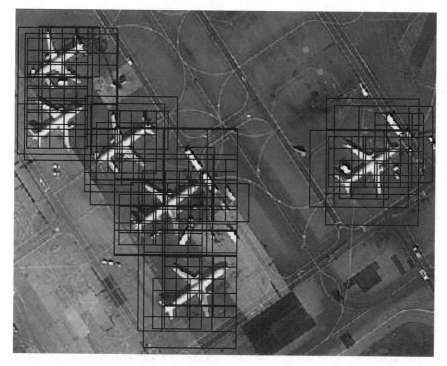

Fig. 5 Results of objects "airplanes" detection

part of aircraft but no one will be contain full aircraft. Moreover each this fragment will be have gray color that can be classified as Road class.

The object detection algorithm can be described as a sequence of steps:

(1) Initialization. In this step doing image *SRC* reading, creating neural network and reading its weights from file, reading *INTERESTED_CLASSES* info about interested classes (i.e. classes to detection on image, default—all classes are presented in this array), reading sliding window array and threshold possibility $P0$ for object classification (default 0).

(2) For each size *ws* (N × N pixels) in sliding window array is performing loop:

1. Considering image fragment *tmp_img* (coordinates $x0$, $y0$, $x0 + N$, $y0 + N$) under the sliding window from original image *SRC*, where $(x0, y0)$ is coordinates of top left point of sliding window.

2. Classifying *tmp_img* by CNN. Result of this classifying is two numbers: calculated class C and possibility P.

3. If calculated class C in *INTERESTED_CLASSES* and $P \geq P0$ then add information $(x0, y0, N, C, P)$ into special vector V.

4. Move sliding window to the next point and continue loop (step for moving is N/2).

(3) Process information from vector V. In the simplest way it's just drawing rectangles $(x0, y0, x0 + N, y0 + N)$ with color that corresponds to class C. Or it can be saving this information to file or sending this vector to another function.

In 3rd step we can paint every fragment by the correspondence color. So will be created segmented image. Image segmented by neural network are described in (Ivanov et al. 2019).

6   Coordinate Rectangles Merging Algorithm

As the coordinate rectangle we will consider a rectangle superimposed on the found object, given by the width and height, which are the dimensions of the desired object, as well as two-dimensional (x, y) coordinates of the upper left corner of the rectangle. It is assumed that the coordinates of the point of the center of such a rectangle coincide, or almost coincide in the permissible error, with the coordinates of the point of the center of the found object.

In addition to the coordinate rectangles, the neural network, after processing the image at the output, gives the probability that is assigned to each of the rectangles. This probability is determined by the possibility of finding a coordinate rectangle over the desired object, it is a floating-point number and takes values from 0 to 1.

The presented algorithm takes as input a list of coordinate rectangles, as well as the associated probabilities. Fusion algorithm consists of two steps: Creating sublists and filtering coordinate rectangles.

Creating sublists from the main list of coordinate rectangles. A rectangle with a zero index (reference rectangle) is selected from the main list. Next, calculate the coordinates of the center of the reference rectangle. After that, the Euclidean distance between the center of the reference rectangle and the centers of the remaining coordinate rectangles is calculated.

All rectangles whose distances from the coordinates of a point of their own center to the center of the reference rectangle are below a certain threshold (the threshold is chosen empirically) are added to the new list with the reference rectangle, and are deleted from the main list. This happens until there are no items in the main list.

Filtering coordinate rectangles by probability and creating a general coordinate rectangle. Each sublist obtained in the previous step is filtered by a certain probability threshold. Consequently, the list contains coordinate rectangles with high probability. Then they form a general coordinate rectangle, the center point of which is the arithmetic average point of the filtered rectangles. Dimensions (width and heights) of the general coordinate rectangle can be calculated in two ways:

(a) The arithmetic average of the dimensions of the filtered rectangles.
(b) Dimensions of the largest coordinate rectangle of the filtered rectangles.

In Fig. 6 on the top side showed an image with plotted coordinate rectangles before the merge algorithm is applied, and the bottom side after. In this example, the object of interest is the aircraft on the runway.

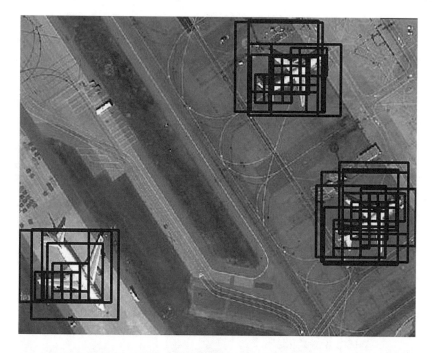

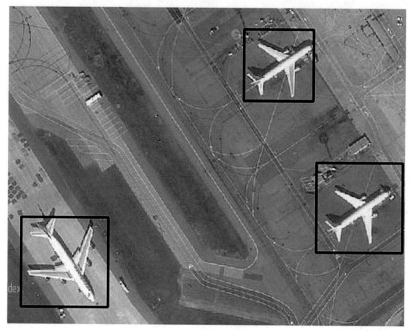

Fig. 6 All rectangles with "airplanes" objects (top) and filtered rectangles (bottom)

Despite the fact that the algorithm consists of only two stages, it is not necessary to wait until the first stage is fully executed. It is allowed to start the second stage immediately after receiving the first sub-list. In this case, you can immediately get a common coordinate rectangle, but the second stage will need to run for each new sublist.

In the experiments, there were some cases when such processing reduced the number of rectangles selected from 167 to 12.

This algorithm can be described as a sequence of steps:

1. Initialization. Reading vector $V <(x0, y0, N, C, P)>$, threshold distention $DIST$, creating V_DST as an empty vector
2. while V is not empty loop:
   V_TMP := empty vector
   a := $V[0]$
   a_cnt := center of rectangle a: $(a.x0 + a.N/2, a.y0 + a.N/2)$
   push a into V_TMP
   delete a from V
   for each b in V loop:
       b_cnt := center of rectangle b: $(b.x0 + b.N/2, b.y0 + b.N/2)$
       if distance between a_cnt and b_cnt is smaller than $DIST$ then
           push b into V_TMP
           delete b from V
       push V_TMP into V_DST
3. for each obj in V_DST loop:
   delete all elements from obj that have low P value // this step requires if value $P0$ wasn't set
   obj_cnt := average center of all not deleted elements from obj
   set obj_size one of the two ways: biggest N for elements in obj or average N-value for not-deleted objects
   push information (obj_cnt, obj_size) into vector RES

4. Draw all rectangles from RES vector

In Fig. 7 and 8 are examples more examples of post-processing object detection algorithm.

7   Adaptive Threshold Changing

Earlier in the article the threshold of entry (T) was specified, which was chosen empirically. The threshold value determined whether the coordinate rectangle is in the group of its neighbors or not.

However, manually setting the threshold for each new image is impractical. To avoid this problem has been developed adaptive threshold change method based on

Fig. 7 Results of objects "airplanes" detection by neural network with filtering coordinate rectangle post-processing

Fig. 8 Results of objects "airplanes" detection by neural network with filtering coordinate rectangle post-processing

the analysis of the sizes of coordinate rectangles and the probability they compare. In fact, the sizes of coordinate rectangles are the sizes of a square scanning window (mask) of a neural network.

The neural network makes several passes through the input image using masks of different sizes. This is done to ensure that objects of interest (in this case, airplanes) of different sizes are also successfully located.

During the experiments with manual selection of the threshold it turned out that the threshold value depends on the size (only the width of the mask can be used, since it is square) of the masks used. Therefore, the adaptive threshold change method is to select masks with the highest probability and their subsequent analysis.

Table 1 Comparing the value of the experimental threshold with the calculated values

T (from experiments)	T1 (AM)	T2 (GM)
150	160	160
150	155	154
140	138	133
120	120	115
100	96	94
80	80	76
75	72	64
60	59	57
45	43	40

As the analysis is used:

1. Calculation of arithmetic mean (AM) across selected widths masks
2. Calculation of the geometric mean (GM) width selected masks

In Table 1 presented a comparison of the value of the selected threshold manually with the obtained adaptive threshold.

The general scheme of searching for objects in an image using filtering coordinate rectangles consists of the sequential execution of modules:

ReadMultiImage. Image reading module.
FindObjects_CNN. A convolutional neural network module launched for detection.
Rectsnglefilter. The module that filters the coordinate rectangles by algorithm described earlier. Output channel for this module is image with filtered rectangles: each rectangle presents a single object.
SaveMultiImage. The module for storing results found using a neural network. Output data is an image with plotted rectangles.

8 Conclusion

At present object detection task is a very popular task in image processing, especially when remotely sensing images or aerial photographs. There are many approaches to finding objects in images.

The article discusses the approach to finding objects in aerial photographs using neural networks by sliding windows to search for aircraft in the photographs. The difference of presented object detection approach is that a full skirting of the image is performed by a sliding window of different sizes, thus all parts of the input image will be classified, instead of individual fragments, found using HOG. However, this approach has a drawback—one object is located in several fragments; thus, in the final image this object can be defined by several coordinate rectangles. To solve this problem, an algorithm was proposed for post-processing of data obtained as a result of the neural network.

An algorithm for post-processing the search results of objects in the image is proposed. This approach allows you to combine a subset of the coordinate rectangles of the corresponding object into one. In some cases, the number of coordinate rectangles was reduced 14 times. The proposed approach was implemented in a distributed data processing system.

Acknowledgements   This work was carried out with the financial support of the state represented by the Ministry of Education and Science of the Russian Federation (unique identifier of the project RFMEFI60419X0236)

References

Cheng G, Han J (2016) A survey on object detection in optical remote sensing images. ISPRS J Photogramm Remote Sens 117:11–28

Chen Q, Sun Q, Heng PA, Xia D (2008) A double-threshold image binarization method based on edge detector. Pattern Recognit 41:1254–1267. https://www.researchgate.net/publication/220604236_A_double-threshold_image_binarization_method_based_on_edge_detector

Cheng G, Zhou P, Han J (2016) Learning rotation-invariant convolutional neural networks for object detection in VHR optical remote sensing images. IEEE Trans Geosci Remote Sens 54(12):7405–7415

Convolutional implementation of sliding windows, object detection, Coursea. https://ru.coursera.org/lecture/convolutional-neural-networks/convolutional-implementation-of-sliding-windows-6UnU4

Convolutional neural network for all, machine learning world. https://medium.com/machine-learning-world/convolutional-neural-networks-for-all-part-ii-b4cb41d424fd

Dorogii YY (2012) Arkhitektura obobshchennykh svertochnykh neironnykh setei (The architecture of generalized convolutional neural networks), Vestnik NTUU KPII, 2012, No. 57, 6 p. Available at: http://www.itvisnyk.kpi.ua/wp-content/uploads/2012/08/54_36.pdf

Face detection using haar cascades, OpenCV. Open source computer vision documentation. http://docs.opencv.org/trunk/d7/d8b/tutorial_py_face_detection.html

Girshick RB, Donahue J, Darrell T, Malik J (2014) Rich feature hierarchies for accurate object detection and semantic segmentation. In: IEEE conference on computer vision and pattern recognition, CVPR 2014. https://www.cv-foundation.org/openaccess/content_cvpr_2014/papers/Girshick_Rich_Feature_Hierarchies_2014_CVPR_paper.pdf

Hinton GE et al (2012a) Improving neural networks by preventing co-adaptation of feature detectors. arXiv:1207.0580v1, 3 July 2012

Hinton GE, Srivastava N et al (2012b) Improving neural networks by preventing co-adaptation of feature detectors. arXiv preprint arXiv:1207.0580. https://pdfs.semanticscholar.org/2116/b2eaaece4af9c28c32af2728f3d49b792cf9.pdf

Ivanov ES, Tishchenko IP, Vinogradov AN (2019) Multispectral image segmentation using neural network. Sovremennye problem DZZ iz kosmosa 16(1):25–34. http://d33.infospace.ru/d33_conf/sb2019t1/25-34.pdf

Kondratyev A, Tishchenko I (2016) Concept of distributed processing system of images flow in terms of π-calculus. In: 2016 18th conference of open innovations association and seminar on information security and protection of information technology (FRUCT-ISPIT), pp 328–334

Kriesel D (2007) A brief introduction to neural networks. ZETA2-EN. http://www.dkriesel.com/_media/science/neuronalenetze-enzeta2-2col-dkrieselcom.pdf

Pitknen J (2001) Individual tree detection in digital aerial images by combining locally adaptive binarization and local maxima methods. Can J For Res 31(5):832–844

Ren S et al (2017) Object detection networks on convolutional feature maps. IEEE Trans Pattern Anal Mach Intell 39(7):1476–1481. https://arxiv.org/pdf/1504.06066

Romanov AA (2018) Svertochnye neironnye seti (Convolutional neural networks), Nauchnye issledovaniya: klyuchevye problemy III tysyacheletiya, 2018, pp 5–9. Available at: https://scientificresearch.ru/images/PDF/2018/21/svertochnye.pdf

Rosebrock A (2015) Histogram of oriented gradients and object detection [Web log post]. Retrieved 31 Aug 2015. http://www.pyimagesearch.com/2014/11/10/histogram-oriented-gradients-object-detection/

Sarangi N, Sekhar C (2015) Tensor deep stacking networks and kernel deep convex networks. In: 4th International conference on pattern recognition: applications and methods, IC RAM, 2015, pp 267–281. https://link.springer.com/chapter/10.1007/978-3-319-27677-=9_17

Ševo I, Avramovic A (2016) Convolutional neural network based automatic object detection on aerial images. IEEE Geosci Remote Sens Lett 13(5):740–744

Modelling and Control of Human Response to a Dynamic Virtual 3D Face

Vytautas Kaminskas and Edgaras Ščiglinskas

Abstract This chapter of the monograph introduces the application of identification and predictor-based control techniques for modelling and design control schemes of human response as reaction to a dynamic virtual 3D face. Two experiment plans were used, the first one—3D face was observed without virtual reality headset and the second one—with virtual reality headset. A human response to the stimulus (virtual 3D face with changing distance-between-eyes) is observed using EEG-based emotion signals (excitement and frustration). Experimental data is obtained when stimulus and response are observed in real time. Cross-correlation analysis of data is demonstrated that exists dynamic linear dependency between stimuli and response signals. The technique of dynamic systems identification which ensures stability and possible higher gain for building a one-step prediction models is applied. Predictor-based control schemes with a minimum variance or a generalized minimum variance controllers and constrained control signal magnitude and change speed are developed. The numerical technique of selection the weight coefficient in the generalized minimum variance control criterion is based on closed-loop stability condition and admissible value of systematic control error. High prediction accuracy and control quality are demonstrated by modelling results.

Keywords Dynamic virtual 3D face · Human response · EEG-based emotion signals · Identification · Prediction · Predictor-based control with constraints · Closed-loop stability · Control error

V. Kaminskas · E. Ščiglinskas (✉)
Faculty of Informatics, Vytautas Magnus University, Vileikos g. 8, 44404 Kaunas, Lithuania
e-mail: edgaras.sciglinskas@vdu.lt

V. Kaminskas
e-mail: vytautas.kaminskas@vdu.lt

© Springer Nature Switzerland AG 2020
G. Dzemyda et al. (eds.), *Data Science: New Issues, Challenges and Applications*, Studies in Computational Intelligence 869,
https://doi.org/10.1007/978-3-030-39250-5_2

1 Introduction

Emotions are very important to human experience because they play an important role in human daily lives—communication, rational decision making and learning (Mikhail et al. 2013).

The most popular way to observe a human emotion in real time is to monitor EEG-based signals as response to stimuli (visual, audio, etc.) (Hondrou and Caridakis 2012; Sari and Nadhira 2009). EEG-based emotion signals (excitement, frustration, engagement/boredom and meditation) are characterized as reliable and quick response signals.

Therefore, it is relevant to construct and investigate methods and models of recognition and estimation dependencies between emotion signals and different stimuli and to design the emotion feedback systems based on these models (Hatamikia et al. 2014; Khushaba et al. 2013; Lin et al. 2010; Sourina and Liu 2011).

The one-step prediction models were proposed for exploring dependencies of the human response to a dynamic virtual 3D face features when a virtual 3D face was observed without (Kaminskas et al. 2014; Kaminskas and Vidugirienė 2016; Vaškevičius et al. 2014a) or using (Kaminskas and Ščiglinskas 2018) virtual reality headset. The technique of dynamic systems identification which ensures stability of the models is applied to build these models (Kaminskas 1982, 1985). Predictive models are necessary in the design of predictor-based control and feedback systems (Clarke et al. 1987; Clarke 1994; Kaminskas 1988, 2007; Peterka 1984). The firsts experimental results of the predictor-based control applications for control of human emotion signals as response to a dynamic virtual 3D face were published in previous papers (Kaminskas et al. 2015; Kaminskas and Ščiglinskas 2016) (3D face was observed without virtual reality headset) and in Kaminskas and Ščiglinskas (2018) (3D face was observed using virtual reality headset).

In this chapter of the monograph the cross-correlation analysis and the technique of identification which ensures stability and possible higher gain of the model for building predictive models of dependencies between EEG-based emotion signals (excitement, frustration) as a human response to a dynamic virtual 3D face features (distance-between-eyes) are developed. Two experiment plans were used, the first one—3D face was observed without virtual reality headset and the second one—with virtual reality headset. Two predictor-based control schemes with a minimum variance or a generalized minimum variance controller and constrained control signal magnitude and changing rate are developed to the control of excitement or frustration as response to a virtual 3D face with changing distance-between-eyes. The numerical technique of computing the weight coefficient in the generalized minimum variance control criterion is proposed. This method is based on stability condition of closed-loop system and admissible value of systematic control error.

2 Experiment Planning and Data Analysis

A virtual 3D face of woman was used as a stimulus for eliciting human reaction. Three types of 3D face features (distance-between-eyes, nose width and chin width) were used as a stimuli for human reaction elicitation and four EEG-based emotion signals (excitement, frustration, engagement/boredom and meditation) were observed and analyzed in previous research (Vaškevičius et al. 2014a; Vidugirienė et al. 2013). Analysis of the results has shown that all three types of the 3D face features has triggered similar human reaction signals, accordingly distance-between-eyes was selected and used as a dynamic 3D face feature in further research (Kaminskas et al. 2014; Kaminskas and Vidugirienė 2016; Kaminskas and Ščiglinskas 2018; Vaškevičius et al. 2014b). From observed four EEG-based emotion signals, excitement and frustration are the most variable signals (Kaminskas et al. 2014; Vaškevičius et al. 2014a; Vidugirienė et al. 2013).

Accordingly, a virtual 3D face of woman with changing distance-between-eyes was used for an input signal and EEG-based excitement and frustration signals were measured as output (Figs. 1 and 2) in this research. The excitement and frustration signals were recorded with *Emotiv Epoc+* device. This device records EEG inputs from 14 channels (in accordance with the international 10–20 locations): AF3, F7, F3, FC5, T7, P7, O1, O2, P8, T8, FC6, F4, F8, AF4 (Mattioli et al. 2015). Values of each output signal vary from 0 to 1. If excitement or frustration is low, the value is close to 0 and if it is high, the value is close to 1.

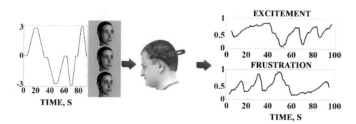

Fig. 1 Input-output experiment plan I

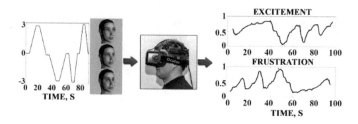

Fig. 2 Input-output experiment plan II

A women virtual 3D face was created with Autodesk Maya and was used as a "neutral" one (Figs. 1 and 2, middle) (Kaminskas et al. 2014; Vaškevičius et al. 2014b). Other 3D faces were formed by changing distance-between-eyes in an extreme manner: large and small distance (Figs. 1 and 2, upper and lower). "Neutral" face has 0 value, largest distance-between-eyes corresponds to value 3 and smallest distance-between-eyes corresponds to value negative 3. The transitions between normal and extreme states were programmed. Ten volunteers (three females and seven males) were tested in case of experiment plan I (Fig. 1) and other ten volunteers (three females and seven males) in case of experiment plan II (Fig. 2). Each volunteer was watching a changing virtual 3D face and each experiment was approximately about 100 s long. EEG-based excitement or frustration and changing distance-between-eyes signals were measured with sampling period $T_0 = 0.5$ s and recorded in a real time.

To estimate the possible relationship between human response signals (excitement or frustration) and virtual 3D face feature (distance-between-eyes) and relationship between excitement and frustration a cross-correlation analysis was performed. The estimates of cross-covariation functions between input (x) and output (y) signals and auto-covariation functions of input and output signals are used for this purpose (Vaškevičius et al. 2014b):

$$R_{yx}[\tau] = \frac{1}{N} \sum_{t=1}^{N-\tau} (y_t - \bar{y})(x_{t+\tau} - \bar{x}),$$

$$R_{xx}[\tau] = \frac{1}{N} \sum_{t=1}^{N-\tau} (x_t - \bar{x})(x_{t+\tau} - \bar{x}),$$

$$R_{yy}[\tau] = \frac{1}{N} \sum_{t=1}^{N-\tau} (y_t - \bar{y})(y_{t+\tau} - \bar{y}), \tag{1}$$

where

$$\bar{x} = \frac{1}{N} \sum_{t=1}^{N} x_t, \quad \bar{y} = \frac{1}{N} \sum_{t=1}^{N} y_t \tag{2}$$

are the averages of input and output, $\tau = 0, \pm 1, \dots$ and $N = 185$.

The maximum cross-correlation function values

$$r_{yx} = \max_{\tau} |r_{yx}[\tau]| = \max_{\tau} \left| \frac{R_{yx}[\tau]}{\sqrt{R_{xy}[0]R_{xx}[0]}} \right| \tag{3}$$

are provided in Table 1. Examples of cross-correlation functions estimates are demonstrated in Figs. 3 and 4.

Table 1 Maximum cross-correlation function values (upper, cross-correlation between excitement and distance-between-eyes, middle—frustration and distance-between-eyes, lower—between excitement and frustration)

No. Volunteer	1 Female	2 Female	3 Female	4 Male	5 Male	6 Male	7 Male	8 Male	9 Male	10 Male
r_{yx}, Experiment plan I	0.66	0.81	0.88	0.51	0.70	0.79	0.46	0.82	0.77	0.69
	0.73	0.84	0.57	0.77	0.70	0.58	0.78	0.56	0.83	0.81
	0.92	0.95	0.61	0.53	0.79	0.74	0.55	0.66	0.82	0.81

No. Volunteer	11 Female	12 Female	13 Female	14 Male	15 Male	16 Male	17 Male	18 Male	19 Male	20 Male
r_{yx}, Experiment plan II	0.61	0.59	0.77	0.86	0.85	0.81	0.82	0.77	0.76	0.70
	0.70	0.60	0.66	0.69	0.77	0.74	0.65	0.69	0.64	0.64
	0.57	0.68	0.59	0.72	0.83	0.91	0.56	0.68	0.62	0.64

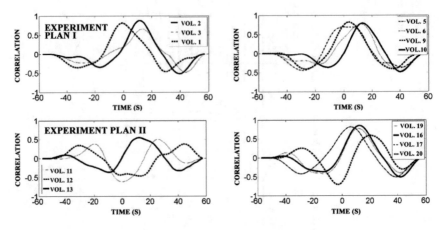

Fig. 3 Estimates of cross-correlation functions of females (left) and of males (right) for excitement

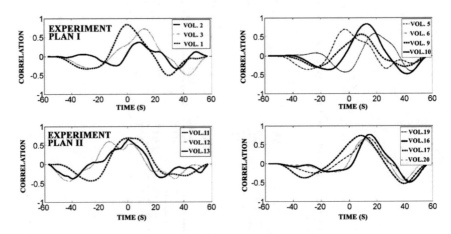

Fig. 4 Estimates of cross-correlation functions of females (left) and of males (right) for frustration

The shift of the maximum values of cross-correlation functions in relation to $R_{yx}[0]$ (Figs. 3 and 4) allows stating that there exists dynamic relationship between 3D face with changing distance-between-eyes and excitement or frustration signals. Maximum cross-correlation values (Table 1) justify a possibility of linear relationship.

3 Input-Output Model Building

Cross-correlation analysis demonstrate that dependency between EEG-based excitement or frustration signal as response to a virtual 3D face with changing distance-between-eyes can be described as linear dynamic input-output model (Kaminskas et al. 2014; Kaminskas and Ščiglinskas 2018; Vaškevičius et al. 2014b)

$$A(z^{-1})y_t = \theta_0 + B(z^{-1})x_t + \varepsilon_t, \tag{4}$$

where

$$A(z^{-1}) = 1 + \sum_{i=1}^{n} a_i z^{-i}, \ B(z^{-1}) = \sum_{j=0}^{m} b_j z^{-j}, \quad m \le n, \tag{5}$$

y_t is an output (excitement, frustration), x_t is an input (distance-between-eyes) signals respectively observed as $y_t = y(tT_0)$, $x_t = x(tT_0)$ with a sampling period T_0, ε_t denotes the equation error of white-noise type, z^{-1} is the backward-shift operator ($z^{-1}x_t = x_{t-1}$) and θ_0 is a permanent component.

Model (4) is obtained from ordinary structure of a dynamic model (Kaminskas 1982, 1985; Kaminskas and Vidugirienė 2016)

$$A(z^{-1})y_t = B_*(z^{-1})f(x_t) + \varepsilon_t, \tag{6}$$

where

$$f(x_t) = f_0 + f_1 x_t, \quad B_*(z^{-1}) = \sum_{j=0}^{m} b_j^* z^{-j}. \tag{7}$$

From (6) is clear, that coefficients of model (4) are

$$\theta_0 = f_0 B_*(1) = f_0 \sum_{j=0}^{m} b_j^*, \quad b_j = f_1 b_j^*, \quad j = 0, 1, \ldots, m. \tag{8}$$

Because excitement and frustration signals are positive, so component $\theta_0 > 0$. Equation (4) can be rewritten as

$$y_t = y_{t|t-1} + \varepsilon_t, \tag{9}$$

where

$$y_{t|t-1} = \theta_0 + L(z^{-1})y_{t-1} + B(z^{-1})x_t, \tag{10}$$

is one-step predictor of the output signal,

$$L(z^{-1}) = z[1 - A(z^{-1})],\tag{11}$$

z is a toward-shift operator ($zy_t = y_{t+1}$).

Parameters (coefficients b_j and a_i, degrees m and n of the polynomials (5) and constant θ_0) of the model (4) are unknown. They must be estimated in the identification process, according to the observations obtained during the experiments with the volunteers.

Basic techniques of system identification and numerical schemes of computing the estimates (Ljung 1987; Soderstrom and Stoica 1989) do not ensure stability of the dynamic models. As a solution of this problem, the techniques and numerical methods of dynamic system identification were developed (Kaminskas 1982, 1985). Applying this technique, the current estimates of the parameters are obtained from the condition

$$\hat{c}_t : \tilde{Q}_t(c) = \frac{1}{t-n} \sum_{k=n+1}^{t} \varepsilon_{k|k-1}^2(c) \to \min_{c \in \Omega_c},\tag{12}$$

where

$$c^T = [\theta_0, b_o, b_1, \ldots, b_m, a_1, a_2, \ldots, a_n]\tag{13}$$

is a vector of the coefficients of the polynomials (5) and constant θ_0,

$$\varepsilon_{t|t-1}(c) = y_t - y_{t|t-1}\tag{14}$$

is one-step output prediction error,

$$\Omega_c = \left\{ a_i : \left| z_i^A \right| < 1, \quad i = 1, 2, \ldots, n \right\}\tag{15}$$

is stability domain (unity disk) for the model (4), z_i^A is the roots of the polynomial

$$z_i^A : A(z) = 0, \quad i = 1, \ldots, n, \quad A(z) = z^n A(z^{-1}),\tag{16}$$

T is a vector transpose sign, sign | | denotes to the absolute value.

The one-step predictor (10) can be rewritten as

$$y_{t+1|t} = \boldsymbol{\beta}_t^T c,\tag{17}$$

where

$$\boldsymbol{\beta}_t^T = \left[1, x_{t+1}, x_t, \ldots, x_{t-m+1}, -y_t, -y_{t-1}, \ldots, -y_{t-n}\right].\tag{18}$$

Considering Eqs. (14) and (17), identification criterion

$$\tilde{Q}_t(\boldsymbol{c}) = \frac{1}{t-n} \sum_{k=n+1}^{t} \left(y_k - \boldsymbol{\beta}_{k-1}^T \boldsymbol{c} \right)^2 \tag{19}$$

is a quadratic function of the vector variable $\boldsymbol{c}$. Accordingly, solution of the minimization problem (12)–(16) is separated into the two stages (Kaminskas et al. 2014; Kaminskas and Ščiglinskas 2018). In the first stage parameter estimates are calculated without evaluation of restrictions (15)

$$\boldsymbol{c}_t = \left[\sum_{k=n+1}^{t} \boldsymbol{\beta}_{k-1} \boldsymbol{\beta}_{k-1}^T \right]^{-1} \left[\sum_{k=n+1}^{t} y_k \boldsymbol{\beta}_{k-1} \right], \quad t > 2n + m + 2 + k_*, \tag{20}$$

where k_* is the first value of discrete time k, when testing signal value is not equal to zero ($x_{k_*} \neq 0$).

In the second stage, vector of the estimates (20) is projected into stability domain (15)

$$\hat{\boldsymbol{c}}_t = \boldsymbol{\Gamma}_t \boldsymbol{c}_t, \tag{21}$$

where

$$\boldsymbol{\Gamma}_t = \begin{pmatrix} 1 & \mathbf{0} & \mathbf{0} \\ \mathbf{0} & \mathbf{I_b} & \mathbf{0} \\ \mathbf{0} & \mathbf{0} & \gamma_t \mathbf{I_a} \end{pmatrix}, \quad 0 < \gamma_t \leq 1 \tag{22}$$

is a $(m + n + 2) \times (m + n + 2)$ diagonal block-matrix of projection, $\mathbf{I_b}$ is a $(m + 1) \times (m + 1)$ unity matrix and $\mathbf{I_a}$ is a $n \times n$ unity matrix.

Factory γ_t in matrix (22) is calculated by

$$\gamma_t = \min\{1, \gamma_{\max} - \gamma_0\}, \tag{23}$$

where $\gamma_{\max} \|\boldsymbol{c}_t\|$ is the distance from the point $\mathbf{0}$ (origin) to the boundary of stability domain Ω_c in the direction of $\boldsymbol{c}_t$, $\|\cdot\|$ is the Euclidean norm sign, γ_0 is a small and positive constant. When model order $n \leq 3$ numerical scheme of computing factor γ_t was given in (Kaminskas et al. 2014; Kaminskas and Vidugirienė 2016)

$$\gamma_t = \min\left\{1, \gamma_{1,t}^{(1)}\right\}, \quad \text{if } n = 1, \tag{24}$$

$$\gamma_{1,t}^{(1)} = \frac{1}{|a_{1,t}|} - \gamma_0, \tag{25}$$

$$\gamma_t = \min\left\{1, \gamma_{1,t}^{(2)}, \gamma_{2,t}^{(2)}, \gamma_{3,t}^{(2)}\right\}, \quad \text{if } n = 2, \tag{26}$$

$$\gamma_{1,t}^{(2)} = \begin{cases} -\dfrac{1}{a_{1,t}+a_{2,t}} - \gamma_0, & \text{if } a_{1,t} + a_{2,t} < 0, \\ 1, & \text{in other cases,} \end{cases} \tag{27}$$

$$\gamma_{2,t}^{(2)} = \begin{cases} -\dfrac{1}{a_{2,t}-a_{1,t}} - \gamma_0, & \text{if } a_{2,t} - a_{1,t} < 0, \\ 1, & \text{in other cases,} \end{cases} \tag{28}$$

$$\gamma_{3,t}^{(2)} = \frac{1}{|a_{2,t}|} - \gamma_0, \tag{29}$$

$$\gamma_t = \min\left\{1, \gamma_{1,t}^{(3)}, \gamma_{2,t}^{(3)}, \gamma_{3,t}^{(3)}, \gamma_{4,t}^{(3)}\right\}, \quad \text{if } n = 3, \tag{30}$$

$$\gamma_{1,t}^{(3)} = \begin{cases} -\dfrac{1}{a_{1,t}+a_{2,t}+a_{3,t}} - \gamma_0, & \text{if } a_{1,t} + a_{2,t} + a_{3,t} < 0, \\ 1, & \text{in other cases,} \end{cases} \tag{31}$$

$$\gamma_{2,t}^{(3)} = \begin{cases} -\dfrac{1}{a_{2,t}-a_{1,t}-a_{3,t}} - \gamma_0, & \text{if } a_{2,t} - a_{1,t} - a_{3,t} < 0, \\ 1, & \text{in other cases,} \end{cases} \tag{32}$$

$$\gamma_{3,t}^{(3)} = \frac{1}{|a_{3,t}|} - \gamma_0, \tag{33}$$

$\gamma_{4,t}^{(3)}$ is a smaller solution (from real positive solutions) of a quadratic equation

$$\left(a_{1,t}a_{3,t} - (a_{3,t})^2\right)\gamma_t^2 - a_{2,t}\gamma_t + 1 = 0. \tag{34}$$

Positive constant $\gamma_0 \in [0.001, 0.01]$.

The model order estimates ($\hat{m}$ and $\hat{n}$) are obtained from conditions (Kaminskas 1982, 1985; Kaminskas et al. 2014; Kaminskas and Vidugirienė 2016)

$$\hat{n} = \min\{\tilde{n}\}, \quad \hat{m} = \min\{\tilde{m}\}, \tag{35}$$

where $\tilde{m}$ and $\tilde{n}$ are polynomial (5) degrees when the following inequalities are correct

$$\left|\frac{\sigma_{\varepsilon,t}[m, n+1] - \sigma_{\varepsilon,t}[m, n]}{\sigma_{\varepsilon,t}[m, n]}\right| \le \delta_\varepsilon, \quad n = 1, 2, \ldots, \tag{36}$$

$$\left|\frac{\sigma_{\varepsilon,t}[m+1, n] - \sigma_{\varepsilon,t}[m, n]}{\sigma_{\varepsilon,t}[m, n]}\right| \le \delta_\varepsilon, \quad m = 1, 2, \ldots, n, \tag{37}$$

$$\sigma_{\varepsilon,t}[m, n] = \sqrt{\frac{1}{N-n}\sum_{t=n+1}^{N}\hat{\varepsilon}_t^2[m, n]} \tag{38}$$

is one-step prediction error standard deviation,

$$\hat{\varepsilon}_t[m, n] = y_t - \hat{y}_{t|t-1}[m, n] \tag{39}$$

is one-step prediction error,

$$\hat{y}_{t|t-1}[m, n] = \hat{\theta}_{0,t} + \hat{L}_t(z^{-1})y_{t-1} + \hat{B}_t(z^{-1})x_t \tag{40}$$

is one-step predictor and $\delta_\varepsilon > 0$ is a chosen constant value. Usually in identification practice $\delta_\varepsilon \in [0.01 - 0.1]$, which corresponds to a relative variation of prediction error standard deviation from 1 to 10%.

Because input signal $x_t^* \in [x_{min}, x_{max}]$, then the largest value of permanent component in output signal is

$$y_{max} = K_t x_{max} + \frac{1}{\hat{A}_t(1)}\hat{\theta}_{0,t}, \quad \text{if } K_t > 0 \tag{41}$$

or

$$y_{max} = K_t x_{min} + \frac{1}{\hat{A}_t(1)}\hat{\theta}_{0,t}, \quad \text{if } K_t < 0, \tag{42}$$

where

$$K_t = \frac{\hat{B}_t(1)}{\hat{A}_t(1)}. \tag{43}$$

From (41) and (42) it's clear that to build predictive model for control system is necessarily to ensure not only the output prediction accuracy, but also highest possible gain (43) of the model (4). Accordingly, we choose predictive model with higher possible gain and admissible value of predictor error

$$\hat{K}_t = \max\left\{\left|\tilde{K}_t\right|\right\}, \quad \tilde{K}_t \in \Omega_K, \tag{44}$$

where

$$\Omega_K = \left\{\tilde{K}_t : \left|\frac{\sigma_{\varepsilon,t}[m, n] - \sigma_{\varepsilon,t-1}[m, n]}{\sigma_{\varepsilon,t-1}[m, n]}\right| \leq \tilde{\delta}_\varepsilon\right\} \tag{45}$$

is a subset of the gain (43).

4 Predictor-Based Control with Constraints

The control law synthesis in predictor-based systems is often based on minimization of minimum variance or generalized minimum variance control criteria (Astrom 1970; Isermann 1981; Astrom and Wittenmark 1984; Clarke 1994; Soeterboek 1992). Basic techniques of a minimum variance or generalized minimum variance control are developed without evaluation of possible control signal constraints. Accordingly, techniques and schemes with constrained control signal magnitude and change rate for linear and nonlinear plants were constructed (Kaminskas 1988, 2007; Kaminskas et al. 1991b, 1993) and applied for control different plants—processes in nuclear reactors (Kaminskas et al. 1990, 1991a, 1992), thermal power plants (Kaminskas et al. 1987, 1989) and etc.

The control law applying these techniques for the input-output plant (4) is obtained from the conditions (Kaminskas 2007)

$$x_{t+1}^* : Q_t(x_{t+1}) \rightarrow \min_{x_{t+1}\in\Omega_x}, \tag{46}$$

$$Q_t(x_{t+1}) = \mathrm{E}\left\{(y_{t+1} - y_{t+1}^*)^2 + q(x_{t+1} - \tilde{x}_{t+1}^*)^2\right\}, \tag{47}$$

$$\Omega_x = \{x_{t+1} : x_{\min} \le x_{t+1} \le x_{\max}, |x_{t+1} - x_t^*| \le \delta_t\}, \tag{48}$$

where E is an expectation operator, y_{t+1}^* is a reference output signal (reference trajectory for excitement or frustration), $\tilde{x}_{t+1}^*$ marks the reference trajectory for the control signal (distance-between-eyes), $x_{\min}$ and $x_{\max}$ are control signal boundaries (smallest and largest distance-between-eyes), $\delta_t > 0$ are the restriction values for the change rate of the control signal and $q \ge 0$ is weight coefficient.

The one-step predictor for control plant (4) may also be constructed as (Kaminskas 1988, 2007)

$$A(z^{-1})y_{t+1|t} = \theta_0 + B(z^{-1})x_{t+1} + L(z^{-1})\varepsilon_{t|t-1}. \tag{49}$$

Then the solution of the minimization problem (46)–(48) for predictor (49) is obtained as

$$x_{t+1}^* = \begin{cases} \min\{x_{\max}, x_t^* + \delta_t, \tilde{x}_{t+1}\}, & \text{if } \tilde{x}_{t+1} \ge x_t^*, \\ \max\{x_{\min}, x_t^* - \delta_t, \tilde{x}_{t+1}\}, & \text{if } \tilde{x}_{t+1} < x_t^*, \end{cases} \tag{50}$$

$$\begin{aligned} D(z^{-1})\tilde{x}_{t+1} &= -L(z^{-1})\varepsilon_{t|t-1} - \theta_0 \\ &\quad + A(z^{-1})(\tilde{y}_{t+1} + \lambda\tilde{x}_{t+1}^*), \end{aligned} \tag{51}$$

$$D(z^{-1}) = B(z^{-1}) + \lambda A(z^{-1}), \quad \lambda = q/b_0, \tag{52}$$

$$\tilde{y}_{k+1} = \begin{cases} y_{k+1}^*, & \text{if } k = t, \\ y_{k+1|k}, & \text{if } k = t - 1, \ldots, t - n. \end{cases} \tag{53}$$

The closed-loop system equation is obtained by inserting (51) into (4)

$$\begin{aligned} D(z^{-1})y_t &= B(z^{-1})\left[y_t^* + \lambda\tilde{x}_t^* - \lambda L(z^{-1})(\tilde{x}_{t-1}^* - \tilde{x}_{t-1})\right] \\ &\quad + \lambda\theta_0 + [B(z^{-1}) + \lambda]\varepsilon_t. \end{aligned} \tag{54}$$

Then stability of the closed-loop system is dependent of characteristic polynomial

$$D(z) = z^n D(z^{-1}) \tag{55}$$

roots—all roots must by inside the unity disk

$$\left| z_i^D \right| \le 1, \quad z_i^D \colon D(z) = 0, \quad i = 1, 2, \ldots, n. \tag{56}$$

The analysis (52) allows state what having a stable model in the process of the identification (12)–(16), stability of a closed-loop system is obtained with any arrangement of roots of the polynomial $B(z^{-1})$, when the weight factor $|\lambda|$ [weight coefficient q in criterion (47)] is increased until all roots of the polynomial (55) rely in the unity disk.

From (54) we get what the permanent component of output signal in stationary regime $\left(y_t^* = y^*, \tilde{x}_{t+1}^* = \tilde{x}^* \right)$ is

$$y = K_p \left(y^* + \tilde{\theta}_0 + \lambda \tilde{x}^* \right), \quad \tilde{\theta}_0 = \frac{\lambda \theta_0}{B(1)}, \tag{57}$$

where

$$K_p = W_p(1), \quad W_p(z^{-1}) = \frac{B(z^{-1})}{D(z^{-1})} \tag{58}$$

is a gain of the transfer function of the reference signal y_t^* in a closed-loop system.
Then the control error can be rewritten as

$$e_p = y^* - y = \left(1 - K_p \right) y^* - K_p (\tilde{\theta}_o + \lambda \tilde{x}^*) \tag{59}$$

and it grows if K_p is significantly different from unit ($|\lambda|$ in (52) or q in (47) is high). Accordingly, the gain K_p is selected from an interval

$$K_p \in [0.8, 1], \text{ if signs of } b_0 \text{ and } K_0 \text{ are equal} \quad \text{or}$$
$$K_p \in [1, 1.2], \text{ if signs of } b_0 \text{ and } K_0 \text{ are different.} \tag{60}$$

Considering expressions (52) and (58), weight factor λ and weight coefficient q are calculated by

$$\lambda = \frac{K_0 (1 - K_p)}{K_p}, \quad q = \lambda b_0. \tag{61}$$

where

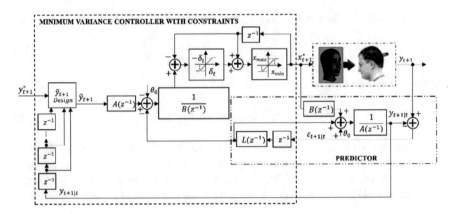

Fig. 5 The scheme of a predictor-based control with minimum variance controller

$$K_0 = W_0(1), \quad W_0(z^{-1}) = \frac{B(z^{-1})}{A(z^{-1})} \tag{62}$$

is a gain of the transfer function of the model (4).

Considering (60), gain K_p values are decreased or increased until calculated λ values (61) ensure stability conditions (56) of closed-loop system.

From (59) and (61) we get what the systematic control error is

$$e_p = (1 - K_p)(y^* - \theta_0^* - K_0\tilde{x}^*), \quad \theta_0^* = \frac{\theta_0}{A(1)}. \tag{63}$$

When gain of closed-loop $K_p = 1$ ($\lambda = 0, q = 0$) we get a minimum variance control quality, in other cases—a generalized minimum variance control quality of a predictor-based control schemes with constraints.

Predictor-based control schemes with minimum variance controller (Fig. 5) and generalized minimum variance controller (Fig. 6), when the reference trajectory for control signal is

$$\tilde{x}_{t+1}^* = x_t^*, \quad t = 1, 2, \ldots \tag{64}$$

are compared in this research.

5 Modelling Results

Modelling experiments consisted of two stages. In the first stage human excitement and frustration signals as response to 3D face with changing distance-between-eyes (testing input, Figs. 1 and 2) were observed. Ten volunteers (three females and seven

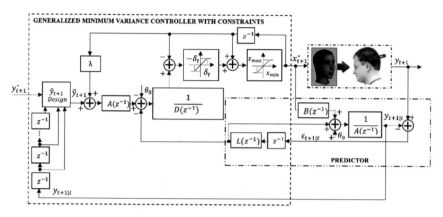

Fig. 6 The scheme of a predictor-based control with generalized minimum variance controller

males) were tested in case of experiment plan I (Fig. 1) and other ten volunteers (three females and seven males) in case of experiment plan II (Fig. 2). Each volunteer was watching a changing virtual 3D face and each experiment was approximately 100 s long. EEG-based excitement or frustration and changing distance-between eyes in virtual 3D face signals were measured with sampling period $T_0 = 0.5s$ and recorded in real time (for each signal, the number of observations is $N = 185$). This duration of the experiment is selected in order to get a sufficient number of input-output measures for parameters identification and model validation. Shorter times (about 60 s) of the experiment were used in previous researches (Vaškevičius et al. 2014b; Vidugirienė et al. 2013).

The results of the input-output model building demonstrated that the dependencies between excitement or frustration and distance-between-eyes can be described using the first order ($\hat{m} = 0, \hat{n} = 1$) models (4). Prediction accuracies were evaluated using the average absolute relative prediction error

$$|\bar{\varepsilon}| = \frac{1}{N-n} \sum_{t=n}^{N-1} \left| \frac{y_{t+1} - \hat{y}_{t+1|t}}{y_{t+1}} \right| \times 100\% \tag{65}$$

Parameter estimates of the models, prediction accuracy measure (64) and discrete time t correspondent to current predictive model with gain (45) are provided in Tables 2 and 3. The examples of one-step excitement prediction results are illustrated in Fig. 7 and frustration prediction results—in Fig. 8.

In the second stage, dynamical virtual 3D is formed according to the control signal (50)–(53). Reference signals were selected to maintain high excitement or frustration levels.

In this case a control efficiency for each volunteer can be evaluated by relative measures:

Table 2 Parameter estimates of the first order models, prediction accuracy measure (65) and the number t of the observations used for the model building in case of input-output experiment plan I

No	Volunteer	$\hat{\theta}_0$		$\hat{b}_0$		$\hat{a}_1$		$|\bar{\varepsilon}|$		t	
		Exc.	Fru.	Exc.	Fru.	Exc.	Fru.	Exc.	Fru.	Exc.	Fru.
1	Female	0.0370	0.0330	0.0160	0.0128	−0.9200	−0.9212	9.3	5.5	60	90
2	Female	0.0373	0.0120	−0.0054	−0.0040	−0.9894	−0.9862	10.0	3.9	100	80
3	Female	0.0202	0.1070	−0.0076	−0.0150	−0.8916	−0.8558	8.6	9.5	20	50
4	Male	0.0307	0.0097	−0.0029	0.0050	−0.9968	−0.9340	10.0	5.2	120	20
5	Male	0.0359	0.0130	0.0038	0.0040	−0.9677	−0.9560	8.4	2.5	100	80
6	Male	0.0436	0.0240	−0.0078	0.0050	−0.9857	−0.9650	5.9	4.3	90	50
7	Male	0.0607	0.0160	−0.0271	0.0083	−0.8035	−0.9441	9.8	3.9	20	50
8	Male	0.0437	0.0380	−0.0083	−0.0100	−0.9678	−0.9580	9.2	4.9	80	50
9	Male	0.0399	0.0745	−0.0139	−0.0100	−0.8607	−0.8240	11.6	4.5	20	20
10	Male	0.0259	0.0090	−0.0037	−0.0040	−0.9876	−0.9726	8.4	3.7	80	60

Table 3 Parameter estimates of the first order models (4), prediction accuracy measure (65) and the number t of the observations used for the model building in case of input-output experiment plan II

No	Volunteer	$\hat{\theta}_0$		$\hat{b}_0$		$\hat{a}_1$		$|\bar{\varepsilon}|$		t	
		Exc.	Fru.	Exc.	Fru.	Exc.	Fru.	Exc.	Fru.	Exc.	Fru.
11	Female	0.0400	0.0370	−0.0067	0.0055	−0.9567	−0.9261	10.1	4.2	100	80
12	Female	0.0300	0.0410	−0.0100	0.0090	−0.9808	−0.9659	6.5	5.1	60	30
13	Female	0.0367	0.2360	−0.0129	0.0127	−0.9633	−0.6683	6.4	7.1	50	20
14	Male	0.0467	−0.0030	−0.0171	0.0035	−0.9764	−0.9750	9.5	3.5	50	50
15	Male	0.0411	0.0810	−0.0089	−0.0060	−0.9896	−0.8400	7.5	3.9	90	60
16	Male	0.0186	0.0209	−0.0049	−0.0030	−0.9512	−0.9586	4.6	2.7	50	90
17	Male	0.0336	0.0110	−0.0064	−0.0080	−0.9766	−0.9904	10.4	4.0	70	70
18	Male	0.0303	0.0240	−0.0090	0.0043	−0.9890	−0.9443	8.1	1.4	70	70
19	Male	0.0399	0.0080	−0.0092	0.0059	−0.8718	−0.9479	9.7	3.4	40	50
20	Male	0.0224	0.0110	−0.0092	−0.0050	−0.9021	−0.9817	7.3	3.8	60	90

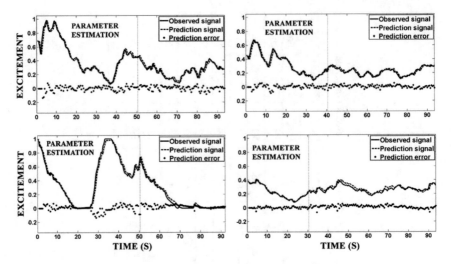

Fig. 7 Examples of one-step excitement prediction results for female (upper left, volunteer No. 2) and for male (upper right, volunteer No. 10), when 3D face is observed without a virtual reality headset (input-output experiment plan I) and results for female (lower left, volunteer No. 11) and for male (lower right, volunteer No. 20), when 3D face is observed using a virtual reality headset (input-output experiment plan II)

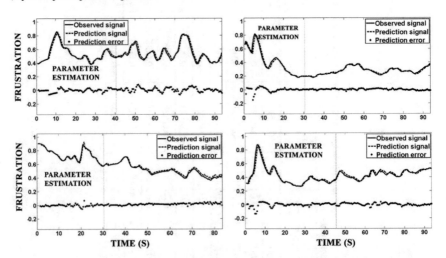

Fig. 8 Examples of one-step frustration prediction results for female (upper left, volunteer No. 2) and for male (upper right, volunteer No. 10), when 3D face is observed without a virtual reality headset (input-output experiment plan I) and results for female (lower left, volunteer No. 12) and for male (lower right, volunteer No. 20), when 3D face is observed using a virtual reality headset (input-output experiment plan II)

$$\Delta y = \frac{|\bar{y}_C - \bar{y}_T|}{\bar{y}_T} \times 100\%, \tag{66}$$

$$|\Delta y| = \frac{1}{N - n} \sum_{t=n}^{N-1} \left| \frac{\hat{y}_{t+1} - y_{t+1}^*}{y_{t+1}^*} \right| \times 100\%, \tag{67}$$

where $\bar{y}_T$ is the average of observed excitement or frustration signal y_t as response to the testing input, $\bar{y}_C$ is the average of output signal $\hat{y}_t$ as response to the control signal (50)–(53). These average measures (65) and (66) are given in Tables 4 and 5.

Results of the excitement and frustration predictor-based control with constraints are demonstrated in Figs. 9, 10, 11 and 12.

The restriction values δ_t in (48) for the change rate of the control signal is selected as: at the highest value $\delta_t = 6(12/s)$, the control signal can pass from minimum x_{min} to maximum value x_{max} with one discrete time step $T_0 = 0.5s$ and vice versa, at the smallest value $\delta_t = 0.15(0.3/s)$ control signal change speed is equivalent to the change rate of the testing signal (Figs. 1 and 2).

6 Conclusions

Experiments data and it's analysis demonstrated that each volunteer reacted to the stimuli individually and response (EEG-based excitement or frustration signal) to dynamic virtual 3D face (with changing distance-between-eyes) can be described by the first order linear dynamic models (4) with different estimates of parameters. The numerical schemes of computing the current estimates of parameters are based on system identification technique which ensures stability and possible higher gain (44) of the input-output models (4). Analysis of the results of a one-step prediction demonstrates that excitement signal can be predicted on average with about 9% average absolute relative prediction error (65) when 3D face is observed without a virtual reality headset or about 8% when it is observed using a virtual reality headset. Frustration signal can be predicted on average with about 5% average absolute relative prediction error when 3D face is observed without a virtual reality headset or about 4% when it is observed using a virtual reality headset. Accordingly, model (4) in the one-step predictor form (10) or (49) can be applied for predictor-based control design in emotion feedback system.

Two schemes of the predictor-based control with constraints were developed for controlling a human excitement or frustration signal as response to a dynamic virtual 3D face with changing distance-between-eyes. Controllers design is based on minimization of minimum variance or generalized minimum variance control criteria in an admissible domain for control signal in these schemes. The numerical technique of computing the weight factor λ in control law (50)–(53) or the weight coefficient q in control criterion (47) is proposed. The technique is based on an admissible value of the systematic control error and stability condition of the closed-loop system.

Table 4 The average of the control efficiency measure (66) (at the same δ_t values, upper row is for excitement control, lower—for frustration control)

$\delta_t\backslash K_p$	Input-output plant I (Fig. 1)					Input-output plant II (Fig. 2)				
	1	0.95	0.9	0.85	0.8	1	0.95	0.9	0.85	0.8
12/s	75.6	74.2	76.3	76.3	76.1	83.3	82.4	82.1	81.9	82.3
	60.8	60.2	61.0	61.0	61.1	45.4	45.3	44.4	43.7	43.6
6/s	75.8	73.7	76.5	76.2	76.0	83.0	82.2	81.9	81.9	82.3
	60.7	60.2	61.0	60.9	61.0	45.3	45.0	44.3	43.6	43.5
2/s	74.1	70.9	73.4	73.9	74.4	81.0	78.4	78.7	79.6	80.2
	59.6	59.4	59.9	59.4	59.7	44.8	43.2	42.1	41.9	41.7
1.2/s	71.1	70.3	72.9	73.3	72.6	66.6	64.4	65.1	63.9	65.5
	59.8	58.9	58.6	58.6	58.9	44.3	41.8	41.2	40.6	43.3
0.6/s	73.6	73.8	71.1	68.6	64.3	67.6	65.6	64.7	64.5	64.8
	59.4	59.3	59.7	56.9	56.6	42.4	40.5	38.4	38.0	37.7
0.3/s	72.9	68.2	68.1	67.4	67.0	66.6	64.3	58.6	57.0	55.3
	59.4	58.2	55.5	53.8	54.1	40.1	36.7	36.0	37.6	37.5

Table 5 The average of the control efficiency measure (67) (at the same δ_t values, upper row is for excitement control, lower—for frustration control)

Input-output plant I (Fig. 1)

$\delta_t \backslash K_p$	1	0.95	0.9	0.85	0.8
12/s	13.8	16.0	17.9	19.4	20.6
	9.6	11.0	11.0	11.9	12.3
6/s	13.9	16.4	18.1	19.6	20.7
	9.8	11.1	11.4	12.0	12.3
2/s	16.3	19.8	21.3	22.2	22.6
	11.1	12.2	12.4	13.1	13.4
1.2/s	19.5	22.0	22.6	23.0	23.9
	11.5	12.8	13.4	13.8	13.9
0.6/s	21.9	23.9	25.9	27.4	28.8
	13.1	13.8	14.0	15.3	16.2
0.3/s	24.4	26.8	27.9	28.6	29.8
	14.9	15.7	17.6	19.2	19.5

Input-output plant II (Fig. 2)

$\delta_t \backslash K_p$	1	0.95	0.9	0.85	0.8
12/s	13.0	14.9	16.4	18.0	18.6
	7.7	8.6	9.8	10.6	11.2
6/s	13.2	15.2	16.6	18.1	18.6
	7.9	8.8	10.0	10.7	11.3
2/s	14.8	17.5	18.8	19.5	20.5
	8.7	10.7	11.7	12.1	12.8
1.2/s	16.0	19.5	21.2	23.2	23.3
	9.8	11.5	12.7	13.8	12.4
0.6/s	20.0	23.1	24.7	25.9	27.2
	11.2	13.4	15.9	16.6	17.2
0.3/s	25.9	28.0	32.0	34.8	36.4
	14.6	17.2	18.2	17.7	17.8

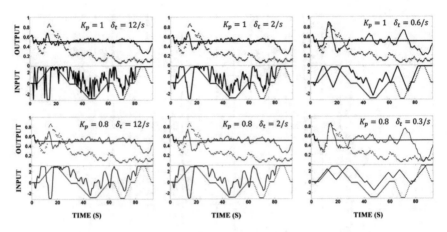

Fig. 9 Examples of excitement control with constraints (upper—minimum variance controller, lower—generalized minimum variance controller for volunteer No. 1 (female) in case of input-output plant I (output: solid line denotes reference signal y_t^*, dotted—output signal $\hat{y}_t$ as response to the control signal and dashed—observed output y_t as response to the testing input; input: solid line denotes control signal x_t^* and dashed—testing input x_t)

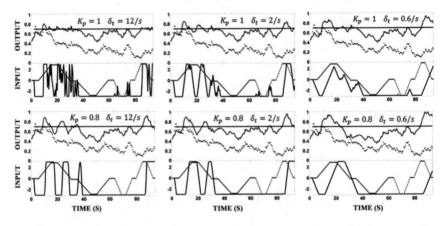

Fig. 10 Examples of excitement control with constraints (upper—minimum variance controller, lower—generalized minimum variance controller) for volunteer No. 17 (male) in case of input-output plant II (output: solid line denotes reference signal y_t^*, dotted—output signal $\hat{y}_t$ as response to the control signal and dashed—observed output y_t as response to the testing input; input: solid line denotes control signal x_t^* and dashed—testing input x_t)

Analysis of the predictor-based control results demonstrated sufficiently good control quality of excitement and frustration signals. Stabilized excitement signal level for input-output plant I is on average 65–75% higher compared to the average of observed response as reaction to the testing input, for input-output plant II—55–80%

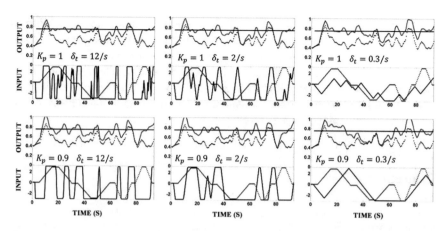

Fig. 11 Examples of frustration predictor-based control with constraints (upper—minimum variance controller, lower—generalized minimum variance controller) for volunteer No. 2 (female) in case of input-output plant I (output: solid line denotes reference signal y_t^*, dotted—output signal $\hat{y}_t$ as response to the control signal and dashed—observed output y_t as response to the testing input; input: solid line denotes control signal x_t^* and dashed—testing input x_t)

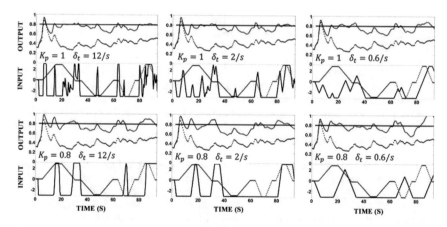

Fig. 12 Examples of frustration control with constraints (upper—minimum variance controller, lower—generalized minimum variance controller) for volunteer No. 20 (male, upper) in case of input-output plant II (output: solid line denotes reference signal y_t^*, dotted—output signal $\hat{y}_t$ as response to the control signal and dashed—observed output y_t as response to the testing input; input: solid line denotes control signal x_t^* and dashed—testing input x_t)

higher. Stabilized frustration signal level for input-output plant I is on average 55–60% higher compared to the average of observed response as reaction on the testing input, for input-output plant II—35–45% higher. Analysis results of control using efficiency measure (67) demonstrated that average absolute relative control error of excitement signals is on average 1.5–2.0 times higher compared to the average absolute relative control error of frustration signals.

Results of the excitement and frustration control demonstrated possibility to decrease variations of the control signal using a limited signal change speed (when constant δ_t of the admissible domain (48) decreases) or a generalized minimum variance control with lower than unit gain of the closed-loop transfer function (58) is used. Variations of the control signals at the same δ_t values are higher in the scheme with a minimum variance controller compared to variations in the scheme with a generalized minimum variance controller.

References

Astrom KJ (1970) Introduction to stochastic control theory. Academic Press, New York

Astrom KJ, Wittenmark B (1984) Computer controlled systems—theory and design (3rd edition 1997). Prentice Hall, Upper Saddle River

Clarke DW (1994) Advances in model predictive control. Oxford Science Publications, UK

Clarke DW, Mohtiedi C, Tuffs PS (1987) Generalized predictive control: parts I and II. Automatica 23:137–160

Hatamikia S, Maghooli K, Nasrabadi AM (2014) The emotion recognition system based on autoregressive model and sequential forward feature selection of electroencephalography signals. J Med Signals Sens 4(3):194–201

Hondrou C, Caridakis G (2012) Affective, natural interaction using EEG: sensors, application and future directions. In: Artificial intelligence: theories and applications, vol 7297. Springer, Berlin, pp 331–338

Isermann R (1981) Digital control systems. Springer, Berlin

Kaminskas V (1982, 1985) Dynamic system identification via discrete-time-observations: Part 1—Statistical method foundations. Estimation in linear systems (1982). Part 2—Estimation in nonlinear systems (1985). Vilnius, Mokslas (in Russian)

Kaminskas V (1988) Predictor-based self-tuning control systems. In: 33 Internationales Wissenschaftliches Kolloquium, Ilmenau, 24–28, 10.1988, Heft 1, Vortragsreiche A1, Technische Kybernetik/Automatisierungstechnik, Ilmenau, Germany, Technische Hochschule Ilmenau, pp 153–156

Kaminskas V (2007) Predictor-based self tuning control with constraints. In: Model and algorithms for global optimization, optimization and its applications, vol 4. Springer, Berlin, pp 333–341

Kaminskas V, Ščiglinskas E (2016) Minimum variance control of human emotion as reactions to a dynamic virtual 3D face. In: AIEEE 2016: Proceedings of the 4th workshop on advances in information, electronic and electrical engineering, Lithuania, Vilnius, pp 1–6

Kaminskas V, Ščiglinskas E (2018) Predictor-based control of human response to a dynamic 3D face using virtual reality. Informatica 29(2):251–264

Kaminskas V, Vidugirienė A (2016) A comparison of Hammerstein-type nonlinear models for identification of human response to virtual 3D face stimuli. Informatica 27(2):283–297

Kaminskas V, Tallat-Kelpša Č, Šidlauskas K (1987) Adaptive minimum variance control of extreme plants. In: Automation and remote control, vol 48, no 9. Consultants Bureu, New York, pp 1188–1195

Kaminskas V, Tallat-Kelpša Č, Šidlauskas K (1989) Self-tuning minimum variance control of nonlinear Wiener-Hammerstein type systems. In: Identification and parameter estimation: selected papers from the 8th IFAC/IFORS symposium, Beijing, PRC, 27–31 Aug 1988. Pergamon Press, Oxford, pp 384–389

Kaminskas V, Janickienė D, Vitkutė D (1990) Self-tuning control of a stochastic nonlinear object. Adaptive systems in control and signal processing: 5th IFAC symposium, Glasgow, 19–21 Apr 1989. Pergamon Press, Oxford, pp 171–175

Kaminskas V, Janickienė D, Vitkutė D (1991a) Self-tuning control of the nuclear reactor power. In: Automatic control in the service of mankind: proceedings of the 11th world congress of the IFAC, Tallinn, Estonia, 13–17 Aug 1990, vol 11. Pergamon Press, Oxford, pp 91–96

Kaminskas V, Šidlauskas K, Tallat-Kelpša Č (1991b) Constrained self-tuning control of stochastic extremal systems. Vilnius: Inst Math Inform, Informatica 2(1):33–51

Kaminskas V, Janickienė D, Vitkutė D (1992) Self-tuning constrained control of a power plant. Control of a power plants and power systems: selected papers from the IFAC symposium, Munich, Germany, 9–11 Mar. Pergamon Press, Oxford, pp 87–92

Kaminskas V, Janickienė D, Šidlauskas K, Vitkutė D (1993) Practical issues in the implementation of predictor-based self-tuning control systems. Vilnius: Inst Math Inform, Informatica 4(1–2):3–20

Kaminskas V, Vaškevičius E, Vidugirienė A (2014) Modeling human emotions as reactions to a dynamical virtual 3D face. Informatica 25(3):425–437

Kaminskas V, Ščiglinskas E, Vidugirienė A (2015) Predictor-based control of human emotions when reacting to a dynamic virtual 3D face stimulus. In: Proceedings of the 12th international conference on informatics in control, automation and robotics, France, Colmar, vol 1, pp 582–587

Khushaba RN, Wise Ch, Kodagoda S, Louviere J, Kahn BE, Townsend C (2013) Consumer neuroscience: assessing the brain response to marketing stimuli using electroencephalogram (EEG) and eye tracking. Expert Syst Appl 40(9):3803–3812

Lin YP, Wang CH, Jung TP (2010) EEG-based emotion recognition in music listening. IEEE Trans Biomed Eng 57(7):1798–1806

Ljung LV (1987) System identification: theory for the user (2nd edition 1999). Prentice-Hall Inc., Upper Saddle River

Mattioli F, Caetano D, Cardoso A, Lamounier E (2015) On the agile development of virtual reality systems. In: Proceedings of the international conference on software engineering research and practice (SERP), pp 10–16

Mikhail M, Allen J, Coan J (2013) Emotion detection using noisy EEG data. Auton Adapt Commun Syst 6(1):80–97

Peterka V (1984) Predictor-based self-tuning control. Automatica 19(5):471–486

Sari L, Nadhira V (2009) Development system for emotion detection based on brain signals and facial images. World Acad Sci Eng Technol Int J Psychol Behav Sci 3(2):13–19

Soderstrom T, Stoica P (1989) System identification. Prentice Hall, Int., London

Soeterboek ARM (1992) Predictive control: a unified approach. Prentice Hall International, London

Sourina O, Liu Y (2011) A fractal-based algorithm of emotion recognition from EEG using arousal valence model. In: Proceedings of biosignals, pp 209–214

Vaškevičius E, Vidugirienė A, Kaminskas V (2014a) Identification of human response to virtual 3D face stimuli. Inf Technol Control 43(1):47–56

Vaškevičius E, Vidugirienė A, Kaminskas V (2014b) Modelling excitement as a reaction to a virtual 3D face. In: Proceedings of the 11th international conference on informatics in control, automation and robotics, Viena, Austria, 1–3 Sept 2014, vol 1. Setubal, Portugal, SCITEPRESS, pp 734–740

Vidugirienė A, Vaškevičius E, Kaminskas V (2013) Modeling of affective state response to a virtual 3D face. In: Proceedings of the 17th European modelling symposium on computer modelling and simulation, 20–22 Nov 2013, Manchester, United Kingdom. Los Alamitos, CA; Washington: IEEE Press, pp 167–172

Knowledge-Based Transformation Algorithms of UML Dynamic Models Generation from Enterprise Model

Ilona Veitaite and Audrius Lopata

Abstract In today's organizations nowadays exists big gap between business and information technologies. Information technology strategy planning is multistage process and information system development relies on implementation of each stage of IS lifecycle. A knowledge-based IS engineering proposed system modelling and decision-making tools and methods, which helps to expand more precise and comprehensive subject area corresponding to the project. Participants of IS project such as developer or programmer is allowed to use not only the knowledge of the project, which is collected in traditional CASE tool storage, but also the knowledge storage, where subject area knowledge is collected according to formal criteria. There have been made many efforts for the analysis of Unified Modelling Language (UML) models generation of diverse knowledge-based models combining frameworks, workflow patterns, modelling languages and natural language specifications. Knowledge-based subsystem as CASE tool component with Enterprise Meta-Model (EMM) and Enterprise Model (EM) within can significantly help with UML dynamic models generation using transformation algorithms. Application of the knowledge-based transformation algorithms grants the possibility to operate additional models validation methods that EMM determines. The main purpose of the paper is to present how EM can be used in UML models generation process. This paper combines results from previous researches and summarizes part of them. There is described importance of knowledge-based IS engineering in business and IT alignment process. There is also described UML models elements roles variations after generation from EM. The most valuable result of this research is presentation of UML Sequence model generation from EM process, by defining transformation algorithm and illustrating it by a particular example what proves EM sufficiency for whole generation process.

I. Veitaite (✉)
Kaunas Faculty, Institute of Social Sciences and Applied Informatics, Vilnius University, Kaunas, Lithuania
e-mail: Ilona.Veitaite@knf.vu.lt

A. Lopata
Faculty of Informatics, Kaunas University of Technology, Studentų g. 50, 51368 Kaunas, Lithuania
e-mail: Audrius.Lopata@ktu.lt

© Springer Nature Switzerland AG 2020
G. Dzemyda et al. (eds.), *Data Science: New Issues, Challenges and Applications*, Studies in Computational Intelligence 869,
https://doi.org/10.1007/978-3-030-39250-5_3

43

Keywords Unified modelling language · Enterprise model · Transformation algorithm · Knowledge-based IS engineering

1 Introduction

Business and IT alignment has significant meaning in many organizations for quite a long time and it still remains as an essential goal. There are many controversies about the term "business-IT alignment". Some would argue that alignment is not necessary and weak goal that can be achieved only by a small number of organizations and some would agree that business and IT alignment ensures suitable function of entire organization. Professionals suggest that organizations need to achieve strategic business and IT alignment to have competitive advantage. Strategic business and IT alignment affect business performance and IT effectiveness (Chen 2008; Chen et al. 2008).

In today's organizations combination of main business purpose and IS development purposes are not related directly. To insure business and IT alignment it is important to create communication between these two parts.

In order to help development of interaction and understanding, many organizations are identifying business-facing roles that have main responsibility for establishing and keeping relationships between IT and the business fields. The balance of IT and business vision, the ability to clarify business and IT challenges with same clarity should be insured. It is often occurrence that understanding how to adopt new technologies to influence business is clear just for one part and talking about this kind of IT application does not always serve well (Henderson and Venkatraman 1999; Peak et al. 2011).

Information systems (IS) become more complex and textual information with primary schemes, diagrams or models is not enough to characterize all business and IT processes. Certain models are applied for computer-based requirements specification for information system modelling. Enterprise modelling has become an integral part of IS development process. Particularly in large and complicated IS, which have to cooperate with other systems and design. By this process, user requirements, knowledge of the business domain, software architecture and other fundamental components are modelled (Jenney 2010; Kerzazi et al. 2013). Recently, the organization's business modelling has become an important stage of modelling design processes (Sommerville 2011). As follows, specialized computer modelling tools have become an integral part of the complex IS development and design processes (Chen et al. 2008).

Traditionally IS engineering stages from modelling to code generation are implemented empirically. Computer-based model is formed by empirical experience of the analyst, while traditional CASE system design models are formed on the basis of the information and data collected on business domain, which is being computerized. In that case, the main role is given to the analyst and improved knowledge of the business domain is not fully controlled by formalized criteria (Sajja and Akerkar 2010).

One of the many disadvantages of computer-based IS engineering methods' is the fact that IS design models are generated only partially, because in the design stage, the designer creates design models based on personal experience, rather than adapting the principles of knowledge generation which are stored in enterprise model. Solution to these type of problems is integration of EM into IS engineering process, as the core knowledge repository (Gudas 2012).

At present, computer-based IS engineering to avoid the empiric, is developing based on new knowledge-based methods. Computer-based IS in knowledge-based IS engineering, is developed using stored enterprise knowledge base of the business domain, i.e., enterprise model, the composition which is defined by formal criteria. This model is based on theory and matches the specific features of all IS used in enterprise. Computer-based IS software development based on that model is known as knowledge-based IS engineering (Gudas 2012).

There are plenty advanced IS development approaches, standards, techniques and business modelling methodologies, which are continually updated (Gudas 2012). UML is one of the most common software specification—standards. It is a universal IS modelling language applied to a number of methodologists and used in the most popular modelling tools. UML is used to describe, visualize and document software, business logic and other systems. Influence of design tools in IS development process is rapidly growing. One of the major results is the generated source code (Skersys 2008). The importance of UML in software development has become more significant since the appearance of model-driven architecture (Butleris et al. 2015; Dunkel and Bruns 2007; Perjons 2011). There are tools for automatic source generating code from the UML models. The main components of IS life cycle design stage models, such as UML, can be generated in semi-automatic mode from knowledge repository—Enterprise model. The method of UML models generation from EM implements a knowledge-based design stage in the IS development cycle. UML dynamic models can be generated through transformation algorithms, when the proper knowledge is collected into knowledge repository, where it is already verified to insure automatically generated design models quality (Eichelberger et al. 2011; Lopata et al. 2016; OMG UML 2019).

Firstly, there will be presented how knowledge-based IS engineering makes positive impact in business and IT alignment process and the significance of EMM and EM usage in this process. Secondly, there will be defined UML models role importance in knowledge-based IS engineering during IS development lifecycle stages. Finally, as the result, transformation algorithm will be explained by steps and illustrated by example of UML Sequence model generation from Enterprise model.

2   Business and IT Alignment Influence to Knowledge-Based IS Engineering

J. C. Henderson and N. Venkatraman defined the interrelationship between business and IT strategies. Strategic Alignment Model (SAM) was created in order to describe these relations. Model consists of four domains: Business Strategy Domain, Business Infrastructure Domain, IT Strategy Domain, IT Infrastructure Domain and relationships between them. Model is based according two parts: strategic integration and functional integration (Henderson and Venkatraman 1999; Plazaola et al. 2008). Strategic integration recognizes that the IT strategy should be expressed in terms of an external domain (how the organization is positioned in the IT environment) and an internal domain (how the IT infrastructure should be managed and configured). There are two types of functional integration: strategic and operational. First relation is between the business strategy (business scope approach) and IT strategy (IT implementation approach). Second relation covers the internal domain and deals with the relation between business infrastructure and IT infrastructure. Functional integration considers how alternative choices made in the IT domain effect those, which made in the business domain and by contraries. Strategic alignment model defines four basic alignment perspectives: Strategy Execution, Technology Transformation, Competitive Potential, and Service Level (Plazaola et al. 2008).

First introduced version of business and IT strategy alignment model was and is conceptual. It does not provide a practical framework to implement this type of alignment, in spite of that, there are alignment variations developed and used in organizations to achieve the business and IT unification, but these variations mostly are diverted for business, not for IT part (Lopata et al. 2014).

There have been done many attempts and researches to simplify and lighten business and IT alignment process using different knowledge-based methods (Perjons 2011; Plazaola et al. 2008). There were many efforts trying to combine different frameworks, modelling languages even natural language specifications for the generation of suitable for IS development process (Sajja and Akerkar 2010; Sommerville 2011). Model driven architecture is related to multiple standards, it develops, compares, aligns, verifies different models and meta-models. It is very important to choose and use particular standard, which is complete and inconsistent. Opportunity to use knowledge-based source for chosen standard's models generation simplifies all process (Dunkel and Bruns 2007; Gailly et al. 2013; Kerzazi et al. 2013).

Knowledge-based CASE Tool subsystem in business and IT model application can be used as the fundamental source for certain frameworks (Fig. 1). These frameworks indicate all important structures inside the organization, including business processes, technological applications, data and relationships between them to perform business. For business and IT alignment process specific data is necessary and it is applied in Knowledge-Based Subsystem. Business strategy domain provides business purposes to knowledge base. Business purposes are defined by the business management and are received from business environment and IT purposes are defined by the IT managements. Business infrastructure domain delivers business

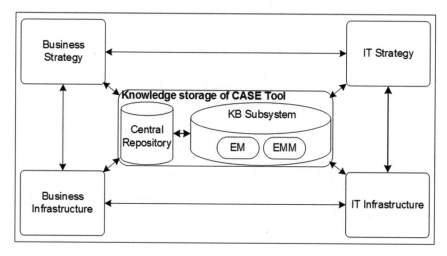

Fig. 1 Business and IT alignment model and Knowledge storage of CASE tool (Lopata et al. 2014)

rules, constraints, processes, functions and other related data and IT infrastructure domain delivers information about IT infrastructure, where describes current software, hardware. All this information is stored in knowledge base and can be used within enterprise model, where is validated regarding to the EMM (Lopata et al. 2011).

Computer-based IS engineering particular methods are developed based on general requirements, which systematizes the chosen methodology. There are some necessary components for knowledge-based IS engineering methodology (Gudas 2009a, b, 2012).

Core of enterprise modelling theoretical basis is the theoretical enterprise model, of which main goal is to identify the necessary and sufficient business components for IS engineering. This model implements enterprise's business management. Theoretical EMis a formalized enterprise management model that establishes business components and its interactions, enterprise management and its interactions (Gudas 2009a, b).

Theoretical enterprise knowledge model is EMM. Based on the IS development life cycle stages the theoretical model components and their interactions are described as EMM. EMM is a structural model, which specifies the necessary and sufficient components of IS engineering enterprise management features and interactions (Gudas 2009a, b; Lopata et al. 2014).

The theoretical basis of knowledge-based IS engineering process is IS engineering process methodology that bases knowledge-based IS development process by using an EM as an additional knowledge source besides the analyst and the user (Gudas 2009a, b; Lopata et al. 2014).

Computer-based IS engineering systems development is based on knowledge-based IS engineering tools. Practical knowledge-based modelling methods are

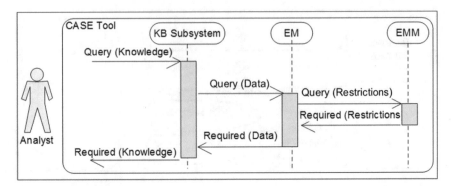

Fig. 2 Representation of interaction between CASE tool's knowledge-based subsystem, EM and EMM (Lopata et al. 2014)

intended for the development of functionality of traditional CASE systems by creating a knowledge-based—CASE intelligent system (Gudas 2009a, b; Lopata et al. 2014).

Computer-based knowledge-based IS engineering project management base is CASE system knowledge-based subsystem. CASE system's knowledge-based subsystem's core component is knowledge base, which essential elements are EMM specification and EM for certain problem domain. Knowledge-based subsystem is interactive member in IS engineering processes (Gudas 2009a, b; Lopata et al. 2014).

Knowledge-based CASE systems containing fundamental components, which arrange knowledge. Knowledge-based subsystem's knowledge base and its essential elements are EMM specification and EM for certain business domain (Fig. 2). Figure 3 presents knowledge-based subsystem connection to the EM and EMM inside CASE tool presented as Sequence diagram (Lopata et al. 2016).

EMM is formally defined EM structure, which consists of a formalized EM in line with the general principles of control theory. EM is the main source of the necessary knowledge of the particular business domain for IS engineering and IS re-engineering processes (Fig. 3) (Gailly et al. 2013; Gudas 2012).

EMM manages EM structure. EM stores knowledge that is necessary for IS development process only and will be used during all stages of IS development life cycle (Fig. 4). EM class model has twenty-three classes. Essential classes are Process, Function and Actor. Class Process, Function, Actor and Objective can have an internal hierarchical structure. These relationships is presented as aggregation relationship. Class Process is linked with the class MaterialFlow as aggregation relationship. Class MaterialFlow is linked with the classes MaterialInputFlow and MaterialOutputFlow as generalization relationship. Class Process is linked with Classes Function, Actor and Event as association relationship. Class Function is linked with classes InformationFlow, InformationActivity, Interpretation, InformationProcessing and Realization as aggregation relationship. These relationships define the internal composition of the Class Function. Class InformationFlow is linked with ProcessOutputAtributes, ProcessInputAtributes, IPInputAttributes and IPOutputAttributs as generalization

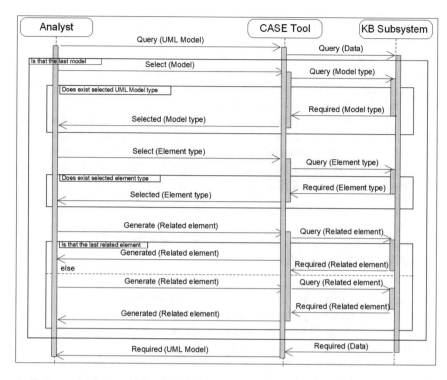

Fig. 3 Sequence diagram of Knowledge-based subsystem connection to the EM and EMM inside CASE tool (Lopata et al. 2016)

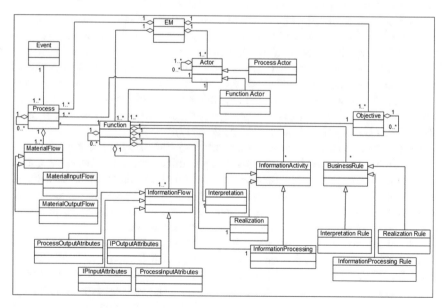

Fig. 4 EMM class diagram (Gudas 2012; Lopata et al. 2014)

relationship. Class InformationActivity is linked with Interpretation, Information-Processing and Realization as generalization relationship. Class Function linked with classes Actor, Objective and BusinessRule as association relationship. Class BusinessRule is linked with Interpretation Rule, Realization Rule, InformationProcessing Rule as generalization relationship. Class Actor is linked with Function Actor and Process Actor as generalization relationship.

EM and EMM are the main components of knowledge-based subsystem of CASE Tool. It would be appropriate to collect necessary knowledge to the knowledge-based subsystem of the particular CASE Tool. This subsystem is the main source of knowledge necessary for design stage models (including UML models) and source code generation process (Gudas 2012; Lopata et al. 2014).

3   Importance of UML Elements Role Variations After Generating from EM

UML model is a partial graphical view of a model of a system under design, implementation, or already in presence. UML model contains graphical elements—UML nodes connected with flows—that represent elements in the UML model of the designed system (Veitaitė and Lopata 2015).

UML specification does not prevent mixing of different kinds of models, for example to unite structural and behavioral elements to express particular case. Therefore, the boundaries between the various kinds of models are not strictly enforced (Jacobson et al. 2005; OMG UML 2019; Diagrams 2012).

UML specification defines two kinds of UML models: structure models and behavior or dynamic models, where Structure models show the static structure of the system and its parts on different abstraction and implementation levels and how they are related to each other. The elements in a structure model represent the meaningful concepts of a system, and may include abstract, real world and implementation concepts; and Behavior models show the dynamic behavior of the objects in a system, which can be described as a series of changes to the system over time (Jacobson et al. 2005; OMG UML 2019; Veitaitė and Lopata 2017).

UML dynamic models can be generated from EM using transformation algorithms. Firstly, certain UML model must be identified for generation process, after this identification, the initial—main element of this UML model must be selected from Enterprise model. Secondly, all related elements must be selected according the initial element and all these related components must be linked between regarding constraints, necessary for UML model type identified earlier.

Information systems design methods specify the arrangement of systems engineering actions, i.e. how, in what order and what UML model to use in the IS development process and how to implement the process (Table 1). Many of them are based on diverse types of models describing differing aspects of the system properties. Sense of each model can be defined individually, but more important is the

Table 1 EM process, function, actor and business rules elements roles variations as different UML dynamic models elements (Eichelberger et al. 2011; Jacobson et al. 2005; OMG UML 2019)

EM	UML model element	UML model
Process/function	Use case	Use case model
	Activity	Activity model
	Behavioural state machine	State machine model
	Protocol state machine	Protocol state machine model
	Message	Sequence model
	Frame	Communication model
	Frame	Interaction overview model
Actor	Actor	Use case model
	Subject	Use case model
	Partition	Activity model
	Lifeline	Sequence model
	Lifeline	Communication model
	Lifeline	Timing model
Business Rules	Extend	Use case model
	Include	Use case model
	Association	Use case model
	Control nodes	Activity model
	Pseudostate	State machine model
	Protocol transition	Protocol state machine model
	Execution specification	Sequence model
	Combined fragment	Sequence model
	Interaction use	Sequence model
	State invariant	Sequence model
	Destruction occurrence	Sequence model
	Duration constraint	Timing model
	Time constraint	Timing model
	Destruction occurrence	Timing model
	Duration constraint	Interaction overview model
	Time constraint	Interaction overview model
	Interaction use	Interaction overview model
	Control nodes	Interaction overview model

fact that each model is the projection of the system. An inexperienced specialist can use UML models inappropriately and the description of the system will possibly be contradictory, insufficient and contentious (Veitaitė and Lopata 2017).

Identifying certain UML model and selecting the initial model element is rather significant, because further generating process depends on it. A lot UML model elements repeats in different UML model, but these elements define various aspects of the system. In example EM elements, such as Process, Function, Actor and Business Rules can be generated into different UML dynamic models elements depending on which UML model is used.

4   UML Models Transformation Algorithms from EM

All UML models: static and dynamic can be generated from EM using transformation algorithms. Certain UML model must be selected for generation process, after this selection, the initial—starting element of this certain UML model must be identified from Enterprise model. Therefore, all related elements must be identified according the initial element and all these related elements must be linked regarding constraints i.e. business rules, obligatory for certain UML model type which was selected in the beginning (Veitaitė and Lopata 2018).

This kind of system definition is rather tangled, because most of the information from the EM overlays in the UML dynamic models and expresses the same matters just in different approaches, as it is explained in the next chapter. Therefore identification of certain UML model for generation process has high significance, because of regarding this selection relies generated element value for system development process (Butleris et al. 2015; OMG UML 2019).

Transformation algorithm of UML model generation from EM process is presented in Fig. 5 and is illustrated by following steps (Veitaitė and Lopata 2018):

- Step 1: Particular UML model for generation from EM process is identified and selected.
- Step 2: If the particular UML model for generation from EM process is selected then algorithm process is continued, else the particular UML model for generation from EM process must be selected.
- Step 3: First element from EM is selected for UML model, identified previously, generation process.
- Step 4: If the selected EM element is initial UML model element, then initial element is generated, else the other EM element must be selected (the selected element must be initial element).
- Step 5: The element related to the initial element is selected from Enterprise model.
- Step 6: The element related to the initial element is generated as UML model element.
- Step 7: The element related to the previous element is selected from Enterprise model.

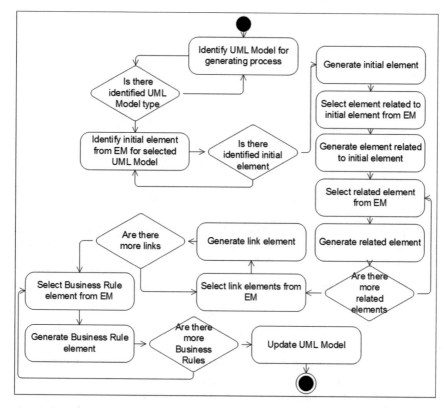

Fig. 5 The top level transformation algorithm of UML models generation from EM process (Veitaitė and Lopata 2018)

- Step 8: The element related to the previous element is generated as UML model element.
- Step 9: If there are more related elements, then they are selected from EM and generated as UML model elements one by one, else the link element is selected from Enterprise model.
- Step 10: The link element is generated as UML model element.
- Step 11: If there are more links, then they are selected from EM and generated as UML model elements one by one, else the Business Rule element is selected from Enterprise model.
- Step 12: The Business Rule element is generated as UML model element.
- Step 13: If there are more Business Rules, then they are selected from EM and generated as UML model elements one by one, else the generated UML model is updated with all elements, links and constraints.
- Step 14: Generation process is finished.

UML Sequence model is most common kind of interaction models which focuses on the message interchange between objects (lifelines). Sequence model shows how

Table 2 UML sequence model elements (Jacobson et al. 2005; OMG UML 2019; Diagrams 2012; Veitaitė and Lopata 2017)

EM element	UML sequence model element
Actor	Lifeline
Process, function	Message
Business rules	Execution specification
	Combined fragment
	Interaction use
	State invariant
	Destruction occurrence

the objects interact with others in a particular scenario of a use case (Plazaola et al. 2008) (Table 2).

UML Sequence model is one of the interaction models group which concentrates on the message interchange between system participants called lifelines (Jacobson et al. 2005; OMG UML 2019; Diagrams 2012).

- Step 1: The initial element Actor from EM for UML Sequence model generation is selected.
- Step 2: Actor element is generated as Lifeline element.
- Step 3: Process element from Enterprise model, which is related with the initial Lifeline element is selected.
- Step 4: If Process element is Message element related to Lifeline, then Message element is generated, else Function element is selected.
- Step 5: Function element is generated as Message element.
- Step 6: There is checking if there are more Process elements in EM related to Lifeline.
- Step 7: Business Rule element as link of Lifeline element from EM which is related with the Message element is selected.
- Step 8: If Business Rule element is UML Sequence model's Execution specification element and serves as link between Lifeline and Message elements then Execution specification element is generated from Enterprise model, else if it is Combined fragment element, then Combined fragment element is generated from Enterprise model, else if it is Interaction Use element, then Interaction use element rom EM is generated, else if it is State invariant element, then State invariant element is generated from Enterprise model, else Destruction occurrence element is generated from Enterprise model.
- Step 9: There is checking if there are more Business Rules in EM related to UML Sequence model. In case, there are, algorithm goes back to step 7.
- Step 10: Lifeline element is updated.
- Step 11: There is checking if there are more Actors in EM related to UML Sequence model. In case, there are, algorithm goes back to step 1.

In UML Sequence model generation from EM (Fig. 6) initial element is lifeline, after generation of this element, follows selection of EM element: process or function and message element is generated. Afterwards the generation of these two types

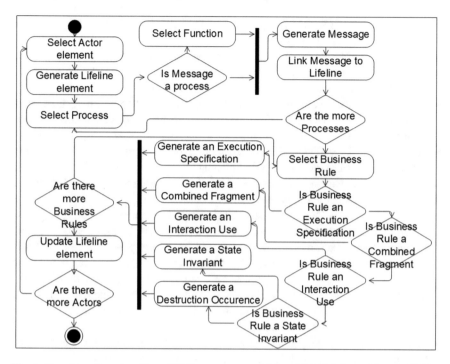

Fig. 6 Transformation algorithm of UML sequence model from EM (Veitaitė and Lopata 2017)

of elements, they have to be linked to each other. In sequence models execution specification, combined fragment, interaction use, state invariant and destruction occurrence are generated from EM business rule elements. Later all these elements are generated, there is update of lifeline element and check are there more actor elements left in enterprise model.

Generation of UML Sequence model is illustrated with the example of Enrolling in a workshop of the conference. Information of this example is stored in Enterprise model. Example shows, how participant enrolls in a Workshop of the Conference, there is checked participants' eligibility and status of enrollment is returned.

Detailed stages of enrolling in workshop of the conference example processes stored in EM are described:

- Stage 1: Participant wants to enroll in the Workshop of the Conference.
- Stage 2: The message of enrollment is sent to check participants' eligibility.
- Stage 3: The workshop history in the Conference is requested.
- Stage 4: The workshop history in the Conference is checked.
- Stage 5: Return message of eligibility status is given.
- Stage 6: Return message of enrollment status is given.

Transformation algorithm of part of UML Sequence model generation Enrolling in a workshop of the conference example from EM process is illustrated by following steps:

- Step 1: The initial element Actor from EM for UML Sequence model generation is selected—aParticipant:Participant.
- Step 2: Actor element is generated as Lifeline element—aParticipant:Participant.
- Step 3: Process element from Enterprise model, which is related with the initial Lifeline element is selected—enrollParticipant(aParticipant).
- Step 4: If Process element is Message element related to Lifeline, then Message element is generated, else Function element is selected—enrollParticipant(aParticipant).
- Step 6: There is checking if there are more Process elements in EM related to Lifeline—isParticipantEligible(aParticipant).
- Step 7: Business Rule element as link of Lifeline element from EM which is related with the Message element is selected.
- Step 8: If Business Rule element is UML Sequence model's Execution specification element and serves as link between Lifeline and Message elements then Execution specification element is generated from Enterprise model, else if it is Combined fragment element, then Combined fragment element is generated from Enterprise model, else if it is Interaction Use element, then Interaction use element rom EM is generated, else if it is State invariant element, then State invariant element is generated from Enterprise model, else Destruction occurrence element is generated from Enterprise model.
- Step 9: There is checking if there are more Business Rules in EM related to UML Sequence model. In case, there are, algorithm goes back to step 7.
- Step 10: Lifeline element is updated.
- Step 11: There is checking if there are more Actors in EM related to UML Sequence model. In case, there are, algorithm goes back to step 1—:Workshop.

After 11 steps of the transformation algorithm generating of Enrolling participant inn workshop of the conference data from EM the 1 stage—Participant wants to enroll in the Workshop of the Conference is illustrated in Fig. 7.

All information of Enrolling in a workshop of the conference example is stored in EM. Each element of the example can be generated through transformation algorithm from EM in UML Sequence model element.

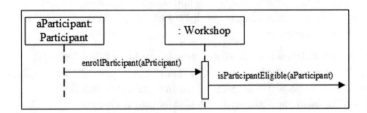

Fig. 7 Stage 1 of UML sequence model example of participant enrolment in workshop of the conference

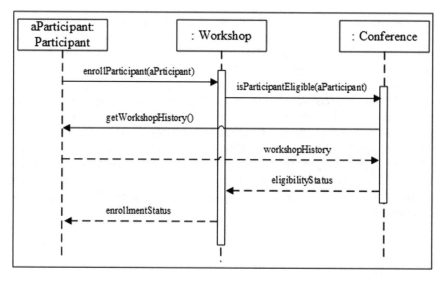

Fig. 8 UML sequence model example of participant enrolment in workshop of the conference

By using transformation algorithm all lifelines: aParticipant: Participant, Workshop, Conference; messages: enrollParticipant(aParticipant), isParticipantEligible(aParticipant), getWorkshopHistory(), eligibilityStatus, enrollmentStatus; businessrules are generated from EM (Fig. 8).

After the implementation of all the steps of UML Sequence model transformation algorithm it can be clearly declared that presented example entirely illustrates accuracy of the UML Sequence model elements generated from Enterprise model.

5 Conclusions

The first part of the article handles with defining the relation between business and IT alignment model usage and importance and how knowledge-based IS engineering can improve the alignment goal. Also there is defined how EMM and EM makes positive impact when these models are integrated into Knowledge-based CASE tool.

The second part handles with UML models elements role variations in knowledge-based IS engineering within all IS development lifecycle stages. This part also handles with explaining how business domain elements stored in EM can be understood and in what different elements of different UML dynamic models they can be generated depending on which model is selected for generation process.

In the third part there is detailed explanation of top level UML model transformation algorithm, which is depicted by steps according previous researches. Also there is presented and depicted by steps UML Sequence model transformation algorithm, where in details each step is explained. This transformation algorithm is illustrated by particular example about Enrolling participant inn workshop of the conference. All data related with this example is already stored in knowledge-based EM. This information is used in UML Sequence model generation process by using transformation algorithm. First stage of generation is described by steps, and further process of full model generation is explained briefly.

In previous researches there were proved that part of UML dynamic models can be generated from EM. Main result of this research together with previous researches proves that UML Sequence model also can be generated from EM. Presented example of Participant enrollment in workshop of the conference shows that data stored in EM is sufficient for generating process and it is possible to declare, that each element of UML models can be generated from the EM. The usage of transformation algorithms in generation process can implement design stage of knowledge-based IS development cycle.

References

Butleris R, Lopata A, Ambraziunas M, Veitaitė I, Masteika S (2015) SysML and UML models usage in knowledge based MDA process. Elektronika ir elektrotechnika 21(2):50–57 (2015). Print ISSN: 1392-1215, Online ISSN: 2029-5731

Chen C-K (2008) Construct model of the knowledge-based economy indicators. Transform Bus Econ 7(2(14))

Chen R, Sun Ch, Helms M, Jihd W (2008) Aligning information technology and business strategy with a dynamic capabilities perspective: a longitudinal study of a Taiwanese Semiconductor Company. Int J Inf Manage 28(2008):366–378

Dunkel J, Bruns R (2007) Model-driven architecture for mobile applications. In: Proceedings of the 10th international conference on business information systems (BIS), vol 4439/2007, pp 464–477

Eichelberger H, Eldogan Y, Schmid KA (2011) Comprehensive analysis of UML tools, their capabilities and compliance. Software Systems Engineering, Universität Hildesheim. Version 2.0

Gailly F, Casteleyn S, Alkhaldi N (2013) On the symbiosis between enterprise modelling and ontology engineering. Ghent University, Universitat Jaume I, Vrije Universiteit Brussel. https://doi.org/10.1007/978-3-642-41924-9_42

Gudas S (2009a) Enterprise knowledge modelling: domains and aspects. Technol Econ Dev Econ Balt J Sustain 281–293

Gudas S (2009b) Architecture of knowledge-based enterprise management systems: a control view. In: Proceedings of the 13th world multiconference on systematics, cybernetics and informatics (WMSCI2009), 10–13 July, Orlando, Florida, USA, vol III, pp 161–266. ISBN 10:1-9934272-61-2 (volume III). ISBN 13:978-1-9934272-61-9

Gudas S (2012) Informacijos sistemų inžinerijos teorijos pagrindai/Fundamentals of Information Systems Engineering Theory. (Lithuanian) Vilnius University. ISBN 978-609-459-075-7

Henderson J, Venkatraman N (1999) Strategic alignment: leveraging information technology for transforming organizations. IBM Syst J 38(2, 3):472–484

Jacobson I, Rumbaugh J, Booch G (2005) Unified modeling language user guide, 3rd edn. Addison-Wesley Professional. ISBN 0321267974

Jenney J (2010) Modern methods of systems engineering: with an introduction to pattern and model based methods. ISBN 13:978-1463777357

Kerzazi N, Lavallée M, Robillard PN (2013) A knowledge-based perspective for software process modeling. e-Inform Softw Eng J 7

Lopata A, Ambraziunas M, Gudas S (2011) Knowledge-based approach to business and IT alignment modelling. Transform Bus Econ 10(2(23)):60–73

Lopata A, Veitaitė I, Gudas S, Butleris R (2014) Case tool component—knowledge-based subsystem UML diagrams generation process. Transform Bus Econ 13(2B(32B)):676–696. Vilnius University, Brno University of Technology, University of Latvia. Brno, Kaunas, Riga, Vilnius, Vilniaus universitetas. ISSN 1648-4460

Lopata A, Veitaitė I, Žemaitytė N (2016) Enterprise model based UML interaction overview model generation process. In: Abramowicz W, Alt R, Bogdan F (eds) Business information systems workshops: BIS 2016 international workshops, Leipzig, Germany, 6–8 July 2016: revised papers. Lecture notes in business information processing, vol 263. Springer International Publishing, Berlin. ISSN 1865-1348

OMG UML (2019) Unified modeling language version 2.5.1. Unified modelling. https://www.omg.org/spec/UML/About-UML/

Peak D, Guynes C, Prybutok V, Xu C (2011) Aligning information technology with business strategy: an action research approach. JITCAR 13(1)

Perjons E (2011) Model-driven process design. Aligning value networks, enterprise goals, services and IT systems. Department of Computer and Systems Sciences, Stockholm University. Sweden by US-AB, Stockholm ISBN 978-91-7447-249-3

Plazaola L, Flores J, Vargas N, Ekstedt M (2008) Strategic business and IT alignment assessment: a case study applying an enterprise architecture-based metamodel. In: Proceedings of the 41st Hawaii international conference on system sciences. 1530-1605/08

Sajja PS, Akerkar R (2010) Knowledge-based systems for development. Adv Knowl Based Syst Model Appl Res 1

Skersys T (2008) Business knowledge-based generation of the system class model Kaunas: Information Systems Department, Kaunas University of Technology. http://itc.ktu.lt/itc372/Skersys372.pdf

Sommerville I (2011) Software engineering, 9th edn. Pearson Education, Inc., Publishing as Addison-Wesley, Boston. ISBN 13:978-0-13-703515-1

UML Diagrams (2012) UML diagrams characteristic. www.uml-diagrams.org

Veitaitė I, Lopata A (2015) Additional knowledge based MOF architecture layer for UML models generation process. In: Abramowicz W, Kokkinaki A (eds) Business information systems: 2015 international workshops: revised papers: proceedings. Lecture notes in business information processing, vol 226. Springer International Publishing, Berlin. ISSN 1865-1348

Veitaitė I, Lopata A (2017) Transformation algorithms of knowledge based UML dynamic models generation. In: Abramowicz W (ed) Business information systems workshops BIS 2017, Poznan, Poland, 28–30 June. Lecture notes in business information processing, vol 303. Springer International Publishing, Cham

Veitaitė I, Lopata A (2018) Problem domain knowledge driven generation of UML models. In: Damaševičius R, Vasiljevienė G (eds) Information and software technologies: 24th international conference, ICIST 2018, Vilnius, Lithuania, 4–6 Oct 2018. Springer, Cham

An Approach for Networking of Wireless Sensors and Embedded Systems Applied for Monitoring of Environment Data

Dalė Dzemydienė and Vytautas Radzevičius

Abstract The aim of this research is to provide an approach for extension of the capacities of wireless sensor network (WSN) by enabling possibilities of monitoring of sea environment parameters. This domain area is related with control of recognition process of the hydro meteorological situations by providing the proper structure of the decision support system (DSS), which can work under specific real-time conditions. The construction issues of such multilayered buoy's infrastructure based on WSN causes some problems: mechanisms of such interconnected devices have to work by properly presented goals and in extreme weather conditions; the modern embedded systems requires the intellectualization possibilities with recognition of adequate evaluation of risky situations; the functionality of WSN has to be based on the technology of the Internet of Things (IoT) with quite limited capacities. Our main findings concern the proposed architectural solution with extended capacities of integrated embedded systems and other gathering sources, by enabling the assessing of water pollution situations. The originality of the proposed approach concerns the issues of network allocation, system resource allocation and the architectural solutions, enabling wireless channels to provide big streams of data. The experimental results show the possibilities of the proposed prototype of multilayer system to work under real conditions of sea with interconnected network of sensors and controllers, based on the principles of other standardized network layers. The prototype method is used for evaluation of functionality of the embedded systems and WSN. The provided experimental results show functional effectiveness of collection and gathering of data under restricted and limited capacities of the embedded systems for operative control needs.

D. Dzemydienė (✉) · V. Radzevičius
Institute of Data Science and Digital Technologies, Vilnius University,
Akademijos str. 4, Vilnius, Lithuania
e-mail: dale.dzemydiene@mif.vu.lt

V. Radzevičius
e-mail: vytautas.radzevicius@mif.vu.lt

© Springer Nature Switzerland AG 2020
G. Dzemyda et al. (eds.), *Data Science: New Issues, Challenges and Applications*, Studies in Computational Intelligence 869,
https://doi.org/10.1007/978-3-030-39250-5_4

Keywords Wireless sensor network (WSN) · Decision support system (DSS) · Internet of things (IoT) · Embedded systems · Monitoring data · Sea water environment data

1   Introduction

Our aim is to ensure WSN integration with embedded systems for monitoring of environment parameters and recognition of situations in sea water. The high concern of environment pollution problems was one of the cornerstones of the Action plans that were set out in the Agenda for Sustainable Development until 2030 (Transforming our World 2015). Frequent outbreaks of contamination, arising algae blooms or other anomalies—each of these factors may be of sufficient severity to cause considerable damage to both: the environment and the quality of the human life.

The large spectrum of required different parameters have to be included in the recognition of multi-component hydro-meteorological situations. For instance, sea wave height, water temperature, wind speed and direction, salinity, turbidity, underwater currents, and similar parameters are participating in real-time decision support processes. Some of these monitoring needs influence the specific constructions of whole system based on the specific WSN adapted for these needs.

The device control systems have to work under conditions of restrict capacities of the embedded components, which monitor environmental parameters and are connected by the wireless network layers with the interoperable structure of data warehouses. Such WSN systems are becoming more and more capable to control other devices based on their real-time values (Belikova et al. 2019; Hart and Martinez 2006; Wang et al. 2018). The proposed in literature operational systems of such type are usually managed by the principles of monitoring of environment parameters, but the identification of situations and control of actions are not solved yet under the conditions of very quick reactions.

Our aim is forwarded for construction of the integrated system of buoys based on the WSN with possibility to work in marine water of the Baltic Sea region. There are some scientific results of our group, which enable us in achievement of the buoy construction for meteorological measurements (Gricius et al. 2014, 2015; Dzemydiene et al. 2016). Described earlier results provide some solutions of one buoy system construction for the monitoring of hydro meteorological data and ship observation. The solutions of construction of the localized multifunction sensor networks, large scale single function networks, biosensor networks are presented in Hart and Martinez (2006), Hoagland and Scatasta (2006), and other works.

In order to ensure even more reliable data collection, it is necessary to develop a distributed information system, with possibilities of predicting of risky situations and supporting of operative decision-making processes. We are trying to propose the system architecture with some of these functions. All of these functions could ensure greater and faster solutions of environment protection and situation recognition of pollution processes.

Mostly, the sources of different information gathering are not concentrated in one area of surroundings, and it is very difficult to collect needful data from different sources, analyze them together as a group. Most of them do not correlate with each other strongly, and during some time intervals only few parameters are provided for recognition of indications. The prediction process requires more widening spectrum of other parameters that values can indicate stronger correlations between groups of environmental and water parameters. For example, the air temperature is assumed to have a slight correlation with the eutrophication process, however the oxygen content is much stronger, and small wave height can increase algal bloom. Therefore, for evaluation of the probability of algal blooms, we need for some special parameters, that have different ties to algal bloom and the correlation is not linear. Thus, we have very different parameters' values, and different correlation strengths when they increase or decrease in a specific range. Furthermore, some of the data from the main meteorological stations is rarely updated and the data is inaccurate, thus, there is an unavoidable need for more accuracy and reliability of concentrated and easily accessible environment information, therefore the needs for WSN increases as well.

The results of this study provide the possibility to construct of WSN under specific environment conditions and adapt this framework for situation recognition in marine water environment of the Baltic Sea. The proposed architecture of constructions helps in the integration of small computing power embedded systems with wireless networking for gathering data in all infrastructure of the monitoring and decision-support system.

By describing the structure of provided material, we shortly analyze the content of sections. The related works are analyzed in Sect. 2, by emphasizing on other research works related with constructions of WSN and development of the DSS in the area of environment protection. Section 3 describes the principles and main components of the proposed architecture of the buoys based of WSNs for monitoring of sea water parameters under restricted capacities. The possibilities of application of other data gathering methods and the means for more clear recognition of situations of water pollution are provided in Sect. 4. Sectin 5 is devoted to working principles of proposed buoys system based on WSN with embedded capacities of recognition of situations. The description of some examples of experimental results of buoys system based on autonomous uninterrupted working regime of WSN system is provided in Sect. 6. In Conclusions we summarized obtained results and present our future tasks.

2   Related Works on the Solutions of Smart WSN

In order to detect and prevent the environmental hazards, a variety of environmental protection means based on monitoring possibilities are used: sensor arrays (Belikova et al. 2019); grab sampling (Baqar et al. 2018); monitoring systems connected to the internet (Wang et al. 2018; Dzemydiene et al. 2016). All these different types of monitoring systems are aimed to predict water quality changes, maintain water quality and protect people from contaminated water, which can cause swimming-associated infections, such as gastrointestinal illnesses, rashes, etc. mentioned in Wade et al. (2008), Hoagland and Scatasta (2006).

Our provided proposal of the buoys system-based on WSN by extended capacities of transformation of big and intensive data streams is based on our continuing research work, which results are published previously in Gricius et al. (2014, 2015), Dzemydiene et al. (2016).

The layers of such type smart systems are working on the IoT technological platform, which integrate multilayered architectural and technological solutions. We would like to mention the works of WSN construction provided by Lewis (2005), in a book edited by Cook and Das in 2005. Interesting proposals of smart environment technologies with applications of WSN are provided in addition with developing of protocols and applications (Cook and Das 2005; Hart and Martinez 2006). The WSN proposed by Zhi-Gang and Cui Cai-Hui in 2009 was designed on the technology and the wireless communication protocol as ZigBee (Zhi-Gang and Cai-Hui 2009). Tasks of such research are forwarded for recognition of air pollution by monitoring of data and using the intellectualization means of geographical information system (GIS) by using WSN. For simulating and analyzing of geographical data under conditions of air pollution they determine the spread, whether the pollution, by relating such data with local or spreader out globally positions.

For working on-line, the structures and methods of adaptable DSS are integrated with smart wireless systems. One direction of research areas concerns the multi-criteria decision-making methods and development of DSS. In this direction of developments we can mention the proposed methodology of multi-dimensional DSS development for sustainable development and especially successful application of such methods in building assessment and certification processes, presented in Medineckiene et al. (2015). The review of multiple criteria decision-making techniques and applications during 1994–2014 are presented in Mardani et al. (2015).

Another direction of research works is concerning with WSNs development which has numerous different applications from few small sensor networks at one location up to big clusters of sensor interpreting big data in the live stream. These applications have different purposes for using in different architectures of WSN (Keshtgari and Deljoo 2012; Jadhav and Deshmukh 2012). The needs of live environment data are increasing, because data gathered from live WSN increases response time for recognition of unexpected accidents in monitoring environments. Furthermore, using intellectualization of the network, we can start to forecast some types of accidents, before they happen.

Some solutions for extension of functionality of WSN are implemented in Jadhav and Deshmukh (2012) for monitoring of forest environment and recognition of situations of fires. Another WSN solutions using ZigBee protocol for monitoring of climatological data are presented in Keshtgari and Deljoo (2012). The successful monitoring and data transferring processes were executed by using ZigBee and extending the power of batteries. Therefore, there was a short lifespan for autonomous data gathering before, having to change batteries of sensor nodes quite frequently (Keshtgari and Deljoo 2012). The power consumption conditions are very import in the WSN constructions, because in order to make the reliable system, the autonomous and long-lasting work of sensor nodes have to be supported.

3 The Architecture of Buoy's System-Based on WSN for Monitoring of Sea Water Parameters

This approach of the adaptable monitoring system is based on continuing research results of our previous works (Gricius et al. 2014, 2015). The main requirements of developing of such smart monitoring system (Fig. 1) with extended capabilities are:

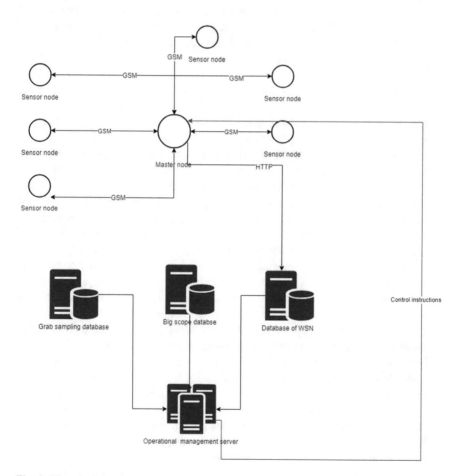

Fig. 1 The principle structure of the system of buoys based on WSN with integration of remote databases and servers

- the structure of the system needs to be with properties of adaptability and working on standard GSM and HTTP protocols;
- the system has to work under on-line conditions and the sea weather conditions;
- the buoy system-based on WSN architecture would have possibilities to work under extreme conditions with restricted capacities of memory and different conditions of transferring of data by big stream flow channels, under the conditions of interruption periods during recharging or replacing the energy resources.

We provide a principle structure of main components of buoys working under WSN system architecture with integration of implement embedded systems. The buoy can incorporate some types of different sensors, communication modules in the own construction. The common components of such architecture are presented in Fig. 1 and more detailed infrastructure of layers is presented in Fig. 2. The types of sensors can be chosen by specific requirements of user needs (e.g., temperature sensor, sensors for measurement of concentration of harmful materials (HM_i), humidity sensors, sensors for atmosphere pressure, etc.).

One buoy can have from 1 sensor up to 20 sensors that gather different data separately. Data from sensors is gathered at time intervals, which are chosen by the system operator. Intervals can vary depending on which parameter is monitored or if some anomalies or a specific situation is recognized, system has a possibility to increase or decrease sampling intervals. The intervals should be extended as long as possible to decrease the battery consumption and increase autonomous working time. All data that are gathered are sending to the Master node through GSM module, which can send information over SMS or GPRS. Mostly such systems are based on the principles of SMS, because the distances in which the nodes can work and distances from the cell tower are high. Usually are chosen the SMS, because it is hard to provide a better alternative that would help to save energy in conditions of the high dispersion of nodes in the high coverage of geographical area.

The main infrastructure represents the layers which are needful for securing requirements of monitoring of hydro meteorological data (Fig. 2). The layers of interoperable functions are represented in left part of the framework and detailed parts show the integrated components which are required for technological solutions, implementation of protocols, and composition of interoperable parts of the system for provision of e-services for sea water monitoring processes.

The sensors have three possibilities to gather the data in short, longer and long time intervals. The short time intervals are in duration of [5–15 min], longer time intervals are in range of 30 min. until 1 h, and the long time intervals can be from 1 h until several days. The data and information is send from buoy (small node) to Master node, which processes the information and saves it to the main database for further analysis. Master node and Operation management subsystem (Fig. 3) are responsible for creation of control instructions for sensor nodes.

Information from grab sampling data is very detailed, but very slow to get, because all of the work is done in the laboratory, therefore, it is mostly used for retrospective analysis and to check the validity of situation recognition made by system. This data

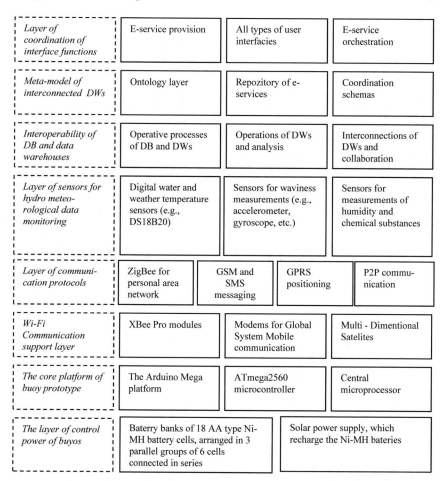

Fig. 2 The multi-layers of infrastructure of integration of components for hydro meteorological data monitoring

most of the time has a delay from 1 day to 2 days, thus it is best to use this data for increasing system forecasting and situation recognition accuracy.

Satellite data is opposite to grab sampling, because it is very inaccurate, it is over big geographical area and has just few parameters. Where grab sampling data has big accuracy, a lot of parameters about a small specific area. The main advantage of satellite data is that we can forecast a lot of information based on data around our monitoring area, because using wind and current direction, we can try to forecast what will come to our monitored area.

So our three main information sources have different database structures and different utilization of that data. Satellite is the most inaccurate, but it is the earliest data, and can be used for forecasting and searching anomalies that might need preparation in advance. For example, in 2018 summer when the Polish coast was hit by the algal

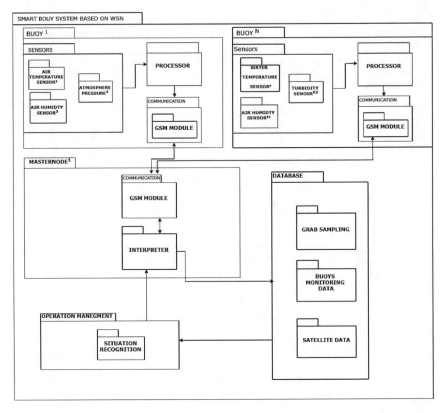

Fig. 3 The communication structure between the components of multi-buoy's system based on WSN

blooms (due to eutrophication), the water mass containing algal blooms could be tracked very well over satellite data. So using satellite data can increase or decrease node sampling data frequency, and the node sampling data is working in real time, and has great accuracy for situation recognition. Grab sampling data is slow, that is why it is used just when system suggests that something wrong might be happening. All of the data from these three sources are firstly gathered at different databases for normalization and are send to the main operational management server.

The more detailed structures of such components are provided in Sects. 3.1 and 3.2.

3.1 Embedded System Deployment in Sensor Node

The embedded systems can be created with attached sensors, such as water temperature at different depths, air temperature, dissolved oxygen matter, atmosphere pressure, wind speed and direction, wave intensity, turbidity, UV intensity and the level of humidity (Fig. 3). The embedded system has the ability to be adapted almost for all environmental parameters, which can be acquired on in-site grab sampling that does not have the needs of high complexity equipment.

The different intellectualization methods are implemented for data forecasting and recognition of situations: such as machine learning, neural networks, and decision tree rules for goal-oriented control. Such methods are used in the server site of the Master node.

One buoy can be constructed from several WSN nodes and can have different sensors that are working for specific purposes in data monitoring process. Every buoy has three parts: a part of sensors, a processor and a communication module. Sensors are responsible for data gathering. The values of monitored parameters are sent to inner database and to the processor where after fixed amount of time data is processed and sent to the communication module where data is sent to the master node.

Data transition from the Sensor node to the Master node is easily accomplishable through embedded systems, using GSM modules, although it not the fastest way, but it is the most reliable way to transfer data in both ways and increase the level of security. It is possible to send near to 140 sensor data points including detailed time, GPS coordinates, without losing sensors' precision or about 140 bytes of information. To achieve data compression and to achieve greater accuracy data is sent not in the normal format, where the full number is sent, but depending on which data parameter is monitored and its' possibility to change in some limits, for example: water temperature in summer varies from 10 to 30 °C, thus, depending on what accuracy is wanted to be achieved, the interval is divided into smaller intervals, where number, which is sent, is the interval number from the beginning, so if the temperature possibility is from 10 to 30, using 5 bits of information we can achieve 0.32 deviation from original value, and adding one bit increase accuracy by 2 times. So there is a possibility to achieve great accuracy with small amount of data transfer.

Autonomy of data gathering nodes

Very important feature of WSN with implemented embedded systems is autonomous working without any psychical interaction after deployment, although it should be capable of controlling needed parameters or functions in sensor nodes after deploy-ment through using wireless network such as GSM. The biggest problem of imple-mentation of sensor nodes is power consumption, because most of the time they are spreaded across river banks or the coast of the sea, where power comes from battery or solar electricity. For decreasing power consumption, there can be different approaches: from mostly working in passive state, where node is turning on and off in periodical time intervals, or using sensors, which use less energy, and decreasing periodicity of those sensors, whose power consumption level is higher. Solar panels, wind generators, waves generators all of these are great for small power creations, which in most cases is all the system would need. Experiment was executed with solar panels, to measure solar radiation at sea level. Experiment is detailed later in the paper.

Other parameter affecting autonomy is robustness of the whole physical system. Most of the time systems are deployed in total wilderness or big open water banks, which make very harsh condition for sensor nodes, because even small imperfections in sealing or the design of the system can lead to corrosion or malfunctions of sensors, therefore, there is a need for human assistance in fixing or replacing a node with another, and that makes hard to have automated and reliable network. Best inexpensive way to increase robustness is to use PVC. It makes sealing easier and there is a lot of space inside the pipe for all electrical components, and batteries. Nevertheless, PVC is very robust to all atmosphere conditions, and when filled with weight at the bottom, it sustains vertical position deployed in water even in high winds and waves.

Flexibility and scalability of system
WSN nodes, created using embedded systems, have great flexibility for constantly changing sensors or parameters which are monitored. Systems have two-way connection with the master node. Sensor node is not only sending monitored data, but it is also receiving data from the Master node, which controls the work of the node, it can increase or decrease time intervals. Because most of WSN nodes have some kind of alternative energy sources, and when energy is low, operational management system can instruct to turn off most of the sensors in the node for power saving, especially when there is no need for constant information, but in opposite situation, if an abnormal situation is discovered, system can increase the monitoring periodicity and make it work until the batteries die.

This topology of proposed system is very easy scalable, because there is the ability to connect up to 100 nodes to 1 Master node, therefore system can be amplified using a rule, that no more than 100 nodes have to connect to one specific Master node, and dividing whole node number by 100 we get our Master nodes number, although it should be mentioned, that we should have at least 2 Master nodes.

3.2   Communication of the Network Sensors and Master Node

Most of WSN communication happens between Sensor and Master nodes, and between Server and Master nodes. They proceed as actors. Our proposed system has to have automated communication that needs very little help of system administrating. Therefore, main communication of the network is shown in Fig. 4.

The Sensor node gathers data and sends in packs of data sets, where data is combined from a few different sensor values at different types, then, when the message size or time is achieved, information is sent to the Master node.

Master node sends data to the Server, where it is checked for integrity and anomalies, then confirmation for correct data is sent to the master node and then confirmation is sent to the sensor node, meaning that sent data can be archived and would not need to be sent again.

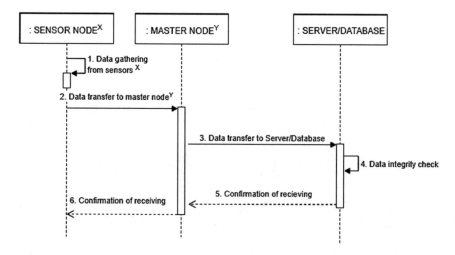

Fig. 4 Representation of scenarios of communication between Sensor and Master nodes

Important part of autonomous system is ensuring of conditions—how the system can withstand an error in communication system or in a hardware failure. System can have the situations of redundancy, if something unplanned can happen, and automatically some additional actions should start in cases of errors. Very important became the detection of such situations when the errors might happen, and anticipate the actions how they should be detected.

For designing of such algorithm of sensor working under abnormal situations (Fig. 5), all possible conditions have to be evaluated in collaboration with other nodes.

If any error occurs in the Sensor node, most of the time it is impossible to see why it is happened. Special activities can be done during the recognition of such situations:

- sending of extra messages for resending of all parts of information or some selected parts of data;
- sampling the marks of the node as faulty and ignore its' parameters.

Errors are detected by the Server, because it is responsible for checking of different parameters during the short time intervals, and controlling of integrity of the whole system. The algorithm of main activities of server working stages and searche phases for detection of errors is shown in Fig. 5.

3.3 Background of Architecture of Buoy—Sensor Node

The proposed system of buoys, as mentioned, consist of a control unit, sensors and a communication module. The control unit is made out of the microcontroller type

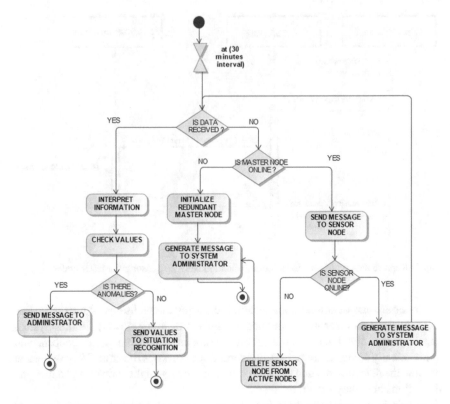

Fig. 5 Example of collaboration algorithm of sensors for recognition of situation of disorderly work of WSN

of Atmega328, mostly used in Arduino Uno R3 microcontroller board. It has 32 KB of flash, 2 KB of SDRAM and 16 MHz operating frequency. Main advantage of such microcontroller is low power consumption (that is 100 mA), and it varies on what frequency and voltage in which it is set up. It can drop even lower. That part of sensors that are connected to the microcontroller are chosen according to the some parameters; the main of them are power consumption and accuracy. As an example, the type of sensors DHT 11 has low power consumption, but the accuracy of measures is in the range of ± 2 °C, which is not acceptable. Therefore, all sensors have to be tested before and because of harsh condition for buoys, expensive sensors are not used.

The communication module is the most important and the most power-needing part of a buoy. Communication can be based on different types of networks such as ZigBee, radio, GSM, 3G or 4G. Our system is based on consumption that our sensor nodes will be placed alongside shore at a few kilometers intervals, so the possibility of having reliable connection is radio, GSM and 2G. Our proposed system is concentrated on utilizing GSM messages that is 140 bytes of information per message, because distances can vary from 1 to 5 or 10 km apart.

Finally, there is a battery of our buoy, battery is used as weight to stabilize the buoy, so it would not flip in high waves case, therefore, it is possible to fit from 20 to 40 batteries in base of the buoy and approximately that might give up to 80,000 mAh of 3.6 V.

4 Possibilities to Apply Data Gathering Methods and Means for Situation Recognition

One of the most significant potential impacts of climate change is the sea level rise and the bacterial pollution, which can cause an inundation of coastal areas and islands, shoreline erosion and destruction of important ecosystems. Satellite altimeter radar measurements can be combined with precisely known spacecraft orbits to measure the sea level on a global basis with unprecedented accuracy. The measurement of long-term changes in global sea level provides a way to test climate models' predictions of global warming and algal blooming.

Most of the time live stream data is not sufficient, we need to combine different types of data gathering methods to gather greater amount of data for better situation recognition.

Usage of satellite data. Information gathered by satellites can reveal some important parameters: ocean bathymetry, sea surface temperature, ocean color, sea and lake ice cover.

Knowing the temperature of the sea surface can tell a lot about current events happening in and around the ocean. Satellite images of sea surface temperature also show patterns of water circulation. Examples include locations of upwelling, characterized by cold waters that rise up from the depths and warm water currents, such as the Gulf Stream.

Satellites also provide information about the color of the water reservoir (river, lake, sea or ocean). For example, color data of the water reservoir helps to determine the impact of floods along the coast, detect river plumes and locate blooms of harmful algae. Water color data from nodes of satellites allows not only to identify where an algal bloom is forming, but also to predict where it might drift in the future.

The data from wireless sensors. The main advantage of the remote wireless sensors is that many data feeds can come into a single base station for storing and analysis, it also enables trigger levels or alert levels to be set for individual monitoring sites and/or parameters so that immediate action can be initiated if a trigger level is exceeded.

For instance, an installed buoy is automated so that it can measure an abundance of different parameters every 15 min as well as collect data on meteorological conditions. The device uses solar power, data logger, GPS and radio. Measured parameters: wind speed and direction, solar radiation, air temperature, pressure, relative air humidity, water temperature and salinity, current velocity and direction, wave height and period, dissolved oxygen content (matter), water turbidity, chlorophyll a concentration.

If a single parameter changes, there is a warning sing to start additional sampling and/or start the decision-making process.

Environmental grab sampling data. In the open seas of marine environments the grab samples can establish a wide range of base-line parameters such as salinity, pH, dissolved organic matter and a range of other significantly important parameters. Water samples can also reveal the water quality, whether the recreational water is suitable for bathers to swim or not.

The grab samples are definitely more time-consuming than the remote ones, but they are crucial, as they can provide scientists with more information. If there are any alerts from the wireless sensor data, then it is easier to take additional grab samples and to compile all of the data together in order to step into the decision-making process.

Therefore, some parameters can only be detected using this particular sampling method. For instance, during the bathing season, there is a certain package of rules, where municipalities are responsible for recreational water management. Still to this day it is impossible to remotely measure the concentration of viruses or, for example, fecal coliform bacteria, so grab samples are the only way to go. In order to save time, the data, which is received from the wireless sensors, determines whether there is a need to do the additional grab-samplings or whether they are unnecessary.

Summarizing, we can state that one way to gather data is not capable to provide needed information, therefore we need to combine data from different sources and different types. Data methods and features are summarized in Table 1.

5 Working Principles of Smart Components Based on WSN System for Recognition of Situations

The main working principles of smart components as embedded DSS are described in such section. The main issues arise in construction of such components:

- How to maintain the interoperability of distributed systems;
- How to create the more adequate construction of decision support models;
- How to support the operative conditions of DSS, working on-line, in conditions of operatively changing extreme environment of sea water.

Such requirements arise in construction of parts, components, communication protocols of the proposed system.

Table 1 Main differences between gathering methods and their parameters

Way of gathering of data	Observed parameters	What can indicate	Accuracy	What used for
Satellite data	Water temperature	Can indicate water temperature at big geographical area	Small, used for localized areas	To indicate common area temperature
Satellite data	Water current	Can indicate big water currents	Small, only for currents in sea or big areas of water as lagoons	Used to indicate water current and to help simulate local and currents brought water mass
Satellite data	Water color	Can indicate big areas of algae bloom	Small, used for massive areas of blooming water	Used to track blooming water, remotely of other countries that can be brought by currents
Wireless sensor network	Water temperature	Indicates water temperature live	Great, but localized are at small intervals	Used to monitor specific places at small time intervals for any changes
Wireless sensor network	Waves height	Indicates intensity and height of waves	Great, but at specific region of shore	Used to determine possible oxygen saturation increase or decrease
Wireless sensor network	Turbidity	Shows water contamination with non-dissolving particles	Great indicator of water contamination start	Used to determine of possibility of algae blooming
Grab sampling	Biological contamination	Shows biological water composition	Slow, but most accurate method	Used to determine full biological composition of water and all bacteria in it
Grab sampling	Chemical contamination	Shows chemical water composition	Slow, but most accurate method	Used to determine full chemical composition and contamination of water

5.1 Data Acquiring and Combining

Data for the operational management system is stored in three different databases, WSN, grab sampling and satellite databases. Most of todays' environment monitoring systems have only one of these databases, thus, it lacks a variety of data inputs or other components for creating automated decision making or abnormality identifications. Therefore, our suggested system architecture combines all databases and tries to evaluate all of the important components through the intellectualization process. Furthermore, there is a capability to expand and add additional databases or data segments from other sources, such as national public health websites or meteorological stations.

Most of the time data combining is based not on combining databases to a big one, but more on combining by some more complex rules. Methods of intellectualization and combining rules are detailed later, before that we need to understand what types of information and parameters are stored.

5.2 Detailing of Monitoring Parameters of Water Reservoir for Evaluation of Situations

There are mainly some types of parameters of water and the environment surrounding it, when we are discussing about monitoring. These parameters need to be monitored differently. The simplest way is to represent the measure parameters that can be monitored by one sensor node, which is anchored with these parameters:

- $Temp^{a/w,S}_{(i,tj,cil,cpl,d)}$—the set of temperature measures i in Celcium [C] at the time moment t_j in coordinates *of cil,cpl*. An index a indicates that it is the air temperature, and w indicates the water temperature measures at different depths d, from water surface;
- $AP^{S}_{(i,tj,cil,cpl,d)}$—the set of atmosphere pressure measures on the conditions of S = [*Spring,Sumer,Authum,Winter*] (i.e. the S indicates the periods of climatic year conditions);
- $OS^{S}_{(i,tj,cil,cpl,d)}$—the set of measures of Oxygen saturation;
- $WS^{S}_{(i,tj,cil,cpl,)}$—the set of wind speed measures and $WD^{S}_{(i,tj,cil,cpl,)}$—the set of wind direction;
- $WAI^{S}_{(i,tj,cil,cpl,)}$—the set of wave intensity measures;
- $Tur^{S}_{(i,tj,cil,cpl,)}$—the set of turbidity measures;
- $UVI^{S}_{(i,tj,cil,cpl,)}$—the set of UV intensity;
- $Hu^{S}_{(i,tj,cil,cpl,)}$—the set of humidity measures.

Not all of these parameters are of the same significance. Temperature of water, atmosphere pressure, wind direction, UV intensity or humidity is slow changing parameters, therefore, monitoring of them can be at longer intervals of time. Turbidity and oxygen saturation are more relevant in our proposed system, thus, it can be monitored more frequently to find the exact moment when the values change, to inform the Master node and start the situation recognition.

Second group can be described as biological and chemical parameters, such as different bacterial contaminations, chemical compositions or finding the exact problem causing the increased level of turbidity, but they cannot be monitored remotely, they can only be monitored by grab sampling.

Third group are most hard to monitor, they could be called geographical data, as water currents, they consist of groups of dots and they have different affects for other parameters, and even to describe them properly, the system has to have some intellectualization systems.

5.3 Requirements for Intellectualization Stage in Environment Monitoring System

For the requirements of identification of monitoring data, all parameters are acquired with specialized intellectualization means. The recognition of situations requires special methods and operable working knowledge base. The spectrum of possible situations is very large. We need some typing and classification of recognition methods.

Water currents on the coastline is a great example, because there is no such sensor that could identify any water current near coast lines, because hot water currents cannot be seen from space. The only way which can determine water currents is the geographical position and to let the buoy be taken by the current and to gather its' GPS coordinates, which could be analyzed later as a whole and not as separate dots on map. In order to track water currents' changes, depending of surrounded environment and changes in the coast-line, we need to create big data storages of information and for WSN nodes to be released in it and carried away along its route. For that type of information, we have to have some type of intellectualization stages and automatizing of finding when we have to gather nodes from the shore and methods with capability to sort out and highlight the patterns in our buoys trajectories.

Furthermore, systems can act on decision trees which can create instructions for better monitoring data. For instance, when monitoring water quality of bathing sites in Lithuanian coast-lines, if turbidity increases or oxygen saturation decreases, water grab sampling has to be done in order to determine the factor (different bacteria or viruses) which caused the outbreak of contamination, the correlation of turbidity and fecal contamination is shown in Fig. 6. Simple rules are not sufficient to indicate the specific situations. Experimental results in dynamics with water samples during appropriate time period are shown in Fig. 6.

Obtained by regression analysis results (in Fig. 6) illustrate situation how the fecal contamination has a strong correlation with turbidity, but its' not only related to this. For example, if a wind is strong, it increases waves height, which increases oxygen saturation, which decreases probability of contamination, therefore, the possibility

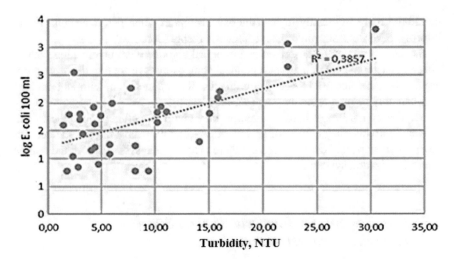

Fig. 6 Example of illustration of obtained results of correlation between turbidity and fecal contamination in samples taken at open source sea water

of turbidity might come from the sand, which is lifted from the seabed, at a small depth where that sensor is, because in high winds and waves there is a possibility that the sensor node is carried away from the original position. Thus, we need complex decision tress or knowledge base for these different scenarios or parameters than can affect values of monitored environment.

6 Experimental Results of Autonomous Uninterrupted Work of WSN System

The proposed system is based on special configuration of sensors which can send information for a long time, but even with the embedded systems, if we use GSM modules, our working time is limited by batteries. Although we mentioned already that battery can be recharged and saved by using system for small time periods for gathering sensor data and then hibernating until next gathering time. We wanted to see how much electricity could be harvested from sun and how fast environment data is changing, so we did an experiment.

Experiment was deployed in the Baltic Sea near shore in 2018 summer from 2018-07-25 to 2018-09-19 to see if energy generated from solar panel could be sufficient to provide power for sensor node. The experiment was executed as follows. At the water level every 15 min solar radiation was measured uninterrupted for 8 weeks, in Tables 2 and 3. It showed how much energy on average was generated for 1, 3, 6, 12 h and day.

Table 2 Minimal values of solar energy generation

Week number	Date	Smallest 1 h W/m^2	Smallest 3 h W/m^2	Smallest 6 h W/m^2	Smallest 12 h W/m^2	Smallest 24 h W/m^2
Week 1	07.25–08.01	2.09	2.28	2.54	8.07	147
Week 2	08.01–08.08	27.27	27.58	28.20	167.76	267.95
Week 3	08.08–08.15	24.44	27.66	27.89	50.25	74.77
Week 4	08.15–08.22	23.61	24.31	24.43	66.89	140.17
Week 5	08.22–08.29	24.58	24.86	25.10	52.43	90.25
Week 6	08.29–09.05	24.40	24.80	25.19	34.09	95.27
Week 7	09.05–09.12	24.40	24.63	24.63	73.54	81.34
Week 8	09.12–09.19	25.30	25.58	25.72	72.53	102.57

Table 3 Average values of solar energy generation

Week number	Date	Average 1 h W/m^2	Average 3 h W/m^2	Average 6 h W/m^2	Average 12 h W/m^2
Week 1	07.25–08.01	224.10	222.77	222.77	222.77
Week 2	08.01–08.08	291.41	289.84	289.84	289.84
Week 3	08.08–08.15	170.95	170.12	170.12	170.12
Week 4	08.15–08.22	226.37	225.19	225.19	225.19
Week 5	08.22–08.29	180.79	179.87	179.87	179.87
Week 6	08.29–09.05	180.07	179.15	179.15	179.15
Week 7	09.05–09.12	164.45	163.62	163.62	163.62
Week 8	09.12–09.19	168.16	167.32	167.32	167.32

The period is divided in 8 weeks. Monitoring and storing such data take background for analysis. The smallest values founded during those weeks are presented in Table 1. It is very important to measure what could be the worst case scenario depending on solar power for recharging buoys' batteries, and that buoy be capable of working without losing power.

The purpose of this experiment is forwarded to the construction of such wireless sensor's communication system with possibilities to withstand in conditions of lacking of solar energy. The experiment results of the smallest energy generation from solar can be similar as showed in Table 2.

Thus, the system should be active during those time intervals when energy generation is less than required for node to operate. Energy should have been accumulated from higher solar intensity periods. Projected energy consumption for system and systems' capabilities should be considered using values showed in Table 3.

Furthermore, we have checked the conditions on how fast monitoring parameters are changing depending on the time period. Data was gathered in intervals of 15 min, and then used to analyze differences between minimal and maximal values in the time period, and showed in Table 4. The smallest time period was 15 min, because for

Table 4 Revealing differences between minimal and maximal values of monitoring data in the time periods

Parameters	15 min max difference	More than 1 in 15 min	Percentage of all	More than 2 in 15 min	Percentage of all
Wind speed	7.78641	624	11.6	148	2.8
Air temperature	4.71	24	0.4	4	0.1
Water temperature at sea surface level	3.02	31	0.6	5	0.1
Water temperature at 1 m depth	1.24	2	0.04	0	0

our experiment we used only one sensor node module that had enlarged battery and was changed periodically, for reassurance, data was monitored in 15 min intervals.

Parameters that were monitored were wind speed, air temperature, water temperature at sea level and at 1 m depth. It was concluded that 15 min intervals are great, but sometimes spikes of value can be very high, but that is very rare and it might be a failure in the sensor itself. Data showed that most of the time values differentiate in 15 min is less than 1 of values (C or km/h).

7 Conclusions

The proposed approach shows the spectrum of problems, which arise in the construction of smart WSN systems with integration of embedded systems for monitoring of environment parameters and the recognition of situations. A huge amount of data of different parameters, which are not connected together, have to be included in recognition of multi-component situations. For instance, sea wave height, water temperature, wind speed and direction, salinity, turbidity and underwater currents are important. Some of these factors influence the specific construction of WSNs.

Remote monitoring technologies can help in supporting of decision making processes, which influence the management of water resources. Some elements of presented architecture of buoys based on WSN show the possibilities to increase capabilities of embedded systems and wireless sensors and provide capabilities of environment monitoring systems of marine type water reservoirs.

Decisions can help in predicting of abnormal situations in water pollution, disturbances of aquatic ecosystems and biodiversity. Our future work is forwarded for more detailed description of the embedded intellectual components for recognition of situations from monitoring systems. But presented principles give us the ability to increase the data precision and help to find some correlations between different environmental parameters.

The experimental results show how the particular parameters have possibilities to exceed the normal limits. Some adequately working online smart monitoring systems can help to predict such situations and show the prognosis, by illustrating and showing what can happen in the next time periods. The approach shows new possibilities of introduction of smart monitoring WSN components. By developing of such systems we have a possibility to make decisions and take actions before these problems occur and cause an abundance of problems.

Acknowledgements Authors would like to express their gratitude to the Marine Research Institute and the Department of Marine Engineering of the Faculty of Marine Engineering and Natural Sciences of Klaipeda University for the supported conditions for experimenting with the system of distant buoys in the Baltic Sea.

References

Baqar M, Sadef Y, Ahmad SR, Mahmood A, Li Y, Zhang G (2018) Organochlorine pesticides across the tributaries of River Ravi, Pakistan: human health risk assessment through dermal exposure, ecological risks, source fingerprints and spatial-temporal distribution. Sci Total Environ 618:291–305

Belikova V, Panchuk V, Legin E, Melenteva A, Kirsanov D, Legin A (2019) Continuous monitoring of water quality at aeration plant with potentiometric sensor array. J Sens Actuators B Chem 282:854–860

Dzemydiene D, Maskeliunas S, Dzemydaite G, Miliauskas A (2016) Semi-automatic service provision based on interaction of data warehouses for evaluation of water resources. Informatica 27(4):709–722

Gricius G, Drungilas D, Andziulis A, Dzemydiene D, Voznak M (2014) SOM based multi-agent hydro meteorological data collection system. In: Nostradamus 2014: prediction, modeling and analysis of complex systems. Springer, Berlin, pp 31–41

Gricius G, Drungilas D, Andziulis A, Dzemydiene D, Voznak M, Kurmis M, Jakovlev S (2015) Advanced approach of multi-agent based buoy communication. Sci World J (2015):1–7

Hart JK, Martinez K (2006) Environmental sensor networks: a revolution in the earth system science? Earth Sci Rev 78:177–191

Hoagland P, Scatasta S (2006) The economic effects of harmful algal blooms. In: Ecology of harmful algae. Springer, Berlin, pp 391–402

Jadhav PS, Deshmukh VU (2012) Forest fire monitoring system based on Zig-Bee wireless. Int J Emerg Technol Adv Eng 2(12):187–191

Keshtgari M, Deljoo A (2012) A wireless sensor network solution for precision agriculture based on ZigBee technology. Wirel Sens Netw 4:25–30

Lewis FL (2005) Wireless sensor networks. In: Cook DJ, Das SK (eds) Smart environments: technologies, protocols, and applications. Wiley, New York, pp 13–46. https://doi.org/10.1002/047168659x

Mardani A, Jusoh A, Zavadskas EK (2015) Fuzzy multiple criteria decision-making techniques and applications–Two decades review from 1994 to 2014. Expert Syst Appl 42(8):4126–4148. https://doi.org/10.1016/j.eswa.2015.01.003

Medineckiene M, Zavadskas EK, Björk F, Turskis Z (2015) Multi-criteria decision-making system for sustainable building assessment/certification. Arch Civ Mech Eng 15(1):11–18

Transforming our World (2015) The 2030 Agenda for Sustainable Development. United Nations, New York

Wade TJ, Calderon LJ, Brenner KP, Sams E, Beach M, Haugland R, Wymer R, Dufour AP (2008) High sensitivity of children to swimming-associated gastrointestinal illness: results using a rapid assay of recreational water quality. Epidemiology 19:375–383

Wang W, Chen J, Hong T (2018) Occupancy prediction through machine learning and data fusion of environmental sensing and Wi-Fi sensing in buildings. Autom Constr 94:233–243

Zhi-Gang H, Cai-Hui C (2009) The application of ZigBee based wireless sensor network and GIS in the air pollution monitoring. In: Proceedings of international conference on environmental science and information application technology, vol 2. Wuhan

Non-standard Distances in High Dimensional Raw Data Stream Classification

Kamil Ząbkiewicz

Abstract In this paper, we present a new approach for classifying high dimensional raw (or close to raw) data streams. It is based on k-nearest neighbour (kNN) classifier. The novelty of the proposed solution is based on non-standard distances, which are computed from compression and hashing methods. We use the term "non-standard" to emphasize the method by which proposed distances are computed. Standard distances, such as Euclidean, Manhattan, Mahalanobis, etc. are calculated from numerical features that describe data. The non-standard approach is not necessarily based on extracted features - we can use raw (not preprocessed) data. The proposed method does not need to select or extract features. Experiments were performed on the datasets having dimensionality larger than 1000 features. Results show that the proposed method in most cases performs better than or similarly to other standard stream classification algorithms. All experiments and comparisons were performed in a Massive Online Analysis (MOA) environment.

Keywords Stream classification · High-dimensional data · KNN classifier · Distance · MOA · Data compression · Hashing

1 Introduction

Data streams are currently an important part of our everyday life. They occur when we are using smart gadgets, when doing an electrocardiogram (ECG) or an electroencephalogram (EEG) in medical facilities. Also, they are used to make weather forecast models shown then on television, or making stock market predictions. We cannot forget about their role in the Internet space; for example, the network packets provide information for intrusion detection systems. Finally, user mouse clicks statistics helps to predict a customer's habits and recommend to buy additional products. All these kinds of data are often transmitted in large quantities. A big challenge for machine learning is the ability to classify data streams as soon as possible, ideally

K. Ząbkiewicz
Faculty of Economics and Informatics in Vilnius, University of Bialystok,
Kalvarijų g. 135, 08221 Vilnius, Lithuania
e-mail: k.zabkiewicz@uwb.edu.pl

© Springer Nature Switzerland AG 2020
G. Dzemyda et al. (eds.), *Data Science: New Issues, Challenges and Applications*, Studies in Computational Intelligence 869,
https://doi.org/10.1007/978-3-030-39250-5_5

83

in real time. We need extracted features to create the classifier model in most cases. Feature extraction/selection can take a significant amount of computational time when dealing with data streams and especially with those having high dimension and/or high volume. The subject of this paper is data stream classification by using non-standard distances. Non-standardness occurs because the distances used do not require numerical features, but instead are computed from the binary structure of the data.

2 The Research Problem

In most cases, classification of data streams is based on the fact that we must build a model from data having features (usually numerical). The problem is more complicated when we are getting raw data, e.g. measurements from medical or smart devices that require feature extraction. Nonetheless, even when features are extracted the number of them can be very large; thus we must perform feature selection. These two procedures can be time-consuming in the case of data stream classification. To solve this issue, a different approach was proposed. It is based on the binary structure of the data. We proposed to implement two non-standard distances based on data compression and hashing. In the next step, we select an appropriate classifier that could deal with the proposed distances. Classification algorithms should be as simple as possible. By simplicity, we mean that the classifier has the minimum number of the parameters to tune. Also, its principle of working must be easy to understand. As a result, we chose the k-nearest neighbour (kNN) classifier. Distance measures are part of its working principle, and it has only one parameter to tune - the number of neighbours.

3 Current Knowledge

A data stream is a potentially unbounded, ordered series of data that continuously arrive with high speed. The main features describing data streams are:

1. Constant flow (elements arrive one following another)
2. The large volume of data (possibly unbounded amount)
3. High data flow speed (processing power is limited)
4. Ability to change in time (data distribution may change on the fly) (Stefanowski and Brzezinski 2016).

3.1 Stream Classification Algorithms

The algorithms for classifying stream can be divided into two groups:

- single classifier
- ensembles.

As a single classifier for data stream classification is a decision-tree-like classifier in most cases, but in other approaches such as Neural Networks, Naive Bayes, and Rule-based classifiers are also used. Ensemble methods are a particularly popular approach for stream classification, because of the fact that classification models can often not be built robustly in a fast data stream. Therefore, the models can be made more robust by using a combination of classifiers. The two most common ensemble methods include bagging and boosting (Aggarwal 2014). A detailed review of current stream classification algorithms is provided in Ditzler et al. (2015), Krawczyk et al. (2017).

3.2 Stream Classification Performance Measures

There are known several approaches on how to measure the performance of data stream classification:

- Prequential accuracy
- Kappa statistics
- Kappa M statistics
- Kappa temporal statistics
- Prequential AUC

Prequential Accuracy may be described as follows. Suppose we have a data stream. Each arriving instance can be used to test the model before it is used for training, and from this, the accuracy can be incrementally updated. When intentionally performed in this order, the model is always being tested on instances it has not seen before. Prequential Accuracy is computed at every moment t as soon as the prediction $\hat{y}_t$ is made

$$Acc(t) = \frac{1}{t} \sum_{i=1}^{t} L(y_i, \hat{y}_i) \tag{1}$$

where $L(y_i, \hat{y}_i) = 1$ if $y_i = \hat{y}_i$ and 0 otherwise.

The Kappa statistics κ is a more sensitive measure for quantifying the predictive performance of streaming classifiers and is defined as follows:

$$\kappa = \frac{p_0 - p_c}{1 - p_c} \tag{2}$$

The quantity p_0 is the classifier's prequential accuracy, and p_c is the probability that a chance classifier–one that randomly assigns to each class the same number of examples as the classifier under consideration–makes a correct prediction. If the classifier is always correct, then $\kappa = 1$. If its predictions coincide with the correct ones as often as those of a chance classifier, then $\kappa = 0$. The Kappa M statistics κ_m are a measure that compares against a majority class classifier instead of a chance classifier:

$$\kappa_m = \frac{p_0 - p_m}{1 - p_m} \tag{3}$$

In cases where the distribution of predicted classes is substantially different from the distribution of the actual classes, a majority class classifier can perform better than a given classifier while the classifier has a positive κ statistics. Kappa temporal statistics considers the presence of temporal dependencies in data streams. It is defined as

$$\kappa_{per} = \frac{p_0 - p'_e}{1 - p'_e} \tag{4}$$

where p'_e is the accuracy of the no-change classifier, the classifier that simply echoes the last label received (Bifet et al. 2015).

The different point of view on measuring the performance of the stream classifier is presented while computing the area under the Receiver Operating Characteristics (ROC) curve AUC (Area Under the Curve). This method was devised rather for static datasets having two classes. More information about it is provided in Fawcett (2006), Majnik and Bosnic (2013). Prequential AUC is dedicated to the streaming data. It is an incremental algorithm that uses a sorted tree structure with a sliding window. More details are provided in Brzezinski and Stefanowski (2017).

3.3   Motivation

In most of the current research, we can notice that streaming data sets have small dimensionality. Instead, the stress is put on their large quantity. In this work, we are considering datasets with a large number of features. The motivation for this arises after analysing the results provided in Zhai et al. (2017). We assume that a 'large' dataset has at least 1000 features. When data sets are rather static, a large number of features becomes a challenge when dealing with the classification task. To overcome this issue the dimensionality reduction methods are often used. There are two approaches to doing this: feature subset selection and new feature extraction from the existing ones. Feature subset selection is burdened by the phenomenon called combinatorial explosion resulting in large computation time practically going to infinity when the number of the features is large. Feature extraction, on the other hand, requires a large amount of memory (e.g. Principal Component Analysis (PCA) requires as much as a square of the number of data vectors) and also computation time (to compute the covariance matrix of high dimensional data). In the case of the streaming data, feature extraction may become less practical or impractical.

4 Classifier Model

An approach proposed in this work will bypass earlier mentioned issues by taking advantage of compression algorithms and hashing functions. This method does not take into account the numeric nature of the features. Most important here is the binary structure of the data. When talking about data compression, we consider its lossless variant. The positive outcome of this method is the fact that we can decompress data any time with no losses. Having such archived data, we can later unpack it and analyse more deeply if it contained some suspicious information. The hashing function also represents some sort of compression, but in this case, we are interested in the generation of the set of hashed numbers. By using this sort of a set and taking advantage of the Jaccard's similarity, we will be able to compute appropriate distance. The algorithm used for data stream classification should be as simple as possible. The nearest neighbours method fulfils this requirement. It requires only a distance measure and the number of neighbours. A model of the classifier is presented in Fig. 1.

As a tool for working with data streams, we will use the Massive Online Analysis (MOA) environment. It will not only generate streams but also provide algorithms for their classification. We have also decided to consider a binary classification problem because most of the research in data stream classification is dedicated to it. In the next chapters, we will present two non-standard distances that were used in the classifier model. Their non-standardness is explained by the fact that we do not need any numerical features to compute them. All that is required is the binary form of data streams, that commonly appears when dealing with measuring devices, e.g. smart gadgets, mine sensors, etc. We will also describe the MOA application and explain why this tool is so vital in the field of data stream classification.

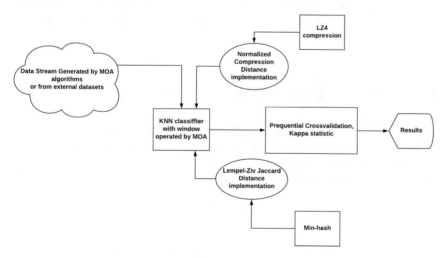

Fig. 1 A scheme of the classifier model

4.1 Normalized Compression Distance

Normalized Compression Distance (NCD) is an approximation of the incomputable Kolmogorov complexity function. The main formula for computing this measure is

$$NCD(A, B) = \frac{Z(AB) - \min(Z(A), Z(B))}{\max(Z(A), Z(B))} \tag{5}$$

where Z(.) is a number of bytes obtained after performing compression of the binary string, AB - a concatenation of the binary strings. NCD acquires values between 0 and 1. Zero distance means that the binary strings are the same; in the case of 1 both strings do not share common information. According to Cilibrasi and Vitanyi (2005) if the compressor is normal, then the NCD is a normalized admissible distance satisfying the metric (in)equalities. Requirements for a normal compressor:

1. Idempotency: $C(xx) = C(x)$, and $C(\lambda) = 0$, where λ is the empty string.
2. Monotonicity: $C(xy) \geq C(x)$.
3. Symmetry: $C(xy) = C(yx)$.
4. Distributivity: $C(xy) + C(z) \leq C(xz) + C(yz)$.

It is also worth noting that for computing NCD we use loss-less data compression. That guarantees us that all compressed information can be restored in the future if we plan to archive it. Of course, there can occur an issue for selecting certain compression algorithm. Dictionary compressor, which is based on the Lempel-Ziv algorithm has limited compression capabilities because of the size of the windows used in the algorithm. When we consider PPM type algorithm the earlier mentioned issue is resolved, but at the price of the compression time - it is slower. An overview of the practical application of this distance can be found in Cilibrasi and Vitanyi (2005), Cilibrasi (2007), Cohen and Vitanyi (2015). Previously it was not applied to stream classification.

4.2 Lempel-Ziv Jaccard Distance

The main approach used in computing this measure proposed in Raff and Nicholas (2017) by Raff and Nicholas is based on the LZ77 algorithm, especially on its part generating a set of all substrings of the given binary strings. Having two sets of substrings the Jaccard similarity index is computed as

$$J(A, B) = \frac{|A \cap B|}{|A \cup B|} \tag{6}$$

Then it is subtracted from 1, and we get Lempel-Ziv Jaccard Distance (LZJD). This theoretical approach was not as fast as desired, so the approximation based on hashing was proposed. It was based on the fact that the Jaccard similarity can be computed approximately from a smaller digest produced from the original sets using the hashing function. The formula for computing the approximation looks like this

$$J(A, B) \approx J\left(\bigcup_{j=1}^{k} h_{\min}^{j}(A), \bigcup_{j=1}^{k} h_{\min}^{j}(B) \right) \tag{7}$$

where $h_{\min}^{j}$ indicates the n'th smallest hash values from the set S.

Computation time for computing LZJD is shorter than in the case of the NCD. This distance also does not suffer from the issue of selecting the appropriate compressor. The proposed measure was successfully used in malware classification. Literature review shows that this measure has not been used in data stream classification before.

4.3 MOA Environment

Massive Online Analysis (MOA) is one of the most popular tools used in the research of data stream mining (Bifet et al. 2010, 2018). Its functionality lets us not only classify streaming data but also perform cluster analysis or detect outliers. An application is written in Java, so it can be used on any operating system that supports Java Virtual Machine (Windows, Linux, MacOS, etc.). It has a user-friendly graphical interface, but we can also use it in command line mode. It can also be imported into the external project. It also has implementation for distributed computing platforms. It is also worth noting that leading researchers not only make experiments with this environment but also update the portfolio of the available algorithms with their proposed approaches. As a result, we do not need to collect all the algorithms from the authors for comparison. These advantages motivated us to perform experiments using MOA functionality.

5 Results of the Experiments

The primary motivation for making experiments with sets having 1000 or more features was the fact that there is a shortage of such research, since most of the research is oriented on streams having large data amounts and not many features. In this chapter, we will present results with streams having various number of features. We analysed three cases:

1. Data streams having feature count from the interval $[10^3, 10^4)$
2. Data streams having feature count from the interval $[10^4, 10^5)$
3. Data streams having feature count from the interval $[10^5, 5 \times 10^5)$

The reason for selecting algorithms for comparison is based on the methodology described in Zhai et al. (2017), Loeffel (2017). Brief descriptions of the nearest neighbour based algorithms are provided below.

KNN with Probabilistic Approximate Window (KNN with PAW) is a modification of kNN described in Bifet et al. (2013). Probabilistic Approximate Window provides a compromise between the relevance of information in the most recent instances and preservation of information carried by older instances. The parameters that can be modified are the same as in the standard kNN.

KNN with Probabilistic Approximate Window and Adaptive Windowing (KNN with PAW and ADWIN) is another modification of kNN described in Bifet et al. (2013). To enable adaptation to concept drift, the ADWIN change detector is added. In this case, the modifiable parameters are the same as in the case of kNN.

Self-Adjusting Memory kNN (SAM kNN) is the newest modification of the kNN classifier proposed in Losing et al. (2018). It includes two memory types to deal with different types of concept drift, called short-term memory (STM) and long-term memory (LTM). The description of other stream classification algorithms is provided in positions (Zhai et al. 2017; Loeffel 2017).

All the data sets were converted to streams by using MOA environment. The parameters used in algorithms were set to their default values. We compared the performance of the kNN and its extensions using proposed non-standard distances with the results of the other algorithms dealing with streams classification. We did not perform any feature reduction, i.e. we took into account all the features. The parameter that was changed during experiments was the number of neighbours. Other variables required for the algorithms were set to their default values. During all experiments, 5-fold prequential cross-validation was performed. As the performance measures we chose prequential accuracy and Kappa statistics.

Data sets selected for the experiments represent different branches of science. A short description of them is provided in Table 1.

In Table 2 we present classification results for the Leukemia data stream. Classification accuracy of the proposed approaches overcomes other stream classification algorithms used in experiments. Kappa statistics is also better in most cases. When comparing between NCD and LZJD, the first approach gives better results. It is also

Table 1 Description of the datasets used in the experiments

Dataset	No. of vectors	Dimensionality	Reference in literature
Leukemia	72	7129	Golub et al. (1999)
Duke breast cancer	86	7129	West et al. (2001)
AnthracyclineTaxaneChemotherapy	159	61,360	Spira et al. (2007)
SMK_CAN_187	187	19,993	Wojnarski et al. (2010)
Physionet_2016	298	114,945	Clifford et al. (2017)

Table 2 Classification results compared with other stream classification algorithms for a stream created from the Leukemia dataset

Leukemia

Classifier	Classification rate, %	Kappa statistics, *100	Number of neighbours	kNN classifier
AccuracyUpdatedEnsemble	65.278 ± 0.000	0.000 ± 0.000	–	–
ADACC	65.278 ± 0.000	0.000 ± 0.000	–	–
DACC	65.278 ± 0.000	0.000 ± 0.000	–	–
DynamicWeightedMajority	65.278 ± 0.000	0.000 ± 0.000	–	–
HoeffdingAdaptiveTree	65.278 ± 0.000	0.000 ± 0.000	–	–
LearnNSE	65.278 ± 0.000	0.000 ± 0.000	–	–
LeveragingBag	65.278 ± 0.000	0.000 ± 0.000	–	–
OzaBag	65.278 ± 0.000	0.000 ± 0.000	–	–
OzaBagAdwin	65.278 ± 0.000	0.000 ± 0.000	–	–
OzaBagASHT	65.278 ± 0.000	0.000 ± 0.000	–	–
OzaBoost	63.333 ± 2.041	27.424 ± 5.281	–	–
OzaBoostAdwin	63.333 ± 2.041	27.424 ± 5.281	–	–
kNN_NCD(1)	**68.611 ± 1.288**	**11.980 ± 4.448**	1	kNN-all[a]
kNN_LZJD(1)	**66.944 ± 1.195**	**10.820 ± 3.687**	1	kNNs[b]
kNN_NCD(best)	**68.611 ± 1.288**	**11.980 ± 4.448**	1	kNN-all
kNN_LZJD(best)	**67.778 ± 0.921**	**9.540 ± 3.195**	3	SAMkNN

[a]kNN-all - means that kNN and its extensions performed in the same way
[b]kNNs - means that kNN and its extensions excluding SAMkNN performed in the same way

worth noticing that we can limit the number of the nearest neighbours to only one to achieve good classification performance.

In Table 3 we provide the result of the Duke Breast Cancer stream classification. Performance measures show the clear superiority of the proposed approaches. The interesting thing is the negative values of Kappa statistics - it means that algorithm performed worse than the chance classifier. What is more, we need only one nearest neighbour to achieve better results in comparison with other stream classification algorithms. kNN classifier using LZJD surpasses its exemplary with NCD. In the case of comparison with other stream classification algorithms, the results of the proposed approaches are significantly better.

To sum up, when considering data streams having more than 1000 features the proposed approach performs better than other stream classification algorithms not related to kNN. We cannot see a clear winner between the two proposed approaches.

The results of the Anthracycline Taxane Chemotherapy stream classification are shown in Table 4. We can notice here similar or slightly better performance rates. Unfortunately, our approach in this case also suffers from negative values of the Kappa statistics. Comparison of the performances between the two proposed approaches shows that in this case, using NCD will give more accurate results.

Table 3 Classification results compared with other stream classification algorithms for a stream created from the Duke breast cancer dataset

Duke breast cancer

Classifier	Classification rate, %	Kappa statistics, *100	Number of neighbours	kNN classifier
AccuracyUpdatedEnsemble	47.674 ± 0.000	0.000 ± 0.000	–	–
ADACC	47.674 ± 0.000	0.000 ± 0.000	–	–
DACC	47.674 ± 0.000	0.000 ± 0.000	–	–
DynamicWeightedMajority	47.674 ± 0.000	0.000 ± 0.000	–	–
HoeffdingAdaptiveTree	47.209 ± 1.078	−9.573 ± 2.022	–	–
LearnNSE	47.674 ± 0.000	0.000 ± 0.000	–	–
LeveragingBag	47.674 ± 1.103	−8.944 ± 2.009	–	–
OzaBag	50.465 ± 0.285	−3.296 ± 0.691	–	–
OzaBagAdwin	50.465 ± 0.285	−3.296 ± 0.691	–	–
OzaBagASHT	50.465 ± 0.285	−3.296 ± 0.691	–	–
OzaBoost	50.233 ± 1.489	−1.995 ± 3.578	–	–
OzaBoostAdwin	50.233 ± 1.489	−1.995 ± 3.578	–	–
kNN_NCD(1)	**58.140 ± 0.973**	**16.386 ± 1.948**	1	kNNs
kNN_LZJD(1)	**68.140 ± 1.014**	**35.457 ± 2.036**	1	kNNs
kNN_NCD(best)	**58.837 ± 0.465**	**17.974 ± 0.934**	7	SAMkNN
kNN_LZJD(best)	**68.140 ± 1.014**	**35.457 ± 2.036**	1	kNNs

In the case of the SMK_CAN_187 stream, results of the experiments are shown in Table 5. The proposed approaches overcome most of the other stream classification algorithms, especially in the case when using LZJD. When comparing performance between classifiers using non-standard distances LZJD confidently overcomes NCD. To sum up, when considering the case of data streams having more than 10000 features, we can see that our approach is better than other stream classification algorithms different than kNN.

The results of the tests made with the Physionet 2016 data set provided in Table 6 show us a significant difference in classification performance when we are using kNN classifier with non-standard distances instead of other stream classification algorithms. We can also observe the fact that using only one nearest neighbour gives us good classification results.

In general, we can see the interesting relationship between the dimensionality of the data and classification performance. In the case of algorithms not based on kNN it is decreasing when the dimensionality is growing and approaches the performance of the chance classifier. Our proposed approach is less vulnerable. We also have noticed that the performance of the non-kNN stream classifiers remains the same for specific data sets. It can be explained by the fact that all these classifiers are the ensembles, and their initialization, feature selection strategy, and base learner setting are the same because we left the parameters of the algorithms set to their default

Table 4 Classification results compared with other stream classification algorithms for a stream created from the Anthracycline Taxane Chemotherapy dataset

Anthracycline Taxane Chemotherapy

Classifier	Classification rate, %	Kappa statistics, *100	Number of neighbours	kNN classifier
AccuracyUpdatedEnsemble	59.748 ± 0.000	0.000 ± 0.000	–	–
ADACC	59.748 ± 0.000	0.000 ± 0.000	–	–
DACC	59.748 ± 0.000	0.000 ± 0.000	–	–
DynamicWeightedMajority	58.616 ± 0.541	−2.121 ± 1.173	–	–
HoeffdingAdaptiveTree	59.623 ± 0.126	−0.251 ± 0.251	–	–
LearnNSE	59.748 ± 0.000	0.000 ± 0.000	–	–
LeveragingBag	57.987 ± 0.728	−2.349 ± 0.955	–	–
OzaBag	56.981 ± 0.706	−3.590 ± 1.433	–	–
OzaBagAdwin	56.981 ± 0.706	−3.590 ± 1.433	–	–
OzaBagASHT	56.981 ± 0.706	−3.590 ± 1.433	–	–
OzaBoost	55.220 ± 0.367	−4.028 ± 0.877	–	–
OzaBoostAdwin	55.220 ± 0.367	−4.028 ± 0.877	–	–
kNN_NCD(1)	58.239 ± 0.760	−0.212 ± 1.412	1	SAMkNN
kNN_LZJD(1)	58.113 ± 0.511	−2.120 ± 0.459	1	SAMkNN
kNN_NCD(best)	59.119 ± 0.199	0.188 ± 0.619	9	SAMkNN
kNN_LZJD(best)	58.742 ± 0.154	−1.511 ± 0.316	7	SAMkNN

Table 5 Classification results compared with other stream classification algorithms for a stream created from the SMK_CAN_187 dataset

SMK_CAN_187

Classifier	Classification rate, %	Kappa statistics, *100	Number of neighbours	kNN classifier
AccuracyUpdatedEnsemble	48.128 ± 0.000	0.000 ± 0.000	–	–
ADACC	48.663 ± 0.239	0.993 ± 0.444	–	–
DACC	48.663 ± 0.239	0.993 ± 0.444	–	–
DynamicWeightedMajority	50.053 ± 0.214	3.578 ± 0.398	–	–
HoeffdingAdaptiveTree	48.877 ± 0.214	1.390 ± 0.397	–	–
LearnNSE	48.128 ± 0.000	0.000 ± 0.000	–	–
LeveragingBag	52.620 ± 0.131	8.363 ± 0.245	–	–
OzaBag	48.235 ± 0.107	0.199 ± 0.199	–	–
OzaBagAdwin	52.406 ± 0.000	7.963 ± 0.000	–	–
OzaBagASHT	48.235 ± 0.107	0.199 ± 0.199	–	–
OzaBoost	55.722 ± 0.262	14.169 ± 0.492	–	–
OzaBoostAdwin	63.316 ± 3.264	28.554 ± 6.181	–	–
kNN_NCD(1)	55.401 ± 2.466	13.600 ± 4.649	1	kNN with PAW and ADWIN
kNN_LZJD(1)	**64.171 ± 1.541**	**30.132 ± 2.930**	1	SAMkNN
kNN_NCD(best)	**61.604 ± 0.393**	**25.253 ± 0.744**	9	SAMkNN
kNN_LZJD(best)	**68.556 ± 1.314**	**38.484 ± 2.510**	5	SAMkNN

Table 6 Classification results compared with other stream classification algorithms for a stream created from the Physionet 2016 dataset

Physionet 2016				
Classifier	Classification rate, %	Kappa statistics, *100	Number of neighbours	kNN classifier
AccuracyUpdatedEnsemble	49.664 ± 0.000	0.000 ± 0.000	–	–
ADACC	49.664 ± 0.000	0.000 ± 0.000	–	–
DACC	49.664 ± 0.000	0.000 ± 0.000	–	–
DynamicWeightedMajority	49.597 ± 0.818	−0.224 ± 1.627	–	–
HoeffdingAdaptiveTree	50.201 ± 0.892	0.155 ± 1.762	–	–
LearnNSE	49.664 ± 0.000	0.000 ± 0.000	–	–
LeveragingBag	56.644 ± 0.658	13.398 ± 1.276	–	–
OzaBag	57.651 ± 1.140	15.554 ± 2.291	–	–
OzaBagAdwin	57.651 ± 1.140	15.554 ± 2.291	–	–
OzaBagASHT	57.651 ± 1.140	15.554 ± 2.291	–	–
OzaBoost	61.611 ± 1.054	23.263 ± 2.101	–	–
OzaBoostAdwin	61.007 ± 0.909	22.049 ± 1.811	–	–
kNN_NCD(1)	**72.886 ± 1.642**	**45.755 ± 3.284**	1	SAMkNN
kNN_LZJD(1)	**77.517 ± 0.773**	**55.024 ± 1.541**	1	kNN
kNN_NCD(best)	**78.389 ± 0.251**	**56.775 ± 0.495**	7	SAMkNN
kNN_LZJD(best)	**79.195 ± 0.978**	**58.373 ± 1.950**	3	SAMkNN

values. We took this step to ensure that algorithms will not be biased by the data set structure. In the future, we plan to implement this approach as an ensemble to make the comparison more objective.

6　Conclusions

We provided here data streams created from data coming from different areas of science: genetics and cardiology. All of them are high dimensional and are almost or are fully raw (in the case of Physionet 2016 data). We showed that the proposed distances could be useful by giving, in most cases, better results than other stream classification algorithms that are not based on kNN. We also have noticed that our classifier achieves good results when performing only 1NN classification. One of the issues that occurred during experiments is a longer computation time. Of course, we can find several causes for that. Firstly, the code written in Java works slower than its implementation in the C language (many compression algorithms are implemented using this language). Secondly, when the k-nearest neighbour classification is performed, distances are computed each time from the beginning. Some caching of known distances could result in faster work of the algorithm. Finally, we did not use parallelisation. Splitting compression task for simultaneous threads could potentially give a speedup. Nonetheless, the results look promising, and after some improvement of the implementation, it may be a useful tool that will help to classify high dimensional raw (or close to raw) data streams.

Acknowledgements We would like to thank the reviewers for their valuable comments and effort to improve this paper. Computations performed as part of the experiments were carried out at the Computer Center of the University of Bialystok.

References

Aggarwal CC (2014) A survey of stream classification algorithms. In: Aggarwal CC (ed) Data classification: algorithms and applications, 25 July 2014. Chapman and Hall/CRC, pp 245–273

Bifet A, Holmes G, Kirkby R, Pfahringer B (2010) MOA: massive online analysis. J Mach Learn Res 11:1601–1604

Bifet A, Pfahringer B, Read J, Holmes G (2013) Efficient data stream classification via probabilistic adaptive windows. In: Proceedings of the 28th annual ACM symposium on applied computing, pp 801–806

Bifet A, de Francisci Morales G, Read J, Holmes G, Pfahringer B (2015) Efficient online evaluation of big data stream classifiers. In: Proceedings of the 21th ACM SIGKDD international conference on knowledge discovery and data mining—KDD '15. Sydney, NSW, Australia, pp 59–68

Bifet A, Gavaldà R, Holmes G, Pfahringer B (2018) Machine learning for data streams with practical examples in MOA. MIT Press

Brzezinski D, Stefanowski J (2017) Prequential AUC: properties of the area under the ROC curve for data streams with concept drift. Knowl Inf Syst 52(2):531–562

Cilibrasi R (2007) Statistical inference through data compression. Ph.D. thesis, Institute for Logic, Language and Computation, University of Amsterdam

Cilibrasi R, Vitanyi PMB (2005) Clustering by compression. IEEE Trans Inf Theory 51(4):1523–1545

Clifford GD, Liu C, Moody B, Millet J, Schmidt S, Li Q, Silva I, Mark RG (2017) Recent advances in heart sound analysis. Physiol Meas 38:E10–E25

Cohen AR, Vitanyi PMB (2015) Normalized compression distance of multisets with applications. IEEE Trans Pattern Anal Mach Intell 37(8):1602–1614

Ditzler G, Roveri G, Alippi MC, Polikar R (2015) Learning in nonstationary environments: a survey. IEEE Comput Intell Mag 10(4):12–25

Fawcett T (2006) An introduction to ROC analysis. Pattern Recogn Lett 27:861–874

Golub TR, Slonim DK, Tamayo P, Huard C, Gaasenbeek M, Mesirov JP, Coller H, Loh ML, Downing JR, Caligiuri MA, Bloomfield CD, Lander ES (1999) Molecular classification of cancer: class discovery and class prediction by gene expression monitoring. Science 286:531

Krawczyk B, Minku LL, Gama J, Stefanowski J, Woźniak M (2017) Ensemble learning for data stream analysis: a survey. Inf Fusion 37:132–156

Loeffel P-X (2017) Adaptive machine learning algorithms for data streams subject to concept drifts. Ph.D. thesis, Université Pierre et Marie Curie, Paris VI

Losing V, Hammer B, Wersing H (2018) Tackling heterogeneous concept drift with the Self-Adjusting Memory (SAM). Knowl Inf Syst 54(1):171–201

Majnik M, Bosnic Z (2013) ROC analysis of classifiers in machine learning: a survey. Intell Data Anal 17(3):531–558

Raff E, Nicholas C (2017) An alternative to NCD for large sequences, Lempel-Ziv Jaccard distance. In: Proceedings of the 23rd ACM SIGKDD international conference on knowledge discovery and data mining, pp 1007–1015

Spira A, Beane JE, Shah V, Steiling K et al (2007) Airway epithelial gene expression in the diagnostic evaluation of smokers with suspect lung cancer. Nat Med 13(3):361–366

Stefanowski J, Brzezinski D (2016) Stream Classification. In: Sammut C, Webb GI (eds) Encyclopedia of machine learning and data mining. Springer, US, Boston, MA

West M, Blanchette C, Dressman H, Huang E, Ishida S, Spang R, Zuzan H, Olson JA, Marks JR, Nevins JR (2001) Predicting the clinical status of human breast cancer by using gene expression profiles. Proc Natl Acad Sci 98(20):11462–11467

Wojnarski M, Janusz A, Nguyen HS, Bazan J, Luo C, Chen Z, Hu F, Wang G, Guan L, Luo H, Gao J, Shen Y, Nikulin V, Huang T-H, McLachlan GJ, Bošnjak M, Gamberger D (2010) RSCTC' 2010 discovery challenge: mining DNA microarray data for medical diagnosis and treatment. In: Szczuka M, Kryszkiewicz M, Ramanna S, Jensen R, Hu Q (eds) Rough sets and current trends in computing. Springer, Berlin, pp 4–19

Zhai T, Gao Y, Wang H, Cao L (2017) Classification of high-dimensional evolving data streams via a resource-efficient online ensemble. Data Min Knowl Disc 31(5):1242–1265

Data Analysis in Setting Action Plans of Telecom Operators

Maria Visan, Angela Ionita and Florin Gheorghe Filip

Abstract In these days it can be noticed a fierce battle of the telecom operators to win the communication service market. The telecom market is currently in full development and it opens opportunities for operators, institutions or for the public, potentially consuming such services. Telecom operators are particularly well positioned to begin to capitalize on this opportunity and build a distinct position in the market for defining suitable value proposals, structuring the right technologies and "go-to-market" partnerships. Fortunately, they possess a huge volume of data that can be analyzed and used in preparing better decisions. For a more in-depth assessment, this paper proposes a detailed analysis of the "battlefield", inlighting the context, the issues of telecom operators that support data for these services, examples of services and the potential users of them, possible solutions to the architectural and methodological implementation. All these aspects will be exemplified by using practical results.

Keywords Big data · Geolocation · Analysis · Methodology · Telecommunication digital transformation

1 Introduction

Excellent service in the information industry means more than having large amounts of data. It is necessary to have the ability to share data in a meaningful, understandable way. Nowadays the need for analyzing data has exploded. Data needs to be pulled

M. Visan (✉)
Department of Engineering, Mechanics, Computers, Romanian Academy, School of Advanced Studies of the Romanian Academy (SCOSAAR), Bucharest, Romania
e-mail: maria.visan@ingr.ro

A. Ionita
Romanian Academy Research Institute for Artificial Intelligence "Mihai Drăgănescu" (ICIA), Bucharest, Romania

F. G. Filip
The Romanian Academy and INCE, Bucharest, Romania

G. Dzemyda et al. (eds.), *Data Science: New Issues, Challenges and Applications*, Studies in Computational Intelligence 869,
https://doi.org/10.1007/978-3-030-39250-5_6

together into one environment to enable the analytics needed for many purposes, to deal with data at a scalable level while reducing cost and complexity (Filip et al. 2017). Public, governmental, other organizations need geolocation information in conjunction with other data from multiple sources, from crowdsourcing to sensors or other contextual Internet of Things (IoT) information, social platform data, or other data sources.

Geolocation, is the intelligent answer to the question "where is the object", using the real-world geographic location. The geolocation refers to smart devices that sense and communicate their geographic position. Capturing this data allows users to consider the added context of a device's location when assessing activity and usage patterns. This technology can be used to track assets, people and even interact with mobile devices to provide more personalized experiences. If the data is available, analysts can incorporate this information to better understand what is happening, where it is happening, and what they should expect to happen. Technology leaders have focused and solutions are easy to use and allow for a low barrier to entry to these types of analytics that manage, analyze and store 2D, 3D information in the database and data infrastructure. With the highest scalability, security and performance, the static data become real-time data when it has been combined with a timeline view for 4D analysis through connections to live sensor feeds as they monitor moving objects, then adds 5D to predict what it will be. The data owners are those that really count in this battlefield and telecom operators have a lot of Big Data collected for their internal workflows. They just need to adopt the vision and define a digital transformation strategy.

The rest of this paper is organized as follows: the problem is described in the next section providing the internal and external origins (characteristics of the telecom operator organizations) that are helpful or harmful in achieving the objective of digital transformation. The existing threats and opportunities in the telecom global market presented in the first section are balanced within the next section, where is inlighted the huge potential hidden in the treasure of information gathered over time by telecom operators, mentioned as a portfolio but untapped at present. The next section is about setting a practical action plan for telecom operators in order to develop organizational change through the digital transformation journey. Using the methodology and the implementation roadmap centered on customer value and the return on investment, the decision for adoption the new technologies will support relevant performance indicators over large volumes of structured and unstructured data, in real-time or near real-time, and will break down the business silos. All these efforts are based on automated systems that are aware of real-time geolocation from mobile devices (managed by telecom operator) combined with events. With the implementation of a proper architecture that includes layers for cross-infrastructure/analytics tools in the current telecommunication market battlefield, the operators with "win-win" mindset will develop new services in order to become an industry driver for a sustainable future.

2 Presentation of the Problem

Understanding the existing situation in the telecom market, at the level of external attributes of the environment and internal attributes of the organizations, gives the helpful and harmful image that supports the telecom operators for the new services, and it drives them to achieve the digital transformation objective. Based on (Ovum 2019) predictions there are several external attributes of the environment identified as opportunities that potentate the digital transformation for telecom operators. We think that the following aspects have the biggest impact:

- the continuous access requests and global broadband services ones due to the increasing number of digital customers and to their intensity of use, that will be growing for the next 10 years;
- for the last two decades, there have been highlighted the fastest technological developments (5 generations over the last 20 years, from 1G to 5G) and the new requests for B2C/B2B (Business to Citizens/Business to Business) IT (Information Technology) disruptive services and infrastructure (digital, broadband, high-speed), generated by the explosion of IoT (Ernst and Young 2018);
- the raising of emerging business models of sharing economy and the world that continuously changed has been generated a huge request for new technologies onto the market;
- the global research and development budgets are now significant and bring these new technologies to life for full development. In the same time, they are opening opportunities in all economic and social sectors, generating new jobs for both operators and their potential customers.

Telecom operators own the essential connectivity infrastructure for the operation and digital growth for all other economic sectors. They collect, on a daily basis, huge volumes of data for internal operations that can be analyzed for providing Big Data as a Service. That's why telecom operators are well positioned to capitalize on their relationship with customers, assets, and to build a distinct position on the market. This is the optimal time for telecom operators to define appropriate value new offers, structuring the right technologies and "go-to-market" partnerships.

The harm to achieving the goal of digital transformation would be that, within organizations, telecom operators would ignore the changing of the moving world and do nothing about it. Loss of revenues, caused by the decrease in demand for traditional services due to the changing of user behavior and lower margins, generate fewer funds available for research and business development just when the rate of needed investment is higher than ever. These organizations are facing with high operating costs with outdated infrastructure. The lack of digital transformation initiatives generates the loss of competitive advantages globally.

From the external perspective, the threat of alignment to the single market rules reduces profit margins (e.g. roaming cost reduction).

The competition from previous customers and partners generate new threats because the global technology giants develop their own infrastructures and move to the telecom market.

For example, taking into account the telecommunications ecosystem, we can look at the dynamics of the value chain and the associated trends. Each player in the value chain has different assets and can be identified as:

- *end-users*—are business users and residents who use and pay, directly or indirectly, for telephone and internet services;
- *telecommunication operators*—who own and provide telecommunication infrastructure and virtually allow any kind of communication;
- *equipment manufacturers*—providing platforms, telephones and other hardware for telecom operators and end-users;
- *content providers*—are companies that primarily provide content consumed over the Internet.

Bearing in mind the explosion of data traffic over the last few years, caused by the ability of consumers to go over the Internet, to view content, use Facebook or access cloud services, we can say that new generations of "virtual service providers" have emerged. These continuous transformations raise some natural questions: how the value chain will be transformed from now on and who will hold the power? Will there continue to be telecommunications operators, such as AT&T and Verizon (historically they have had a customer relationship)? Or will be the new comers like the content providers (such as Google and Amazon) that until yesterday were just users infrastructure or partners with the operators? Or will the phone makers be winning (such as Apple or Google both providing content and equipment)?

At present, one of the relevant trends in the telecom world is to increase the power of these kind of providers. For instance, Facebook has more users than any telecom operator while Apple and Google have their own content systems. That is why operators have become more and more transparent, and the user's concern does no longer geared toward who owns connectivity.

If we analyze where the telecom industry's profits have been collected over the past two decades, we find out there are not traditional operators, but technology owners like Apple, Amazon, Facebook, Google, or Netflix (to name only the well-known). Moreover, the global extensions of the technology, media and telecommunications sectors increase annually the profit of giants (Apple, Google, Microsoft, etc.). These companies, including other players in the service market, not only redefine customer experience, but also are now competing with operators to take responsibility for customer relationships. While the technology giants benefit from scalability in global economies, national boundaries fragmented the traditional operators.

As a result, it is not clear who will be at the forefront of the value chain in these dynamics of the telecom market. However, several scenarios can occur. Content providers may become new telecom operators—owning customers who already communicate via Facebook and other applications, while the traditional operators can adapt to become content providers. To become more complex, cloud computing and machine-to-machine penetration services are other market forces that influence this path. The traditional operators will have to decide their positioning strategically by defining who their real competitors are in the future. Imagining the scenario where AT&T (the world's number 1, based on Forbes statistics) would win 1% or 2% of

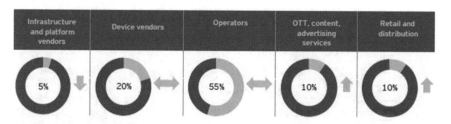

Fig. 1 Telecommunications industry value chain—2015 share of revenues by segment (EY, Global telecommunications study: navigating the road to 2020 (EY 2015)

Verizon's mobile market share (the world's number 2) or vice versa, the question would be how Google or Facebook can influence the AT&T or Verizon customers in their final decision. Some operators look at these threats and make strategic moves to compete better against technology players. In order to reposition itself in the telecoms value chain, DirecTV from AT&T purchased the Time Warner Cable and Verizon acquired AOL and Yahoo. In the same time, they have to share distribution networks via alliances with other traditional operators in order to have a chance against those technology giants.

The fast pace of digital technology evolution and new disruptive business models (new cloud apps and communication services, some of them almost free) jeopardize current investments, erode service revenues and put significant pressure on the core costs of telecom operators.

The ascending trend regarding the focus on telecom content services mentioned since 2015 in data source by Ey analyses in Fig. 1, is confirmed by Statista after few years in Fig. 2 in its prognosis for 2019 for Global telecommunication services market value all over the world.

In order to remain relevant in the digital market, as we have seen above, telecom operators should consider at least two strategic moves and take immediate action for reinventing their business models:

- upgrading the business, becoming cost-effective and more agile, and
- Identifying new growth areas (new innovative products and services), in combining the great potential of digitization and their core competencies.

3 Main Telecom Operators Portfolio

The telecom operator's portfolio includes a huge amount of databases collected over the years for their internal or external operations. These databases should be harnessed as valuable assets for digital transformation without any other effort.

For example, Network Infrastructure data are currently used for internal optimization processes such as capacity planning, infrastructure upgrades, proactive maintenance, performance management, network traffic shaping. Service and Security

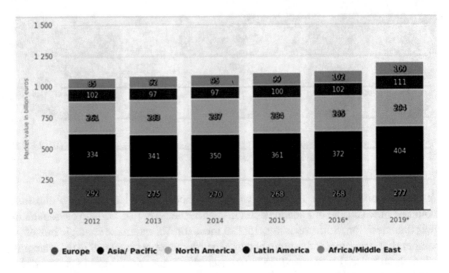

Fig. 2 Global telecommunication services market value from 2012 to 2019 by region (in billion Euros) (Statista 2019)

data are currently used to deliver better service and security results, minimizing total cost serving and improving customer experience, reducing quitting operator and ensuring compliance with security regulations and company policies: customer experience analysis, contact center productivity, field service productivity, data protection compliance, data protection, user devices security, financial management, fraud prevention and detection and localization of information or sales and marketing data are used for business outcomes development in sales and marketing through 360° view of customer value, smart/custom marketing campaigns, anticipation of customer needs, upselling and cross-selling, churn reduction. Based on these data, there are Next Product To Buy (Nptb) recommendations for other services (Knott et al. 2002).

The external data that could generate new products/services are based on geolocation information that is able to be combined with various city data insights and offer a new strong web services platform, such as the ones for smart cities or for delivering new and adjacent lines of business to telecom operators. Combining the citywide network of data from both human movement and technical sensors, it could constantly inform the city leaders of what is going on in their communities. New powerful-targeted marketing and geolocation intelligence products/services, based on location insights and location intelligence, become actionable intelligence for advertisers, merchants, payment processors, or for citizens.

4   Methodological Aspects

When the decision to introduce and create an information system that leads the digital transformation of the target enterprise is taken, a successful implementation requires fulfillment of the main influence factors mentioned in Fig. 3.

First of all, the management and also the system designer should understand the context. This includes analysis of the influencing factors (market trends, legislation) for the decision-making process and definition of a strong business case. Then the main success factor is human resource, the allocated team (key and non-key resources) for this transition, putting in balance their previous experience. Another influencing factor is the one giving for each team member a very good definition and understanding of the orientation and purpose of the project, the organizational settlement. Then the other influence covers the methods and standards adopted and the proper selection of ICT (Information and Communications Technology) products/tools. Selecting the most appropriate alternatives approaches and methods, that include the standards, such as ISO 9241 ("Ergonomics of Human-System Interaction")—recommended for obtaining a user-centered solution, ISO 9241-171.2008 ("Guidance in Software Accessibility"), ISO 9241-151.2008 ("Guidance WWW User Interface"), ISO 37120 ("Sustainable development of communities—Indicators for city services and quality of life"). The information system purchase option could be one of the following: IT on premises, IT as a service (SaaIT), or software as a service (Saas). The decision to purchase will take into consideration the advantages and disadvantages of each of these options. Defining the roadmap with detailed steps for implementation will cover the necessary time, money and resources for the defined results.

Fig. 3 Main influence factors for a successful information system implementation (Filip 2012)

Due to the fact everything is measured and monitored, there will be necessary to define principles for evaluating implementation results: defining successful implementation evaluation criteria, selecting methods used for multiple evaluation (benefits/cost analysis, the "net present value" (NPV) of the investments, value analysis, or "rating and scoring", event logging, and so on), in terms of expected results for each phase, quality, time and available budget.

During the entire implementation process, strong attention should be allocated for monitoring constantly the external factors as market trends, and the movements of competitors in the industry, to changes in the legislation, and to all other specific technical and non-technical aspects related to the integration of the new system into the target enterprise environment.

The decisional perspective for the digital transformation of telecom operators and alignment to the methodological aspects is the successful key combined with permanent management involvement, permanent monitoring, analyses and decision.

5 Proposed Implementation Roadmap

Based on previous experience in the implementation of Information Technology (IT) systems and in accordance with practical criteria, there are various approaches for designing, building and implementing an information system in the targeted telecom operator enterprise that decide the digital transformation as a project. Each of the following steps implies methodological aspects presented above, procedures and checkpoints that are strictly monitored and well documented.

Step 1 Initial efforts centered on customer value and return on investment (ROI): Improving understanding, forecasting and generating the real value for customers, and relevant as new experiences.

Step 2 Big Data project developed across the entire organization. Digital transformation requires a strong IT foundation (data, software and hardware infrastructure) flexible and scalable architecture that generates sustainable business value, meeting business requirements and providing value to customers. Big Data strategy for collection, organizing, validation and controlling redesigning databases by new criteria depending on request eliminating the data silos to allow faster and more innovative daily activities circulation of data across the organization.

Step 3 Quick results over existing data. Take advantage of data stored in existing workspaces while expanding data repositories by needs to manage larger volumes and types of data. Only after successful completion of the previous step it could be realized an extension to other data sources.

Step 4 Analytics centered on customer value. Transforming data perspective by providing relevant, integrated, timely and actionable information for immediate decision and execution. Adoption of new technologies (such as query and reporting, dashboard control and data score control) to support relevant performance indicators, and customer experience management. The predictive analysis offers patterns, analytics over large volumes of structured and unstructured data in real-time or near real-time. Intelligent analysis and autonomy, automating prescriptive analyzes based on automated systems that are aware of real-time events (contextual IoT information and social platform data). Example: when a customer searches for a specific bistro, automatically receive a discount on a drink in the same bistro. A call center uses Watson from IBM for providing accurate to customer inquiries more quickly (less than 3 s could parse 1 million books or 200 million pages), more efficiently and accurately.

Step 5 Develop a strong business case: active management engagement, strong and continuous collaboration between IT and business.

6 Proposed/Recommended Architecture for Information System

The digital transformation of the big data information system implementation is not an easy task. The sheer volume, velocity, and variety of data make it difficult to extract information and business insight. As tools for working with big data sets advance, so does the meaning of big data. More and more, this architecture relates to the value you can extract from your data sets through advanced analytics, rather than strictly the size of the data, although in these cases they tend to be quite large. In a simplified schema (Fig. 4), presented a high-level logical architecture applicable for any digital transformation process that involves big data solutions. Next, every

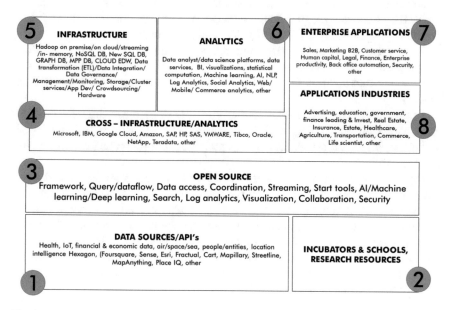

Fig. 4 Functional architecture (El Kaim 2018)

pattern from various industries will be customized and the necessary tools/products that offer the relevant functionality for problem solving will be implemented.

Data sources/API The first component is the one used primarily to collect data from different sources and formats directly from project partners or via API's. This component add any data from IoT sensor feeds, demographic, health, financial and economic data, from air/field/space/sea, people/any type of entities, location intelligence (using different technology formats as: *Hexagon, Esri, Foursquare, Sense, Fractual, Cart, Mapillary, Streetline, MapAnything, Place IQ*, and so on).

Data Resources The second component, is the external one used for bridging a longstanding gap between different industry partners, ambitious students, world-class education, and passionate founders under one community. Their main purpose is managing information to innovate and support the ecosystems and any other company that has been created through a private-public collaboration between incubators, schools, universities or research centers.

Organizations can choose to preprocess unstructured data before analyzing it.

Open Source, Cross—Infrastructure/Analytics, Infrastructure These components are tools for data access, coordination, streaming, query/dataflow, preprocessing data format, search, log analytics, analytics preprocess, AI (Artificial Intelligence)/machine learning/deep learning, query, visualization, collaboration, storing, security tools that allow data to be consumed by the Analytics component, that include different technologies platforms for data analyzing such as data analyst, data science, BI (Business Intelligence) visualizations, vertical analytics, statistical computation, data services, **machine** learning, AI, NLP (Natural Language Processing), search, log analytics, social analytics, web/mobile/commerce analyst.

Analytics The analytics include Data analyst/data science platforms, data services, BI, visualizations, statistical computation, Machine learning, AI, NLP, Log Analytics, Social Analytics, Web/Mobile/Commerce analytics or other.

Enterprise Applications The enterprise applications are designed for different purposes such as sales, marketing B2B (Business to Business), customer service, human capital, legal, finance, enterprise productivity, back office automation, security.

Applications Industries The applications industries mean different business applications for advertising, education, government, finance leading and invest, real estate, insurance, estate, healthcare, agriculture, transportation, commerce, life scientist or other.

Decisions regarding architectures or software components for data analysis and business intelligence platforms requires a deeply analyses base on chosen Big Data strategy and the development of the telecommunication market. According to the functional architecture, the system that assures the digital transformation of the telecom operator will consist of several functional components that interact with each other to optimize the process of collecting, processing, analyzing and providing information according to the architecture Fig. 5, using the same components numbered as it has been presented in Fig. 4.

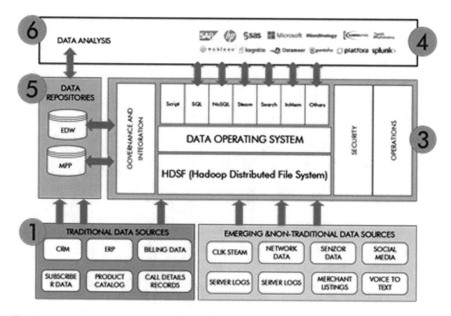

Fig. 5 Physical architecture (Visan and Mone 2018)

7 Hub Services Based on Location Positions

There are three levels for the digital transformation solution for a telecom operator, the collect/storage of the data, application, and client web-based, as it has been presented in Fig. 6.

The first level cover the data and API's data connectors from different sources and data storage that allow using different data types static 2D or 3D data, 4D sensor feeds, or 5D data resulted from prediction. The second layer includes application layer designed for extracting the location positioning from telecom operational data, real-time data processing acquisition and anonymization—that handle big data, different visualization for different applications domains like traffic speed, urban planning, environment, agriculture, health, tourism, other. The result is an Open Geospatial Consortium/Simple Object Access.

Protocol (OGC/SOAP) web services hub based, but not limited to geography, and open interfaces (REST, SOAP, JDBC, SQL, …). The client web services are available in a web OGC portal that publishes the services in order to be consumed by different users.

For each of these layers, there are security interfaces and open standards implemented in order to allow consumers an easy access to these services. For example, the practical approach for a Big Data traffic solution, it analyzes multiple types of data (network technical data, customer calls, road fingerprints, real-time car positions and speeds, traffic history) for real-world traffic analysis and information. By combining location with customers call patterns, operators can identify traffic loads

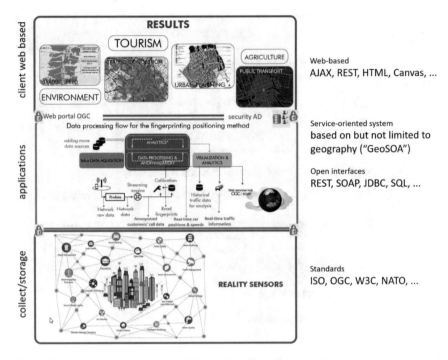

Fig. 6 New services based on geolocation (intelligent location)

and traffic jams to separate travel and stationary models (Visan and Mone 2018). The client's location allows managing technical network and user experience, identify customers in locations and lead real marketing information to these locations, increasing user experience and value for customers (Filip 2007). The benefits from such project implementation are: road sensor precision for lower price, virtual traffic sensing, for example at every 500 m, near real-time traffic information with maximum delay of, for example, 5 min, performance monitoring of the national road network, active traffic control possible, base for the future-simple Google Maps to professional traffic operation centers (Visan and Mone 2018).

Each business scenarios could be mapped to the appropriate atomic and composite patterns that make up the solution pattern. Architects and designers can apply the solution pattern to define the high-level solution and functional components of the appropriate big data solution.

8 Conclusion

Worldwide around us, real life is continuously changing, altering the existing processes, and, in many cases, becoming disruptive. For many companies, the ones that are not sufficient agile, the impact of this disruption generates decision delays for changing and failing to adapt fast enough in order to avoid the negative impact.

The modern analytic tools, machine learning, and artificial intelligence solve with better results most of the existing sensitive points, but nowadays these tools should become part of the transformative digital daily processes.

The focal purpose of this paper was to summarize the journey to the implementation of a successful information system, that offers a business alternative for telecom operators and allows them to develop a new line of business. This is materialized as a completely different one from today's operator services offered. The new services (Ernst and Young 2018) optimize both the internal workflows and benefit from the existing databases, and also provide answers to the market's requests enabling customers to shape smart change across diverse business and industry landscapes. Technology adoption is not enough. A smart approach based on a methodology and a customized roadmap is needed. The implementation of the proposed methodology can be extrapolated to any of the companies seeking to improve their operations and provide a better quality of their services to the community.

Setting a clear objective and methodology is the starting point. Any company, and in our paper any telecom operator, should start developing and strengthening their position in the market, and should multiply cross-selling activities by predicting the product each customer would be most likely to buy next (Knott et al. 2002), doing all these by structuring the right technologies and "go-to-market" partnerships. In the first section, it has been presented the existing threats and opportunities raising in telecom global market, followed by a description of the problem and the main telecom operator's portfolio. The next section provides the methodological aspects of the main influence factors for a successful information system implementation. The roadmap and the architecture for the proposed information system follow in the next sections, in order to obtain the final results, the hub services based on geolocation.

Combining analytical tools with geolocation, the telecom operator answers to the market requests and offers actionable information for executive decision-making. The vision for structuring the right technologies, and developing new services will solve the threatening growing competition. The telecom operators become appropriate business partners for developing smart applications (smart mobility, smart safety, smart healthcare, smart agriculture, smart retailer, smart tourism or smart "any") and convenient administrative services, or that support the emerging business models like the sharing economy. Sometimes, the customer of such apps or services becomes itself a partner, that is using or sharing his own data and is developing new services for the community.

In this equation of digitalization, the operator becomes an enabler, integrating solutions to anticipate and answer to the ever-growing needs of connectivity. In a growing digitalized world, to lead the digital transformation and accompany its own customers through digital care, interactive apps and digital communities become a real necessity (Visan and Mone 2018).

To stay ahead of competitors, managers from telecom operators, and not only, do they need to completely change their present approach, including what they do and how they do it, but also to reinvent their business to improve the quality of the user experience.

References

El Kaim W (2018) Enterprise architecture and governance. Available at https://www.slideshare.net/welkaim/big-data-architecture-hadoop-and-data-lake-part-1. Accessed on 20 Mar 2019

Ernst & Young (2018) Global telecommunications study: navigating the road to 2020. Available at https://www.ey.com/Publication/vwLUAssets/ey-global-telecommunications-study-navigating-the-road-to-2020/$FILE/ey-global-telecommunications-study-navigating-the-road-to-2020.pdf. Accessed on 25 May 2019

Filip FG (2007) Sisteme suport pentru decizii, Ed. TEHNICA, Bucuresti, ISBN978-973-31-2308-8 (in Romanian)

Filip FG (2012) A decision-making perspective for designing and building information systems, Int J Comput Commun 7(2):264–272. ISSN 1841-9836

Filip FG, Zamfirescu CB, Ciurea C (2017) Computer-supported collaborative decision-making. Springer International Publishing, Cham. ISBN: 978-3-319-47219-5

ISO 9241—Ergonomics of human-system interaction. Available at https://www.iso.org/standard/52075.html. Accessed on 20 Mar 2019

Knott A, Hayes A, Neslin SA (2002) Next-product-to-buy models for cross-selling applications. https://doi.org/10.1002/dir.10038. Accessed 12 Mar 2019

Ovum (2019) 10 predictions for 2019. https://www.digitaltveurope.com/comment/ovum-10-predictions-for-2019. Accessed 12 Mar 2019

Statista (2019) The global telecommunications services market: 2012–2019 data available at 15. The global telecommunications services market: 2012–2019 data. Available at https://www.faistgroup.com/news/the-global-telecommunications-services-market-2012-2019-data/. Accessed 25 May 2019

Visan M, Mone F (2018) Big data services based on mobile data and their strategic importance. https://ieeexplore.ieee.org/xpl/mostRecentIssue.jsp?punumber=8383963. Accessed 12 Mar 2019

Extending Model-Driven Development Process with Causal Modeling Approach

Saulius Gudas and Andrius Valatavičius

Abstract The model-driven development is most promising methodology for cyber-social systems (CSS), cyber-enterprise systems (CES), cyber-physical systems (CPS), and some other types of complex systems. Causality is an important concept in modeling; it helps to reveal the properties of the domain hidden from the outside observer. Great results of CPS engineering based on the perceived causality of specific domain—physical system. The subject domain of the CES as well as of CSS is a complex system type named "an enterprise". The aim of the article is to enhance a model-based development (MDD) process with a causal modeling approach. The causal modeling aims to reveal the causality inherent to the specific domain type and to represent this deep knowledge on CIM layer. To do this, you need to add a new layer of MDA—a layer of domain knowledge discovery. Traditional MDA/MDD process use the external observation-based domain modeling on CIM layer. Such models assigned to empirical as they based on the notations that do not include causal dependencies, inherent to the domain type. From the causal modeling viewpoint, an enterprise considered to be a self-managed system driven by the internal needs. The specific need creates a particular causal dependence of activities—a management functional dependence (MFD). Concept of the MFD denotes some meaningful collaboration of activities—the causal interactions required by the definite internal need. The first step is conceptualization of the perceived domain causality on CIM layer. A top level conceptual causal model of MFD is defined as a management transaction (MT). The next step is the detailed MT modeling when an elementary control cycle (EMC) is created for each MT. EMC reveals the internal structure of MT and goal-driven interactions between MT internal elements: a workflow of data/knowledge transformations. The results of this study help to better understand that the content of the CIM layer should be aligned with the domain causality as close as reasonable. The main contribution is the extended MDA scheme with a new layer of the domain knowledge discovery and the causal knowledge discovery (CKD) technique tailored for enterprise domain. Technique uses twofold decomposition of management transaction: a control view-based and self-managing

S. Gudas (✉) · A. Valatavičius
Institute of Data Science and Digital Technologies, Vilnius, Lithuania
e-mail: saulius.gudas@mif.vu.lt

© Springer Nature Switzerland AG 2020
G. Dzemyda et al. (eds.), *Data Science: New Issues, Challenges and Applications*, Studies in Computational Intelligence 869,
https://doi.org/10.1007/978-3-030-39250-5_7

view based. The outcome of technique is hierarchy of management transactions and their internal components: lower level management functions and processes, goals, knowledge and information flows. Causal knowledge discovery technique is illustrated using the study programme renewal domain.

Keywords Causality · MDA · Enterprise domain · Causal model · Management transaction · Knowledge discovery

1 Introduction

One of the most promising enterprise application software (EAS) development methodologies is the model-driven enterprise software development approach [Model Driven Architecture (MDA) and Model Driven Development (MDD)] (OMG 2019; Trask and Roman 2009). The MDA/MDD process of the enterprise application software engineering mostly based on OMG specifications (OMG 2019). From the system analysis perspective the application software development MDA/MDD techniques (OMG specifications for MDA layers CIM, PIM, and PSM), belong to black box approach, that corresponds to the external modeling paradigm (Gudas et al. 2019; Valatavičius and Gudas 2017).

The capabilities of enterprise modeling (EM) and business process modeling (BPM) techniques become important in applying MDD methodology. A wide variety of EM and BPM methods used for problem domain modeling on CIM layer.

However, traditional methods are not sufficient to identify the deep features of domain, i.e. causal dependencies between management activities, goal-driven data/knowledge transformation logic, which essential in the development of enterprise information systems software. Causal modeling approach is aimed to shift the traditional MDD process of the EAS engineering to internal modeling paradigm, relying on the use of the causal model of enterprise.

Causality is an important concept in modern science (Bunge 2011); it helps to reveal the domain properties hidden from the outside observer. Integration the causal modeling with the application software development is aimed to shift the traditional MDA/MDD process to internal modeling paradigm, thus creating causal knowledge based development techniques.

Great results of CPS engineering based on the perceived causality of specific physical system, which is a causal model, the specification of internal domain regularities. Concepts *causality, causation, causal knowledge, cause-effect relationship, causal dependence* used in various fields of research: social and economic process management, health care, educational, production management. Great results of CPS engineering based on the perceived causality of specific physical system, which is a causal model, the specification of internal domain regularities.

Viable results today are achieved with the development of smart systems, autonomic systems and other types of the cyber-physical systems (CPS). CPS engineering uses the inherent properties of real world domain (i.e. physical system) that are defined by the term *causality*. In other words, CPS engineering methods based on the causal knowledge (scientific law, scientific explanation) of the subject domain (Bunge 2011). From the system analysis perspective such methods belong to the internal modeling paradigm, that corresponds a white box/grey box approach (Gudas and Valatavicius 2017).

The real world domain of the cyber-enterprise systems as well as of cyber-social systems is defined using concept "an enterprise". An enterprise is a common name of a subject domain in the IS engineering methodologies. An enterprise defined as one or more organizations sharing a definite mission, goals and objectives to offer an output such as a product or a service in (ISO 15704:2000). An enterprise refers to different type of the complex systems compared to the physical systems, which is called an organizational systems. The features of organizational systems revealed by second-order cybernetics are as follows: goal-driven (purposeful) behavior, self-organization, self-management and some other (Heylighen and Joslyn 2001; Scott 1996). These features help explain and model the system's internal interactions, causality. Another characteristic feature of this system type is *a circular causality* that provides a control feedback loop inside such systems. An importance of a circular causality in understanding the essential ...has been noticed yet by Foerster: "one-dimensional cause-and-effect chains was replaced by the two-dimensional notion of a circular process" (von Foerster 1953). The circular causality is a feature common to the cyber-systems (CPS, CES, CSS, and etc.), it includes criteria of behavior (goals, purposes) as a component of structure. Challenge is discovery of the real world domain causality (i.e. the inherent causal dependencies enterprise), and usage of the perceived causality for engineering.

MDD process of the enterprise application software (EAS) development starts from the subject domain modeling on CIM layer. The results of this study help us understand that the content of the CIM layer should be as close as possible to the causality of the domain. Traditional MDA/MDD process use the external observation-based domain modeling on CIM layer using a wide range of languages such as DFD, BPMN, IDEF, ARIS, UPDM, or other modeling specifications. Models created in this way are attributed to empirical models as they do not uncover causality inherent to the domain type, e.g. do not include causal knowledge of the enterprise domain.

The purpose of the paper is to shift a model-based development (MDD) to the internal modeling paradigm (Gudas et al. 2019; Valatavičius and Gudas 2017) supported with a causal modeling approach that requires the addition of a new MDA layer—domain knowledge discovery layer. The causal knowledge discovery (CKD) technique presented.

The precondition of the causal modeling approach is prior knowledge of causality inherent to the specific domain type. We focus on the definite subject domain—a complex system type named "an enterprise". From the causal modeling viewpoint, an enterprise is a subject domain, considered as a self-managed system driven by internal needs (strategy, management system goals).

Causal modeling approach enabled the disclosure of gap between the domain causality and content of CIM layer in traditional MDA/MDD methods (OMG 2019).

The difference between the empirical model and the causal model is essential, qualitative. An empirical model is the external observation based. A causal model is a formal representation of a subject domain causality (regularities). Causal model is constructed using deep knowledge of intrinsic properties (inherent properties) of the domain type, e.g. using causal knowledge that have already been discovered (Bunge 2011). An analogy could be the science of physics, which finds the regularities of nature, defines them formally, and applies them to engineering solutions.

We rely on the causal modeling methodology of enterprise described in Gudas (2012a, 2016). The main causal modeling concepts are the management functional dependence (MFD) and the management transaction (MT) (Gudas 2012a). The reason to appear of some particular MFD (in the managers' heads (thoughts) or documents) is the strategic/tactic/operational need of enterprise. MFD predetermines the interdependencies of required activities and management information (data, knowledge, goals) flows among these activities (Gudas 2012a). The implementation of MFD creates an appropriate management transaction (MT) in the subject area (organizational environment). Standard modeling notations (e.g. BPMN, UML, DFD, IDEF0, IDEF3 etc.) are suitable for disclosing the content of MT: management information flows (data flows, knowledge structures, goal structures) and sequence of transformations inside MT.

The causal MDA/MDD process supplemented with the domain knowledge discovery layer, and supported with causality modeling frameworks. The causal MDA/MDD process will help to trace the causal dependencies across the MDA layers, and to follow the domain causality within application software, to trace the influence of the domain causality to orchestration and choreography of the application software components.

The remainder of the paper structured as follows: Sect. 2 reviews related works on the concept of causality, circular causality, and causal modeling of enterprise. Section 3 describes the internal modeling perspective of enterprise modeling based on the concepts of management functional dependence and management transaction. System analysis of MDA/MDD process in Sect. 4 reveals the gaps of MDA/MDD process from the internal modeling paradigm perspective. The causality driven MDA/MDD process described in Sect. 5. The causal modeling frameworks the management transaction (MT), and the elementary management cycle (EMC) depicted and discussed in Sect. 6. Section 7 discusses the causal knowledge discovery technique: the principal schema of causal knowledge discovery, and the main stages of causal domain modeling technique. The causal modeling technique illustrated in Sect. 8. The results and further work required discussed in the conclusion section.

2   Causality and Causal Modeling of Enterprise

Evolution of complex systems entail an development methodology shift away from the empirical methods toward a more systemic methods, based on the causality concept and termed herein *causal modeling*. There are many different definitions of causality, causal model, and causal modeling in the literature (SEP 2018; Bunge 2011; Frosini 2006; Khoo et al. 2002). The definition of "causal model" content depends on the interpretation of "causality" concept (viewpoint of modeling).

The concepts of causality (causation) is considered to be fundamental to all natural science, especially physics, it is complex and difficult to define. A comprehensive analysis of causality, the place of the causal principle in modern science is presented by Bunge (2011). Causality is studied in philosophy and mathematics. Philosophers D. Hume and J. S. Mill contributed a great deal to our understanding of causation (Khoo et al. 2002).

Causality is an abstraction that indicates how connects one process, event or state (the cause) with another process, event or state (the effect). There are different types of causality (causation) in various real world domains: relational causality, probabilistic causality, distributed causality, Granger causality (Granger 1969), and other.

Causality is related to physical processes, events or states, mental activity (human thinking), psychological processes, teleology cause (purpose), and statistics (Khoo et al. 2002).

The related concepts such as causal dependence, causal model, cause-effect relationship, causal knowledge are applied in all areas of science and technology: social and economic systems, health care, education and research, production management, security systems, data science, etc.

Stanford Encyclopedia of Philosophy provides the following definition of causal model: "*Causal models are mathematical models representing causal relationships within an individual systems or population. They facilitate inferences about causal relationships from statistical data.*" (SEP 2018). Thus, (as indicated here) a source for reasoning about causal relationships is collected data (statistics or measurements), i.e. a source is the external observation based information about real world.

According to the *semiotic tetrahedron* defined in FRISCO report (Falkenberg et al. 1996) a conception (data, information, model) is a result of interpretation of domain (referent), perceived by interpreter (an actor or device). Thus, information or data is a second-order source (perceived and interpreted reality) for developing the causal model of domain.

The primary source of knowledge should be the real world itself, perceived as a real world regularity (phenomenon as in physics).

Since application software development is focused on the support of enterprise activities, a primary source of causal knowledge should be a regularity inherent to enterprise domain. Causal model is a conceptual model representing the inherent causal dependence of activities within specific domain. The causal dependence of activities facilitate inferences about causal relationships (functional dependence) of data.

There are different definitions of the notion causality and causal dependence in the different fields of science. We present some definitions the notion causality and causal dependence which are close to our research area.

A counterfactual definitions of causality in Halpern (2016) gives a sufficient condition for causality: A is a cause of B if there is counterfactual dependence between A and B: if A hadn't occurred (although it did), then B would not have occurred.

　　　The definition of the notion of causal dependence and causation proposed in Lewis (1973): "An event E *causally depends* on C if, and only if, (i) if C had occurred, then E would have occurred, and (ii) if C had not occurred, then E would not have occurred." Causation then defined as a chain of causal dependence.

　　　In our approach the notion of *a causality (causation)* considered as a chain of cause-effect relationships relevant to the real world regularity in the specific domain.

　　　A cause-effect relationship is a conceptualisation of the *causal dependence of real world:* it is a relationship in which one event (process, activity, i.e. *the cause*) makes another event (process, activity) happen (*the effect*). One cause can have several effects. A separate cause-effect relationship is not yet "causality" in full sense (scientific law, scientific explanation), it is only a fragment of the chain, an item of the causal knowledge.

　　　A cause-effect relationship is one-dimensional cause-and-effect, named "linear causality" in the early cybernetics literature in the Macy conference proceedings, 1953 (Scott 1996). An important idea was that a proper modeling of systems with self-organization (i.e. CSS, CES) require "the simple notion of one-dimensional cause-and-effect chains" to be replaced by "the two-dimensional notion of a circular process." This conclusion captures the importance of circular causality concept understanding, avoiding defied analysis of the enterprise domain.

　　　A *circular causality* is a key feature of some well-known business management models: Deming's PDCA cycle (Plan, Do, Check, Act) (Deming 1993), interactions of primary activities and Support Activities in Porter's Value Chain Model (Porter 1985), Harmon business management model (Harmon 2010), and Rummler-Brache model of enterprise management (Rummler 2010).

　　　A *circular causality* is very basic concept in the science and research of real world phenomenons. In control theory a *circular causality* is formalized and represented as a feedback control loop of control system. Circular causality is abstracted in concept "transaction" in the systems engineering and software engineering. "A transaction in the context of a database, is a logical unit that is independently executed for data retrieval or updates" (Techopedia 2019).

　　　A transaction is a key concept for discovering of deep properties (causality) of the subject domain. A management transaction in the context of an enterprise causal modeling, is a logical unit which is an independent (management and control) cycle of activities interaction. The conceptual structure of the generalized transaction corresponds to the conceptual structure of the control system with the feedback loop (Fig. 1):

$$T(Q) = \{(S_1, \ldots, S_n), (M_1, \ldots, M_n), Rs, Feedback\} \tag{1}$$

　　　Here: T—transaction, a single indivisible logical unit; Q—goal, objective, criterias, requirements, rules, etc. (depends on the layer or stage); S_i—process, activity, information transformation, application, procedure, and other (depends on the layer or stage); M_j—flow, data, information, message, Rs—a sequence relationship of S; Feedback—a constraint to establish a closed loop of S, and in this way to create a single unit.

Fig. 1 Topology of the generalized transaction is a wheel graph

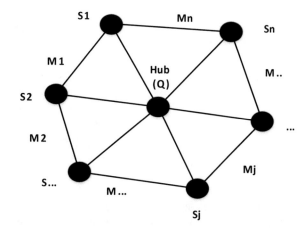

The topology of the generalized transaction is a wheel graph (Fig. 1). In the graph theory, a wheel graph obtained from a cycle graph C_{n-1} by adding a new vertex called *a Hub* that connected to all the vertices of cycle graph C_n (Bondy and Murty 2008). The premise is that a conceptual structure of the transactions in all MDA layers or systems development life cycle stages is the same—transaction is a single indivisible logical unit of work, a closed loop sequence of steps (processes, actions, activities, works, transformations, procedures, and other) (Gudas and Valatavicius 2017; Gudas and Lopata 2016).

3 Causal Model of Enterprise

Circular causality is common feature of the management and control processes in complex systems such as CPS, CES, CSS, and other types. One more example is the autonomic computing approach of IBM based on the circular causality (Kephart and Chess 2003).

The content of transactions on the Enterprise Architecture (EA) layers or MDA layers is different; it corresponds to the viewpoint (semantics) of the definite layer. On business management layer of EA frameworks, there are several interpretations for transactions. However, in the business management frameworks (e.g. Action Workflow approach, Deming's PDCA cycle, transactional workflows) a transaction is a closed loop sequence of goal-driven activities (i.e. a value-oriented transaction) as in Medina-Mora et al. (1992), Deming (1993), Porter (1985), Georgakopoulos et al. (1995), Rummler 2010). On business process layer, the enterprise transaction in Dietz (2006), Papazoglou (2003) or the management transaction in Gudas and Lopata (2016) is a single indivisible logical unit of work (however, it is a complex process) comprising a closed loop sequence of information transformation steps.

Fig. 2 Management
transaction is a circular
causality inherent to
enterprise

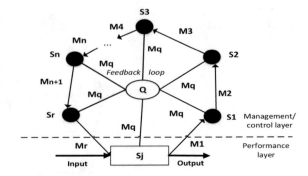

A primary reason of MT emerging is a management functional dependency (MFD); it causes collaboration of activities that needed for achieving some enterprise goal (Gudas 2012a). MT is a closed loop sequence of goal-driven information transformations (comprising a management function Fj) focused on the control of enterprise process Pi (Gudas 2012a). On application layer, the transaction is defined a closed loop sequence of information exchange that is treated as a unit for the purposes of satisfying a request.

A notation of management transaction reveals the information content of self-managed activities: data and knowledge flows M1, …, Mr, and internal transformations S1, …, Sr, including an impact of goals/purposes (Q) (Fig. 2).

Thus, causal modeling is not limited to studies of cause-and-effect relationship, seeks to discover and use deep knowledge. A deep knowledge in Wang et al. (2015) is divided into two types—causal knowledge and first law knowledge. The definition of the first law knowledge is important: "knowledge related to physical properties, …including theory, law, formula and rule" (Wang et al. 2015). We will continue to use the concept of causal knowledge in the sense that includes first law knowledge.

This approach to the notion of causality and causal modeling relies on the premises as follows:

- The modeling methodology belongs to internal modeling paradigm (white box approach);
- A source of knowledge is real world activities, a pre-defined knowledge is regularity of domain type (causal knowledge), and a perceived causality includes the circular causality (transaction) and causal dependence or real world activities.

Summarizing the content of the causality-related concepts we have such types of models by causality modeling aspects in Table 1.

From the internal modeling perspective an enterprise domain considered as an entirety of self-managed activities. The strategic or tactical needs that determines the causal links between some set of activities is called here a management functional dependence (Gudas 2012a). Causal model of the abstract enterprise is an entirety of collaborating MFD in Fig. 3. The required by MFD set of activities forms a complex unit—a self-managed activity (a management transaction). Thus, MFD predefines the appearance of some aggregated activity and predefines the causal

Table 1 Types of models by causality modeling aspects

Aspects and factors	Options	Causal model	Knowledge-based model	Empirical model
		A white box	A grey box	A black box
Modeling paradigms	External modeling (black box approach)		X	X
	Internal modeling (white box approach)	X		
Source of knowledge	Primary: real world activities	X	X	X
	Secondary (derivative): data		X	X
	Third row: process model or data model		X	X
Predefined knowledge	Empirical knowledge: observation-based, and experience-based		X	
	Causal knowledge: regularity, scientific law	X		
Analytical horizon (a perceived causality)	Separate cause-and-effect relationship, a structure of cause-and-effect relationships		X	X
	Sequence of cause-and-effect relationships (a workflow)		X	X
	Closed loop of cause-and-effect relationships (a circular causality, transaction)	X		

(continued)

Table 1 (continued)

Aspects and factors	Options	Causal model	Knowledge-based model	Empirical model
		A white box	A grey box	A black box
Type of dependence (a perceived causality)	Primary dependence: causal dependence of real world activities	X		
	Secondary dependence (derivative): functional dependence of data		X	X

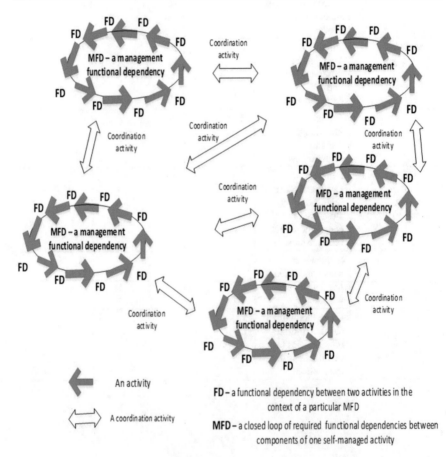

Fig. 3 Causal model of the abstract enterprise

interdependencies of the internal elements of aggregated activity (i.e. predefines a management transaction) (Gudas 2012a).

A summary of the main features of the causal model of enterprise: (a) Enterprise considered a set of management functional dependencies; (b) MFD is implemented as a self-managed activity; (c) A conceptual causal model of a self-managed activity is a management transaction. The specific conceptual frameworks for enterprise causal modeling developed in previous works (Gudas 2012a, b, 2016) presented in the section below describing the knowledge discovery layer of the extended MDA schema.

4 Evaluation of MDA/MDD Process Gaps

Model-Driven Architecture (MDA) is a software development approach proposed by the OMG (in 2001). The MDA layers of abstraction Computation Independent Model (CIM), Platform Independent Model (PIM), and Platform Specific Model (PSM), and model transformations (inside each layer and between these layers) define steps of development, which lead to the last step—specification of the Implementation Code. CIM layer contains of conceptual domain model: business process model. PIM layer contains of system specifications on (platform independent) languages without any technical or technological details. Platform-specific (PSM) layer is detailed specification of all technical and technological aspects. Models on the CIM, PIM and PSM layers must be compliant with relevant meta-models.

Model-Driven Development (MDD) is a system development methodology based on the Model-Driven Architecture (MDA). MDA/MDD process relies on the use of OMG standards. MDD uses model transformation process to generate (semi-automatically) the target model from the source model (Kleppe et al. 2003).

MDA/MDD transformations are meta-model driven, i.e. the pre-defined knowledge is required for transformations: Meta-model MM2 for transformation TR2: CIM to PIM, Meta-model MM3 for transformation TR3: PIM to PSM, and Meta-model MM4 for transformation TR4: PSM to code (Fig. 4).

When examining MDA from a causal modeling perspective, some uncertainties observed. The issue is how to form CIM layer. The uncertainty lies in the interaction between the CIM layer and the request to develop application (Fig. 4). What kind of knowledge does an analyst use for conducting transformation TR1? Finding answers helps to analyze MDA/MDD process at CIM layer.

Regarding real world domain modeling and transformation to CIM layer, the focus is on the analyst's relationship with the real world. Analysts have prior knowledge of the modeling techniques: business process modeling notations (e.g. BPMN, SBVR, DMN, UML, SysML, IDEF, DFD), and enterprise architecture frameworks (MODAF, UPDM, UAF). In addition, analysts have experience in some enterprises; however have limited deep knowledge of the enterprise domain regularities. The reason is that there is no formal theory examining regularities of enterprise management from cybernetics perspective (focus on the informational content of management

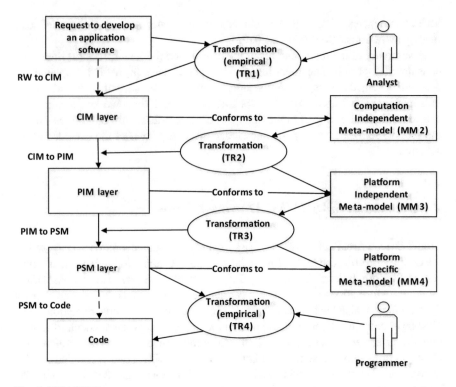

Fig. 4 MDA/MDD process

activities). One can only mention FRISCO Report as an attempt to create such a theory of enterprise information systems (Falkenberg et al. 1996).

Thus, a real world domain modeling and transformation to CIM layer belongs to external modeling paradigm. Therefore, model on CIM layer considered as black box of real world domain. Understanding of domain activities is more observation based. Analyst has no prior-knowledge of the underlying dependencies of that domain type, i.e. there is a lack of enterprise causality knowledge.

Summarizing, real world domain modeling at CIM layer is constructing of black boxes, because no prior-knowledge of RW domain causality was used (Fig. 5). Regarding the analyst's relationship with models at CIM layer, these models are white boxes with regard to analyst, because a prior-knowledge of modeling notations used to represent observations.

The analysis of the evolution of OMG modeling specifications confirms that so far little attention paid to real world domain analysis (Fig. 6). OMG modeling specifications for software engineering (UML 1.1) appeared in 1997. Notation for real world domain modeling came later: *Business Process Model and Notation* (BPMN 1.0) appeared only in 2007 (revised version BPMN 2.0.2—in 2014). The first specification for real world domain knowledge modeling—*Business Motivation Model*—appeared in 2008 (2015—BMM 1.3.), Ontology Definition Metamodel (ODM 1.0)

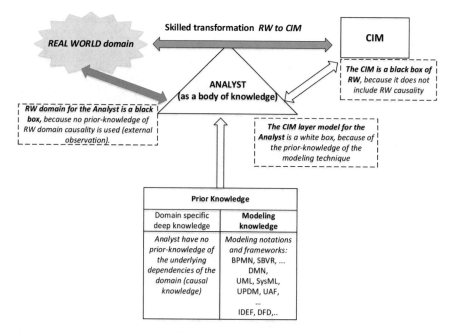

Fig. 5 System analysis of MDA/MDD process at CIM layer

appeared in 2009 (2014—ODM 1.1). Notation for deeper real world domain modeling *Decision Model and Notation* (DMN) 1.0 appeared only in 2016 (2018—DMN 1.2).

Analyzing the evolution of OMG's modeling specifications, the following trend is emerging from software engineering to process modeling, and knowledge (decision) modeling. However, all these notations attributed to the external modeling paradigm, because they do not rely on deep knowledge. This highlights a drawback of advancement in real world modeling techniques for software engineering needs.

The uncertainty in the interaction between the CIM layer and the request to develop application depicted in Fig. 5. This indeterminacy solved if the MDD process adjusted as shown in Fig. 7.

We make the basic assumption that the causal modeling analyst has firmly mastered the theoretical fundamentals of domain causality, i.e. has knowledge about the regularities of specific RW domain type. This is a fundamental difference in the causal modeling approach from traditional (empirical) modeling techniques (see Fig. 5). Several theoretical works have already been published on the issues of mapping causal knowledge about real world domain to software development process (Osis and Asnina 2011; Osis and Donins 2017). Topological modeling method for domain analysis and software development presented in Osis and Asnina (2011). The topological model of functioning uses the cause–effect relations for the modeling of subject domain in the framework of MDA. Method allows constructing a Computation Independent Model (CIM) in a form of Topological Model that is transformable

MDA / MDD approach	Process modeling & Software engineering	Knowledge modeling	Causal knowledge modeling (discovery)
	OMG specifications	OMG specifications	
Real World domain	Category Business modeling (14)	Category Business modeling	< *Category Causal knowledge discovery>*
(Types of complex systems: Business Enterprise, Cyber-physical System, Cyber-Enterprise system, Cyber-Social systems, …)	BPMN 2.0.2 SBVR 1.4 SysML 1.5 UML 2.5.1 …	DMN 1.2 ODM 1.1	< *Required for causal knowledge discovery>*
CIM layer *(Computation Independent Models)* **PIM layer** *(Platform Independent Models)*	Category Business modeling (14) BPMN 2.0.2 SBVR 1.4 SysML 1.5 UML 2.5.1 … Category Enterprise modeling (2) UPDM 2.1 UAF 1.0	Category Business modeling (14) BPDM 1.0 BMM 1.3 BPMM 1.0 OCL 2.4 ODM 1.1 DMN 1.2 QTV 1.3 …	
PSM layer *(Platform Specific Models)*	Category Software engineering (9) UML SysML SoaML …	Category Software engineering (9) KDM 1.4 RTF	
Code Layer	Programming Languages		

Fig. 6 Clusters of MDA/MDD modeling specifications

to the Platform Independent Model (PIM) (Osis 2004). Application of cybernetics and control theory to the subject domain analysis provides a strong foundation for the knowledge-based software engineering theory, connecting the enterprise causality modeling and model-driven approach to software engineering (Gudas 2012a). Thus, in performing causal modeling, the analyst must have prior knowledge of the causality of the domain. In addition, the analyst have prior knowledge of the standard modeling techniques on CIM layer of representation techniques of discovered causality. That is why the analyst's relationship with both RW domain modeling and CIM layer models classified as a white box. Causal modeling and model-driven development processes integration details discussed in the next section.

5 The Causality Driven MDA/MDD Process

A prior knowledge is an important concept in causality driven MDA/MDD process. We focused on the prior causal knowledge (deep knowledge).

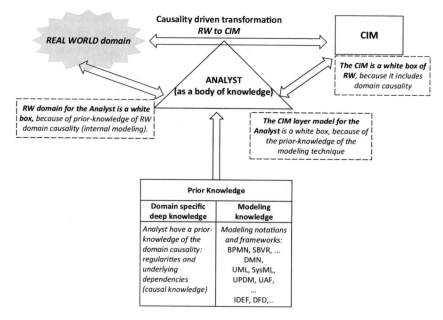

Fig. 7 System analysis of causal modeling at CIM layer

Different things (in a scientific sense) is having a prior causal knowledge of mechanical system (physical system), and having a prior causal knowledge of goal-driven, self-organized organizational system (such as enterprise, CES or CSS or CBS).

"An engineer working with a mechanical system, on the other hand, almost always know its internal structure and behavior to a high degree of accuracy, and therefore tends to de-emphasize the system/model distinction, acting as if the model is the system" (Heylighen and Joslyn 2001). The *"high degree of accuracy"* is because of the known theory of operation and control of this system type, i.e. a prior causal knowledge is available.

In contrast, an engineer working with organizational system (manager, business process analyst, enterprise application developer…), usually don't know internal structure and behavior "to a high degree of accuracy", since there is no theory of causal information/knowledge interactions of management and control, i.e. a prior causal knowledge is not available. Therefore, an enterprise causal knowledge discovery method needed. Knowledge discovery often described as deriving knowledge from the input data. Our approach focused on the definite type of knowledge—a causal knowledge. Causal knowledge discovery considered as an analysis process of real world domain, when using a prior knowledge—the theoretical reliable knowledge of inherent causal dependencies (regularity) of particular domain type.

The proposed causal MDA/MDD process of enterprise application software development based on the assumptions as follow (Gudas and Valatavicius 2017):

Assumption 1 The software development methods should be based on the domain causality (a deep knowledge-oriented), and this means shifting of MDA/MDD process to the internal modeling paradigm, i.e. causal modeling approach.

Assumption 2 The modified MDA/MDD process should include the Knowledge Discovery Layer (KDL), and the cross-layer transformations based on the internal model (IM) principle. The internal model (IM) principle expresses qualitative requirements for the validity of models on MDA layers. The cross-layer transformations, based on the internal model (IM) principle ensure top-down transferring of domain causality:

$$IM(1) \rightarrow IM(2) \rightarrow IM(3) \rightarrow IM(4) \tag{2}$$

Here: the internal model IM(1) is a causal model of enterprise domain represented on the Knowledge Discovery Layer; the internal model IM(2) is an enterprise/business process model (a causal CIM*) with inherited causality from KDL; the internal model IM(3) is an application architecture model (a causal PIM*) with inherited causality from CIM* layer; the internal model IM(4) is a detailed application software model (a causal PSM*) with inherited causality from PIM* layer.

Assumption 3 The essential features of the real world domain which accumulate in the internal model IM(1) are transformed and remain in the lower layer internal models IM(2), IM(3), and IM(4).

The causality driven MDA schema in Fig. 8 supplemented with the domain knowledge discovery layer (KDL). The model transformations in the causality driven MDA are internal-model-based transformations (IM-based transformations).

One of the milestones of the causal modeling approach is a concept of internal model (IM) related with a good regulator theorem (Conant and Ross Ashby 1970; Francis and Wonham 1976). The internal model (IM) is a knowledge model of the problem domain inside the control system, and is used to maintain the stable behavior. IM also could be considered as a white-box of the problem domain (causal model of domain) which specifies the essential elements and their causal dependencies (laws of behavior inside a domain). Analysis of the role of the internal model (IM) in the control systems concludes that adapting of the internal model (IM) in the context of software development is a relevant topic for enhancing intelligent MDD technologies (Gudas and Valatavicius 2017).

The internal model was defined in 1970 as a good regulator theorem (Conant and Ross Ashby 1970). The regulator theorem is the following idea: "*any regulator (if it conforms to the qualifications given) must model what it regulates*" (Conant and Ross Ashby 1970). Internal modeling can be used for enhancement of intelligent software technologies (i.e. utility-based or intelligent software agents) and serve as a background of knowledge-based software systems. The internal model first was articulated as the internal model principle of control theory in 1976 (Francis and Wonham 1976). The internal model approach emerged in the control theory, whose problem domain is a device, object or open system in general. The purpose

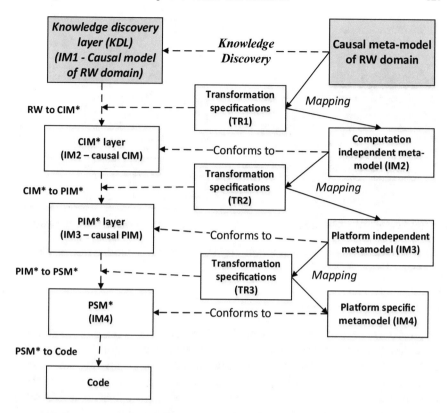

Fig. 8 Causality driven MDA/MDD process

of the Internal Model is to supply the closed loop of the Control System with control signals which maintain the stable behavior. The Internal Model (IM) is a model of the problem domain created in advance using prior knowledge. Thus, IM is a predefined model, based on knowledge of the essential properties of the domain. Due to IM, an important, intelligent feature of prediction occurs in IM control systems because the control is based not only on the measurements or evaluation of the state (Kumar 2012).

6 A Prior Knowledge and Causal Modeling of Enterprise

A comprehensive description of the enterprise causal modeling approach presented in Gudas (2012a), Gudas and Lopata (2016). Here we discuss the key concepts and frameworks, and provide the basics of the causal modeling technique—a prior knowledge for causal modeling of enterprise.

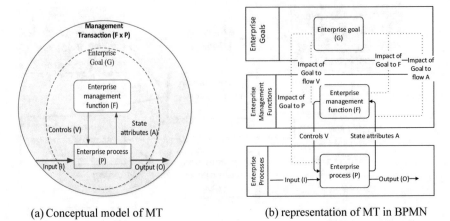

(a) Conceptual model of MT   (b) representation of MT in BPMN

Fig. 9 Management transaction is a conceptual representation of real world causality (self-managed activity)

- **Management Transaction**

The main concept of enterprise causality discovery is a *management functional dependency*. An enterprise defined as a set of interrelated self-managed activities, wherein each self-managed activity is implementation of the specific management functional dependency (Gudas 2012a). *A management functional dependency (MFD)* is the primary cause of dependency between a set of activities that are collaborating to achieve a definite goal G related with a definite enterprise process P. The content of a management functional dependency is information/knowledge transformations between activities, and specified using the concept of *management transaction*. The first step of enterprise causality modeling is discovering of MFD within the problem domain and conceptual representation of perceived MFD as the management transaction (MT) (see Fig. 9).

We draw attention to what the MFD and MT are the examples of the circular causality in the organizational systems (e.g. in enterprises, CES and CSS). The difference between MFD and MT: MFD denotes a circular causality of domain activities perceived in the mind of enterprise staff (in the human brain), and here MT is representation of the perceived causality (e.g. identified MFD) in some modeling notation (e.g. BPMN, DFD, IDEF, etc.), aimed to define the information content of MFD.

A management transaction defined as a closed loop of goal-driven information transformations between the enterprise management function F and the enterprise process P (Fig. 9a). Enterprise goal (G) affects all actions (interactions of elements) inside management transaction (Fig. 9b).

An example of a specific MT: *MTij ((Order fulfillment) = ((Fj—Order fulfillment management) × (Pi—Build and ship product), A—Orders received, V—Product shipment invoice)*.

Management transaction is a conceptual representation of the real world causality perceived as MFD (see Fig. 9). A top level management transaction $MT^* = ((F \times P), A, V)$ comprises the enterprise goal G, the enterprise management function F, the enterprise process P, the management information flows "State attributes" (A)" and "Controls" (V) integrated by the feedback loop of management and control (Gudas 2012a).

The reason to represent the proposed domain modeling frameworks in two different ways—conceptual model and de facto *standard* BPMN—is to examine and evaluate the capabilities of BPMN to specify the closed loop processes. As discussed above, the closed loop processes are a common type of constructs for depicting the essential features of business management activities.

The next step reveals deep knowledge of target domain—getting the internal structure of the MT considered as a self-managed system. Here is the deep structure of management transaction MT defined as the elementary management cycle (EMC) (Fig. 10).

- **Elementary Management Cycle**

The elementary management cycle (EMC) is the internal model of MT, which reveals detailed structure of MT (Fig. 10). The steps {Tp} of EMC specify transformations of management information (data gathering rules, data interpretation and processing rules, decision making rules etc.) required for implementing the management function Fj closed loop interaction with process Pi (Gudas 2012a). The components of EMC are goal-driven items, because the enterprise management goal (G) makes

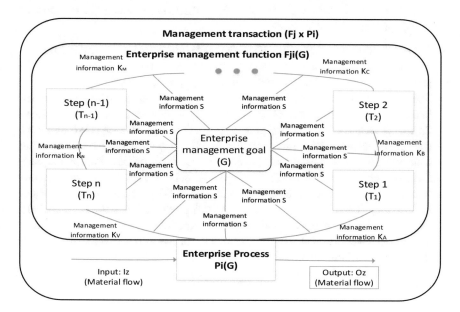

Fig. 10 The framework of elementary management cycle (EMC)

an impact on the EMC steps $(T_1, ..., T_n)$ as well as on management information flows $(K_A, K_B, ..., K_V)$ between EMC steps. This impact of the goal G denoted by management information flow (S).

The elementary management cycle (EMC) considered as a causal model of the self-managed activity (Gudas 2012a) which includes enterprise management goal (G), enterprise process P(G), and the management function F(G). F(G) has a predefined internal structure which comprises: the information transformation steps $T(G) = (T_1, ..., T_n)$, the management information flows $K(G) = (K_A, ..., K_V)$ between the steps T(G), and a set of impacts S(G) focused on the process P(G), steps T(G), and flows K(G) (Fig. 10).

The elementary management cycle (EMC) defined as follows:

$$EMC = (G, P(G), F(T(G), K(G), S(G)));$$ (3)

The concept "management information" in Fig. 10 denotes all types of information flows: data, knowledge, information, goals, rules, directives, constraints, etc. The management information flows S indicate an impact of goal (G) on the EMC steps T as well as on flows Ki between steps.

In addition to the management transaction, the elementary management cycle (EMC) is a key element of the enterprise causal model. In parallel with definitions of atoms and molecules of organizations in the ontological approach towards enterprise modeling (Dietz 2003), EMC could be considered as an atomic component of the enterprise management model.

Therefore, the generalized EMC was adapted for the application software engineering in Gudas (2012a), Gudas et al. (2005). Particular semantics of components of the adapted EMC is as follows:

- The enterprise process Pi of the EMC denotes some primary activity Pi (transformation of materials, energy, products). Inputs (I) and outputs (O) of the enterprise process Pi are material flows (materials, energy, products, services);
- The management function Fj of the EMC includes four major typical management information transformations (steps T) required to control the enterprise process Pi: Interpretation (IN), Data Processing (DP), Decision Making (DM) and Realization of decisions (RE); Interpretation IN(G) is a goal-driven process of fixing, identification and gathering data (state attributes A) of process Pj. The output of IN is a systematized data flow B. Data processing DP(G) is a goal-driven data processing that depends on the enterprise goals (G). A DP prepares data for a decision making step (DM). Decision making DM (G) carried out according to the rules that depend on the enterprise goals (G). The output of the Decision Making step is a management decision D intended to direct the process Pj. Realization of decision RE (G) is carried out according to the enterprise goal (G) requirements. The output of RE is controls V—an impact required by decision D on the enterprise process Pj (Gudas 2012a).

7   The Causal Knowledge Discovery Technique

The principal schema of the knowledge-based (KB) domain analysis and causal knowledge discovery (CKD) technique presented in Fig. 12. The premise is that experts understand the concepts of enterprise domain causality such as MFD, MT and EMC. The realization of some need (some goal or capability) in the field of activity is a specific MFD covering appropriate causal interactions between specific activities. This MFD is conceptualized and depicted as a management transaction $MT = ((F \times P), A, V)$ (see Fig. 6).

The technique integrates two top-down decomposition perspectives: a control view, and a self-managing view. A control view implies that top level MT* considered to be an aggregate of lower level MT's. A control view driven decomposition means that: (a) the MT* function F subdivided into sub-functions F_1, Fj, ..., Fn; (b) MT* process P decomposed into sub-processes P_1, ..., Pi, ..., Pm; (c) the lower-level management transactions $MTij = ((Fj \times Pi), Ai, Vj)$ are created by linking relevant lower-level functions Fj with processes Pi. Experts determine sub-functions and sub-processes as well as new MT's. Making a new MTij has sense if required by enterprise need (defined as some subgoal or capability). The resulting set of management transactions forms DSCM—the Detailed Supply Chain Model (Gudas 2012a, 2016).

Self-managing view implies that MT* considered to be a self-managed system having the predefined internal structure represented as EMC framework (see Fig. 10).

- **The main stages of causal domain modeling technique**

The main stages of causal domain modeling technique are as follows:
Stage 1. Conceptualization of the management functional dependence
   Top level. Experts name the most important (strategic) needs (enterprise goals or capabilities). Specific needs (goal or capability) can be realized by creating a specific MFD* linking the required business activities. Such perceived MFD* is conceptualized as a top-level management transaction $MT* = ((F \times P), A, V)$ (see Fig. 9).
   Level 1. The top-down MT* decomposition is twofold and performed in the following steps:

Step 1.   Control view: MT* divided into a set of lower level management trans-actions {MTji} on the level 1. The resulting set of collaborating MTs comprises the Detailed Supply Chain Model: $DSCM^{(1)} = \{MTij\}$ (see Fig. 11).
Step 2.   Self-managing view: the internal components of MT* are determined by experts and in this way a top-level management transaction MT* is specified as $EMC^{(1)}$ on the level 1.
Step 3.   Verification on the level 1: comparison of two different internal models of MT*: the structure of $DSCM^{(1)}$ compared with the structure of $EMC^{(1)}$, and the correspondence and mismatch of elements is determined; required model adjustments made.

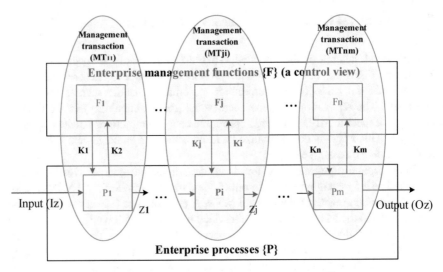

Fig. 11 The top level management transaction MT* specified as DSCM$^{(1)}$

Step 4. Validation on the level 1: evaluation of two different internal models of the management transaction MT*: a control view-based model DSCM$^{(1)}$ = {MTij}, and a self-managing view-based ECM$^{(1)}$ assessed individually against the needs of the enterprise.

Stage 2. Decomposition of management transactions {MTji}

Level 2. Step 1. Control view: decomposition of each management transaction MTji$^{(1)}$ (level 1), into a set of lower level management transactions {MTji$^{(2)}$}, and finally, composition of a lower level DSCM$^{(2)}$;

Step 2. Self-managing view: decomposition of each management transaction MTji$^{(1)}$ (level 1) into a relevant lower level EMCji$^{(2)}$ (level 2). Thus, a lower level EMCji created for each management transaction MTji, identified at a higher level.

Step 3. Verification of two representations of the same (higher level) management transaction MTji, i.e. comparison of all pairs of DSCMji and ECMji on the same level of knowledge discovery: Verification-1 of all the developed DSCMs$^{(2)}$ against relevant EMC$^{(2)}$; Verification-2 of all the developed EMCs$^{(2)}$ against relevant DSCM$^{(2)}$;

Step 4. Validation: evaluation of two representations of the same (higher level) management transaction MTji, i.e. validation-1 of all DSCMji and validation-2 of ECMji on the same level of modeling: Validation-1 of the developed set of DSCM$^{(2)}$; Validation-2 of the developed set of EMC$^{(2)}$.

Stage 3. Lower level causal modeling

A causal modeling at level 3 and at lower levels include the same steps as defined in Stage 2. The end level of decomposing hierarchy depends on the enterprise needs and experts.

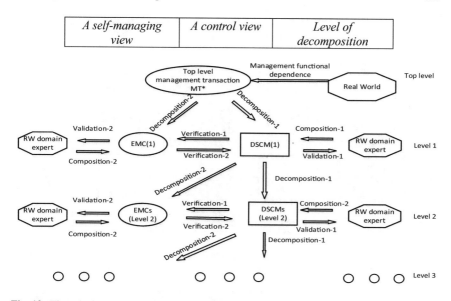

A self-managing view	A control view	Level of decomposition

Fig. 12 The principal schema of causal knowledge discovery

To go further into potentiality of the internal modeling, the causal knowledge discovery (CKD) technique specified in detail according to principal schema in Fig. 12 (i.e. using this CKD technique), and using the prior causal knowledge frameworks (MT and EMC).

Stage 1. Top-level decomposition.

The proposed causal knowledge discovery technique defined as the top-level management transaction $MT^* = (F =$ "Causal modeling of the subject domain", $P =$ "Prior knowledge based analysis of subject domain", $Ki =$ "Perceived causal dependencies", $Kj =$ "Verified and validated causal model"). The outcome of the *Stage 1. Top level decomposition* depicted in Figs. 13 and 14. A control view of the causal knowledge discovery process depicted as $DSCM^{(1)}$ in Fig. 13: $DSCM(1) = \{((Fj \times Pi), Ki, Kj)\}$, here: Fj is the enterprise management function ($Fj \in F$, F is a set of management functions); Pi is the enterprise process ($Pi \in P$, P is a set of enterprise processes), $(Fj \times Pi)$ is the Cartesian product of F and P, Ki—process Pi state attributes, Kj—controls of process Pi. Four specific MTs identified $MT_{11} = ((F_1 \times P_1), K_1, K_2)$, $MT_{22} = ((F_2 \times P_2), K_3, K_4)$, $MT_{33} = ((F_3 \times P_3), K_5, K_6)$, and $MT_{44} = ((F_4 \times P_4), K_7, K_8)$. The management transaction MT_{11} named "Development of the DSCM of subject domain" and used to illustrate this method below. A self-managing view of causal knowledge discovery technique represented in Fig. 14: $EMCji^{(1)} = ((Gji, Pi(G), Fj(Tjn(G), Kj(G), Sj(G)))$, here: Gji—an enterprise goal co-related with the process Pi and management function Fj. Fj is a complex structure which includes: $Kj(G)$—knowledge/data flows, $Tjn(G)$—transformations of K flows (steps), $Sj(G)$—impacts of goal Gj.

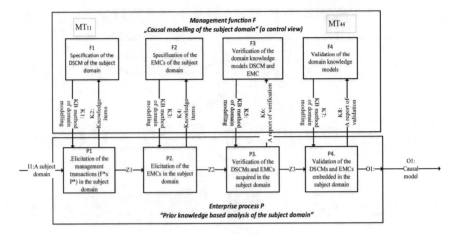

Fig. 13 The causal knowledge discovery process represented as DSCM (control view, level 1)

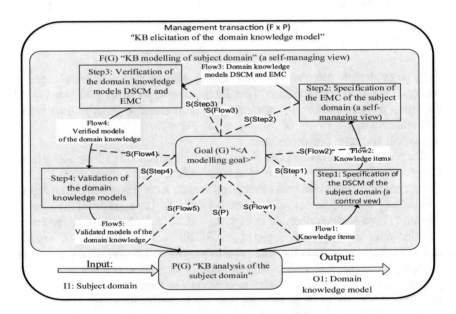

Fig. 14 The causal knowledge discovery process represented as $EMC^{(1)}$ (self-managing view, level 1)

The causal modeling on the level 2 is illustrated by the control view-based decomposition of $MT_{11}^{(1)}$ "Development of the DSCM of subject domain" (see Fig. 13). The detailed internal structure of $MT_{11}^{(1)}$ represented as $DSCM_{11}^{(2)} = \{(F_{11} \times P_{11}), (F_{12} \times P_{12}), (F1_{13} \times P_{13}), F_{14} \times P_{14})\}$ in Fig. 15.

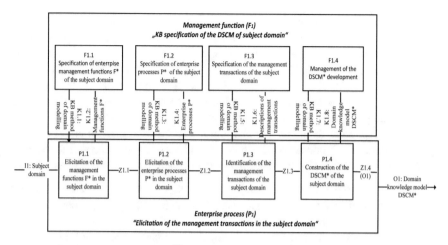

Fig. 15 The detailed structure of $MT_{11}^{(1)}$ "development of DSCM of the subject domain" represented as $DSCM_{11}^{(2)}$ (control view, level 2)

The self-managing view-based causal domain modeling at level 2 is illustrated by decomposition of the same $MT_{11}^{(1)}$ "Development of the DSCM of subject domain" (see Fig. 13). $MT_{11}^{(1)}$ presented as $EMC_{11}^{(2)}$ in Fig. 16.

- **Verification of the domain knowledge model**

We follow this definition of verification: "*Verification: The process of determining that a model implementation and its associated data accurately represents the developer's conceptual description and specifications*" (DoD 2009). Verification of

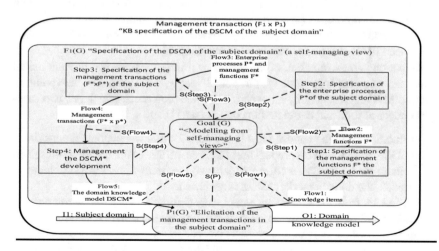

Fig. 16 Lower level $EMC_{11}^{(2)}$ of the management transaction MT_{11} "development of DSCM of the subject domain" (level 2, a self-managing view)

the domain model includes bidirectional comparison of control view based model with self-managing view-based model. Thus, the causal domain modeling technique includes a twofold verification of the developed domain models DSCM and EMC at each level of modeling. Verification-1 is comparison of DSCM content against EMC on the same level of modeling. Verification-2 is comparison of EMC content against DSCM on the same level of modeling. Informal rules of the verification-1 of DSCM mapping to EMC are as follows: the management transactions MT within DSCM correspond to the steps of EMC, i.e. in DSCM each pair (Fj × Pi) of management functions Fj and enterprise processes Pi matches to some definite step of EMC. Informal rules of the verification-2 of EMC mapping to DSCM are as follows:

(a) the steps of the EMC correspond to the management transactions MT within DSCM, i.e. the step of EMC matches to the pair (F × P) of management function F and enterprise process P of some MT;

(b) The information flows of the EMC and MT are partly compliant. Due to difference of modeling viewpoints, a part of the information flows of MT, namely feedback information flows between F and P, are hidden in EMC framework, i.e. the internal flows of the management transaction (F × P) are not represented in EMC model of the same domain.

- **Validation of the domain knowledge model**

Validation of the domain knowledge model is aimed for determining the degree to which the model is an accurate representation of the real world (subject domain. We follow such definition of validation: "*Validation. The process of determining the degree to which a model or simulation and its associated data are an accurate representation of the real world from the perspective of the intended uses of the model*" (DoD 2009).

This technique of KB domain modeling includes a twofold validation at each level of domain modeling: (a) validation-1: evaluation of control view–based DSCM of the domain (or sub-domain), and (b) validation-2: evaluation of self-managing view-based EMC of the same domain (or sub-domain).

Validation is the evaluation of *the degree to which a* represented domain model is an *accurate representation of* the actual real system from the viewpoint of required MFD represented by an expert. The validation process includes two cases independent of time and correlated in content (see Fig. 9):

(a) Validation from the control view (validation-1): real world domain considered as DSCM. An expert (propagating required MFD) evaluates the represented domain model DSCM, a correspondence to the needs of real world domain. If needed, the domain model DSCM is reviewed.

(b) Validation from the self-management view (validation-2): a real world domain considered as EMC. An expert (propagating required MFD) determines the degree to which the internal components of represented EMC (Fj × Pi) correspond to the activities within a real world domain. If needed, the domain model EMC is reviewed.

Validation from the control view (validation-1) defines the accuracy of the subject domain model represented by DSCM. The core informal rules of validation-1 are as follows:

- Are some management functions F* omitted or insufficiently defined in the model?
- Are some enterprise processes P* omitted or insufficiently defined?
- Are elements of the feedback loop (information flows) between some F and P omitted or insufficiently defined?
- Is the domain model (represented as DSCM) quite adequate to required MFD or not?

The validation from the self-managed view (validation-2) defines the accuracy of the subject domain model represented by EMC. The core informal rules of validation-2 are as follows:

- Are some steps of management function Fx omitted or insufficiently defined?
- Are some flows between the steps of the management function Fx omitted or insufficiently defined?
- Are the enterprise process Px inputs and outputs omitted or insufficiently defined?
- Are impacts of the enterprise goal G on the steps and flows omitted or insufficiently defined?
- Is the domain model (represented as EMC) quite adequate to required MFD or not?

8 Experimental Research

The causal modeling technique illustrated for the education enterprise of Study programme (SP) development (Gudas et al. 2019; Tekutov et al. 2012). The education enterprise "Study programme development and renewal" is considered as the management functional dependence MFDSP aimed to maintain the quality of study programmes. A top-level model of this MFDSP is the management transaction $MT_{SP} = ((F_{SP} \times P_{SP}), K1, K2))$ "Study programme development and renewal" presented in Fig. 17.

A top down decomposition of MT_{SP} on level 1 is twofold: Step 1: a control-view based (Fig. 18), and Step 2: a self-managing view-based (Fig. 19).

A control view-based causal model of MT_{SP} "Study programme management and renewal" defined as $DSCM_{SP}$ in Fig. 18. $DSCM_{SP}$ includes management transactions as follows: $MT_{11} = ((F1 \times P1), (K1, K2))$—constructing of the encapsulated SP knowledge model by mining the SP content; $MT_{22} = ((F2 \times P2), (K3, K4))$ elicitation of actual domain model; $MT_{33} = ((F3 \times P3), (K5, K6))$—renewal of (old) SP knowledge model (using actual domain knowledge); $MT_{44} = ((F4 \times P4), (K7, K8))$—development of renewed SP based on the causal domain knowledge. It has become clear, that the study programme management function F_{SP} is a complex

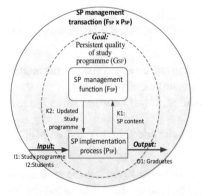 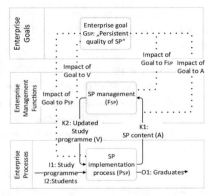

(a) The top level conceptual model of MFD$_{SP}$ (b) The top level model of MFD$_{SP}$
in BPMN

Fig. 17 MFD$_{SP}$ "study programme development and renewal" depicted as the management transaction MT$_{SP}$: **a** conceptual model, and **b** BPMN notation

component and includes a set of lower level management functions: $F_{SP} = \{F_1, F_2, F_3, F_4\}$. Note that a control view-based decomposition of MT$_{SP}$ does not include goal analysis.

MT$_{SP}$ "Study programme development and renewal" specified from the self-managing viewpoint as EMC$_{SP}$ (at level 1) in the Fig. 19. The causal model of management function $F_{SP}(G)$ "Study programme management" includes the steps as follows:

- Step 1. Knowledge-based modelling of the existing SP content: analysis of the existing SP content aimed to reveal knowledge items of the SP domain, that are encapsulated in the existing (old) study programme;
- Step 2. Knowledge-based elicitation of the actual domain model: investigation of the SP related real world domain, using internal modelling methodology;
- Step 3. Renewal of the SP domain model stored in the knowledge base;
- Step 4. Knowledge-based renewal of the study programme content in accordance with the actual SP domain model.

9 Conclusions

Causal modeling becomes a fundamental issue of the enterprise application software engineering. The prevailing now external modeling paradigm in MDD is not sufficient to justify the requirements of EAS development. Evaluating from the system analysis perspective, the domain modeling methods on CIM layer belong to external modeling paradigm (or black box approach). Such modeling notations as DFD, BPMN, IDEF, ARIS, UPDM assigned to empirical as they based on the notations that do not include

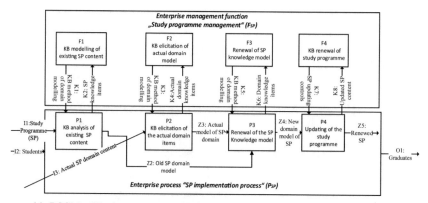

(a) DSCM$_{SP}$ "Study programme management and renewal" (Level 1, a control view)

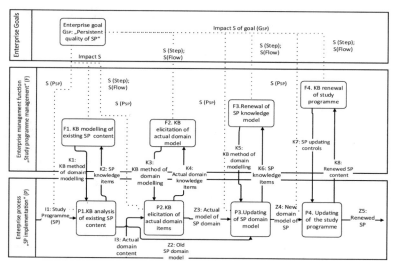

(b) DSCM$_{SP}$ "Study programme management and renewal" in BPMN notation (Level 1, a control view)

Fig. 18 The internal model of management transaction MT$_{SP}$ (level 1, a control view)

causal dependencies, inherent to the domain type. The causality-based approach is a shift of domain modeling to an internal modeling paradigm that transcends the boundaries of traditional enterprise and business process modeling methodologies. Causal modeling requires intellectual readiness—to have prior knowledge of the specific causality inherent to the particular domain type.

Domain causality modeling at CIM level are promising to figure out deep properties of the enterprise—causal dependencies of goal-driven activities, and data/knowledge transformations. The results of this study help us understand that

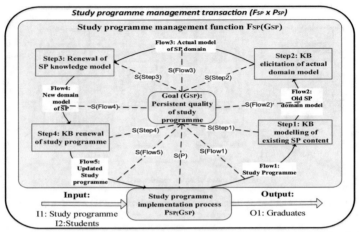

(a) The internal model of management transaction MT$_{SP}$ depicted as
EMC$_{SP}$ "Study programme management and renewal" (Level 1, a self-
managing view)

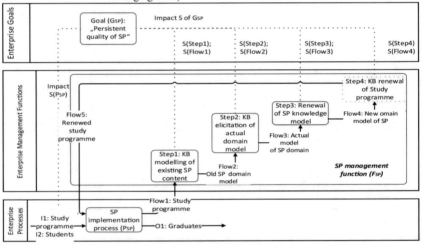

(b) EMC $_{SP}$ "Study programme management and renewal" in BPMN
notation (Level 1, a self-managing view)

Fig. 19 EMC$_{SP}$ of MFD$_{SP}$ "study programme development and renewal" notation (level 1, self-
managing view)

the content of the CIM layer should be as close as reasonable to the causality of the
domain.

The main contribution is causal modeling approach which enabled the disclosure
of gap in conceptualization of the subject domain on CIM layer of MDA and to
develop the expanded MDA schema with a new layer of the domain knowledge
discovery. The causal knowledge discovery (CKD) technique presented, it is tailored

for enterprise domain causal dependencies capturing and modeling. In this approach, a notion of circular causality is important.

The proposed technique of domain modeling described using the previously introduced concepts: a perceived domain causality MFD—management functional dependence, MT framework (a control view-based causal model of MFD) and EMC framework (a self-managing view-based internal model of MT). A top level causal model is defined as a management transaction (MT). The next step is the detailed MT modeling when an elementary management cycle (EMC) is created for each MT. EMC reveals the internal structure of MT and goal-driven interactions of MT elements: internal transformations and workflow of data/knowledge.

The extended MDA framework and causal knowledge discovery (CKD) technique will help to capture the domain causal dependencies on CIM layer, and next, to develop transformation specifications to map to PIM layer. This is the first step in transferring the causality of the domain through other MDA layers to the application software. The influence of the causal domain model on the PIM and PSM layers and the choreography of the application should be further investigated.

References

Bondy A, Murty USR (2008) Graph theory. Springer, Berlin

Bunge M (2011) Causality and modern science, 3rd revised edn. Courier Corporation, DOVER Publications, Inc., Mineola

Conant RC, Ross Ashby W (1970) Every good regulator of a system must be a model of that system. Int J Syst Sci 89–97

Deming WE (1993) The new economics for industry, government and education. Massachusetts Institute of Technology, Center for Advanced Engineering Study, Cambridge

Dietz JLG (2003) The atoms, molecules and fibers of organizations. Data Knowl Eng 47(3):301–325

Dietz JLG (2006) The deep structure of business processes. Commun ACM 49(5):58–64

DoD (2009) Department of Defense, Instruction 5000.61, M&S VV&A. www.dtic.mil/whs/directives/corres/pdf/500061p.pdf

Falkenberg ED et al (1996) FRISCO: a framework of information system concepts. The IFIP WG 8.1 Task Group FRISCO, Dec 1996

Francis BA, Wonham WM (1976) The internal model principle of control theory. Automatica 12(5):457–465

Frosini BV (2006) Causality and causal models: a conceptual perspective. Int Stat Rev 74(3):305–334

Georgakopoulos D, Hornick M, Sheth A (1995) An overview of workflow management: from process modeling to workflow automation infrastructure. In: Distributed and parallel databases, vol 3. Kluwer Academic Publishers, Boston, pp 119–153

Granger CWJ (1969) Investigating causal relations by econometric models and cross-spectral methods. Econometrica 37(3):424–438

Gudas S (2012a) Foundations of the information systems' engineering theory. Vilnius University Press, Vilnius

Gudas S (2012b) Knowledge-based enterprise framework: a management control view. New research on knowledge management models and method. http://www.intechopen.com/books/new-research-on-knowledge-management-models-and-methods/knowledge-based-enterprise-framework-a-management-control-view

Gudas S, Lopata A (2016) Towards internal modeling of the information systems application domain. Informatica 27(1):1–29

Gudas S et al (2005) Approach to enterprise modeling for information systems engineering. Informatics 16(2):175–192

Gudas S, Tekutov J, Butleris R, Denisovas V (2019) Modelling subject domain causality for learning content renewal. Informatica 30(3):455–480

Halpern JY (2016) Sufficient conditions for causality to be transitive. Philos Sci 83(2):213–226

Harmon P (2010) The scope and evaluation of business process management. In: von Brocke J, Rosemann M (eds) International handbooks on information systems, handbook on business process management, vol 1. Springer, Berlin, pp 37–81

Heylighen F, Joslyn C. (2001) Cybernetics, and second-order cybernetics. In: Meyers RA (ed) Encyclopedia of physical science & technology, 3rd edn. Academic Press, New York. Accessed 02 Apr 2019. http://pcp.vub.ac.be/Papers/Cybernetics-EPST.pdf

ISO 15704:2000 Industrial automation systems—requirements for enterprise-reference architectures and methodologies

Kephart JO, Chess DM (2003). The vision of autonomic computing. Computer 36(1):41–50

Khoo C et al (2002) The many facets of the cause-effect relation. In: Green R, Bean CA, Myaeng SH (eds) The semantics of relationships: an interdisciplinary perspective. Kluwer, Dordrecht, pp 51–70

Kleppe A et al (2003) MDA explained, the model-driven architecture: practice and promise. Addison Wesley, Boston

Kumar S (2012) Kac-Moody groups, their flag varieties, and representation theory. Springer Science & Business Media, Berlin

Lewis D (1973) Causation. J Phil 70(17). Seventieth annual meeting of the american philosophical association eastern division, 11 Oct 1973, pp 556–567

Medina-Mora R, Winograd T, Flores R, Flores F (1992) The action workflow approach to workflow management technology. In: CSCW 92 Proceedings, pp 281–288

OMG (2019) Business modeling category—specifications associated. https://www.omg.org/spec/category/business-modeling

Osis J (2004) Software development with topological model in the framework of MDA. In: Proceedings of the 9th CaiSE/IFIP8.1/EUNO international workshop on evaluation of modeling methods in systems analysis and design (EMMSAD'2004) in connection with the CaiSE'2004, 1, Riga, Latvia: RTU, pp 211–220

Osis J, Asnina E (2011) Model-driven domain analysis and software development: architectures and functions. IGI Global, Hershey

Osis J, Donins U (2017) Topological UML modeling: an improved approach for domain modeling and software development. Elsevier, Amsterdam

Papazoglou MP (2003) Web services and business transactions (2003). In: World wide web: internet and web information systems, vol 6. Kluwer Academic Publishers, Dordrecht, pp 49–91

Porter ME (1985) Competitive advantage. The Free Press, New York

Rummler GA (2010) White space revisited: creating value through process. Wiley, San Francisco

Scott B (1996) Second-order cybernetics as cognitive methodology. Syst Res 13(3):393–406

SEP (2018) Stanford encyclopedia of philosophy. https://plato.stanford.edu/entries/causal-models/#CausDiscInte. Accessed 15 Mar 2019

Techopedia (2019) Transaction. https://www.techopedia.com/definition/16455/transaction. Accessed 04 Apr 2019

Tekutov J et al (2012) The refinement of study program content based on a problem domain model. Transform Bus Econ 11(1(25)):199–212

Trask B, Roman A (2009) Introduction to model driven development with examples using eclipse frameworks. In: ACM conference on object-oriented programming, systems, languages and applications, Orlando

Valatavičius A, Gudas S (2017) Toward the deep, knowledge-based interoperability of applications. Inf Sci 79:83–113 von Foerster et al (1953) The Macy conference proceedings note

Wang L et al (2015) Research on fault diagnosis system with causal relationship in equipment technical manual's deep knowledge. In: Yang (ed) Advances in future manufacturing engineering. Taylor & Francis Group, London, pp 28–31

Discrete Competitive Facility Location by Ranking Candidate Locations

Algirdas Lančinskas, Pascual Fernández, Blas Pelegrín and Julius Žilinskas

Abstract Competitive facility location is a strategic decision for firms providing goods or services and competing for the market share in a geographical area. There are different facility location models and solution procedures proposed in the literature which vary on their ingredients, such as location space, customer behavior, objective function(s), etc. In this paper we focus on two discrete competitive facility location problems: a single objective discrete facility location problem for an entering firm and a bi-objective discrete facility location problem for firm expansion. Two random search algorithms for discrete facility location based on ranking of candidate locations are described and the results of their performance investigation are discussed. It is shown that the ranking of candidate locations is a suitable strategy for discrete facility location as the algorithms are able to determine the optimal solution for different instances of the facility location problem or approximate the optimal solution with a reasonable accuracy.

Keywords Facility location · Combinatorial optimization · Multi-objective optimization · Random search algorithms

A. Lančinskas (✉) · J. Žilinskas
Institute of Data Science and Digital Technologies, Vilnius University,
Akademijos 4, 08412 Vilnius, Lithuania
e-mail: algirdas.lancinskas@mii.vu.lt

J. Žilinskas
e-mail: julius.zilinskas@mii.vu.lt

P. Fernández · B. Pelegrín
Department of Statistics and Operations Research, University of Murcia Campus Espinardo,
30071 Murcia, Spain
e-mail: pfdez@um.es

B. Pelegrín
e-mail: pelegrin@um.es

© Springer Nature Switzerland AG 2020
G. Dzemyda et al. (eds.), *Data Science: New Issues, Challenges and Applications*, Studies in Computational Intelligence 869,
https://doi.org/10.1007/978-3-030-39250-5_8

145

1 Introduction

Facility Location (FL) deals with finding locations for facilities with respect to minimization of the production costs or maximization of the profit of a firm. The right location for a facility depends on different factors such us type of service or product that the firm provides, production and delivery costs, market environment, customer behavior when choosing a facility, etc. Customers are ready to travel up to a certain distance which depends on the type of a service they want to buy; e.g. to buy a car or to reach an airport customers can travel very far, but for daily shopping they usually choose local market places. Cristaller (1950), Lösch (1940) and von Thünen (1910) were among the first authors in developing theory of central places, explaining how different types of businesses locate in a region (Fischer 2011). The goal for facility location is to find the optimal locations for a given set of new facilities with respect to maximize utility of the new facilities and minimize any unwanted effect caused by locating the new facilities. Although the location as an object of papers and books was from the beginning of the 20th century, continuous research on facility location started in the 1960s from the works of Balinski (1965), Hakimi (1964, 1965), ReVelle and Swain (1970), etc.

Some FL problems, which have attracted a lot of attention, are Competitive Facility Location Problems (CFLPs), where two or more firms compete for market share by choosing appropriate locations, competitive prices, high service quality, and other properties to attract as many customers as possible (Ashtiani 2016; Drezner 2014; Eiselt et al. 2015). Some real-world competitive facility location problems require to take into account several criteria when determining optimal locations for the new facilities (e.g. market share, costs for establishment and maintenance, any undesirable effects, etc.), thus making the decision maker to face a multi-objective optimization problem. See Doerner et al. (2009), Farahani et al. (2010) for instances of multi-objective facility location problems and Chinchuluun and Pardalos (2007), Chinchuluun et al. (2008), Montibeller and Franco (2010) for the concept and developments in solution of multi-objective optimization problems.

Solving a real-world CFLP is usually complex and computationally expensive due to reasons such as complex objective function(s) or need of complex analysis of a large amount of data, e.g. population of prospective customers, their current and expected behavior when choosing the facility for a service, etc. Due to these and similar reasons it can be impossible to find the optimal solution(s) within a reasonable time. Therefore, heuristic methods, which can be applied to approximate the optimal solution(s) of a specific optimization problem, are often used to tackle a real-world CFLP. One of the most popular heuristic algorithms for single-objective optimization is the Genetic Algorithm (GA) which has been proposed by Holland (1975) in 1975 and described by Goldberg (1989) in 1989. GA has been successfully applied to solve various optimization problems including facility location problems; see Jaramillo et al. (2002) for examples of application of GA to FLPs.

Srinivas and Deb (1994) proposed a Non-Dominated Sorting Genetic Algorithm (NSGA) which is suitable for multi-objective optimization and was one of the first multi-objective optimization algorithms applied to various problems. Deb et al. (2002) proposed the updated version of the algorithm – NSGA-II. NSGA-II has been applied to solve various multi-objective FLPs. Villegas et al. (2006) utilized NSGA-II to solve a bi-objective FLP by minimizing operational cost of Colombian Coffee supply network and maximizing the demand; Liao and Hsieh (2009) used NSGA-II to optimize the location for distribution centers with respect to two objectives: maximization of customer service and minimization of the total cost; Medaglia et al. (2009) utilized hybrid NSGA-II and mixed-integer programming approach to solve a bi-objective obnoxious FLP related to the hospital waste management network.

In this work we will focus on two CFLPs: the single-objective CFLP for an entering firm which is used by firms planning to enter the market and the multi-objective CFLP for firm expansion which is used by firms which are already in the market but want to expand their market share by locating new facilities. Formulation of the problems, which will be presented in Sect. 2, includes three different customer behavior models: proportional, binary, and partially binary, all of which will be introduced in the same section. Two random search algorithms to solve the formulated facility location problems will be described and discussion on their performance will be presented in Sects. 3 and 4. Finally, conclusions will be formulated in the last section.

2 Facility Location Problems

Consider a set of firms

$$K = \{1, 2, \ldots, k\}, \tag{1}$$

providing services or goods in a certain geographical region. The i-th firm owns a set J_i of preexisting facilities which satisfy customer needs in the region. A firm wants to establish a set X of s new facilities in order to enter the market or expand its current market share if the firm is already in the market. Locations for the new facilities can be chosen from a discrete set L of candidate locations taking into account the competition for the market share with other market participants. Then the entering or expanding firm faces a Discrete Competitive Facility Location Problem (DCFLP) aimed at selecting a subset $X \subset L$ of location candidates for the new facilities with respect to optimize the objective function.

In order to make the problem computationally tractable, all customer demand in every spatially separated area is aggregated to a single demand point. Let I be the set of demand points, assuming that demand at each demand point is fixed and known as w_i, where $i = 1, 2, \ldots, |I|$. See Francis et al. (2002) for demand aggregation.

2.1 Customer Choice Rules

The rules describing attractiveness of a facility to customers are a remarkable part of CFLP. Some methods to estimate the customer attractiveness have been proposed by Huff (1964), where attractiveness of a facility is estimated by dividing facility quality indicator by a positive non-descending function of the distance between the customer and the facility. One of the expressions of the attractiveness that the i-th customer feels to the j-th facility, used in Fernández et al. (2017), Lančinskas et al. (2017), is

$$a_{ij} = \frac{q_j}{1 + d_{ij}},\tag{2}$$

where q_j is the quality of the j-th facility and d_{ij} is the distance between the i-th customer and the j-th facility. The quality of a facility is expressed by a numerical value which depends on properties of the facility such as location, size, variety of services and additional attractions, leisure areas, parking lots, etc.

The attractiveness of a facility is the base for a key ingredient of CFLPs – the rules which customers follow when choosing a facility to satisfy their needs. The most popular customer choice rules are proportional (Drezner and Drezner, 2004; Peeters and Plastria, 1998; Serra and Colomé, 2001) and binary (Hakimi, 1995; Suárez-Vega et al., 2004, 2007). Further we will review the latter customer choice rules together with the partially binary customer choice rule, which has been presented and investigated in Fernández et al. (2017).

Proportional customer choice. In the case of the proportional customer choice rule the buying power of a demand point is divided among all facilities in proportion to their attractiveness. Then the total market share of the set X of s new facilities is

$$M(X) = \sum_{i \in I} w_i \frac{\sum_{j \in X} a_{ij}}{\sum_{j \in X} a_{ij} + \sum_{k \in K} \sum_{j \in J_k} a_{ij}},\tag{3}$$

where w_i is the demand of the i-th demand points.

Binary Customer Choice Rule. In the case of the binary customer choice rule the whole buying power of a single demand point is satisfied by a single facility – the most attractive one. It may occur that there are more than one facility with the maximum attractiveness for a single demand point. If all tied facilities belong to the firm which locates new facilities, then the firm captures the whole buying power of the demand point. If some of the tied facilities belong to the competitors, it is assumed that the firm locating the new facilities captures a fixed proportion $\theta \in [0, 1]$ of customer's demand. If none of the tied facilities are owned by the entering firm, then no demand is captured from the customer.

Let us define the set of demand points, which maximal attractiveness is for a facility in X,

$$I^> = \{i \in I : a_i(X) > \max\{a_i(J_k) : k \in K\}\}\tag{4}$$

and the set of demand points which maximal attractiveness for a facility in X is equal to a competitor's facility from J_k, $k \in K$

$$I^= = \{i \in I : a_i(X) = \max\{a_i(J_k) : k \in K\}\}, \tag{5}$$

where $a_i(\cdot)$ is the maximum attractiveness that customers from the i-th demand point feel for the facilities in a given set. Then the market share captured by the facilities in X is

$$M(X) = \sum_{i \in I^>} w_i + \sum_{i \in I^=} \theta w_i. \tag{6}$$

Partially Binary Customer Choice Rule. In the case of the partially binary customer choice rule the buying power of a single demand point is satisfied by all firms, but the customers patronize one facility per firm – the most attractive one. Then the demand is split between these facilities in proportion with their attractiveness. In this model it is not necessary to consider ties between more than one facility because they are irrelevant when it comes to obtain the total market share captured by the entering firm.

The market share captured by the entering firm for the partially binary rule is

$$M(X) = \sum_{i \in I} w_i \frac{a_i(X)}{a_i(X) + \sum_{k \in K} a_i(J_k)}. \tag{7}$$

2.2 Discrete Competitive Facility Location Problems

DCFLP for an Entering Firm (DCFLP/EF) is one of the most popular FLPs in the literature (Aboolian et al. 2008; Ashtiani 2016; Drezner 2014). It is important for firms which want to enter the market and, therefore, must compete for the market share with other firms which are already in the market. The goal of the entering firm is to find optimal locations for the new facilities with respect to maximization of their market share or profit, taking into account the customer behavior when choosing a facility or facilities to serve their demand.

Mathematically, the DCFLP/EF can be described as to choose set of locations X for s new facilities from the a set L of all candidate locations with respect to maximize the total market share of the new facilities:

$$\max\{M(X) : |X| = s, X \subset L\}, \tag{8}$$

where $M(X)$ is the total market share of the new facilities located in X, calculated taking into account the customer choice rules.

The **DCFLP for Firm Expansion (DCFLP/FE)** is another DCFLP important for firms which are already in the market, but want to expand their market share by establishing new facilities. The firm faces a bi-objective optimization problem with the following objectives: (1) maximize the total market share of the new facilities in X and (2) minimize the negative effect of the loss of customers of the preexisting facilities belonging to the expanding firm caused by the entrance of the new facilities. Such an undesirable effect is called *cannibalism*, which was studied for the first time in franchise systems in Ghosh and Craig (1991).

Assume that a firm $k \in K$, which owns the set J_k of facilities, wants to extends its market share by establishing a set X of s new facilities. Lets denote by $M^b(J_k)$ and $M^a(J_k)$ the market share captured by the facilities in J_k before and after the expansion, respectively. Since the new facilities can attract customers from preexisting facilities including those in J_k the following equality must be valid:

$$M^a(J_k) \le M^b(J_k).$$ (9)

The cannibalized market share can be expressed by the loss of them market share of the facilities in J_k caused by the expansion:

$$C(X) = M^b(J_k) - M^a(J_k).$$ (10)

Then the bi-objective DCFLP/FE can be formulated as to find a subset $X \subset L$ of s elements to represent locations for the new facilities with respect to maximization of the market share $M(X)$ and simultaneous minimization of the effect of cannibalism $C(X)$.

It is natural that a solution that is better by one objective can be worse or even the worst by another objective. Two solutions can be compared with each other by dominance relation: it is said that a solution X dominates another solution X' if X is not worse than X' by all objectives and X is strictly better than X' by at least one objective. Then the solution X is called a dominator of X'. A solution which has no dominators is called non-dominated or Pareto-optimal. A set of Pareto-optimal solutions is known as Pareto set and the corresponding set of objective values is known as Pareto front. Determination of the Pareto front is the main goal of the multi-objective optimization. See Fig. 1 for the example of the Pareto front of a DCFLP/FE in the context of a dominated point. The filled points in the picture stand for the Pareto-optimal points and the empty point is dominated by three filled points all of which are better by both objectives.

3 Solution of Single-Objective DCFLP/EF

This section aims to the description of the procedure of solution of the single-objective DCFLP/EF. The heuristic algorithm based on the ranking of candidate locations is described in the next section and performance of the algorithm is discussed in Sect. 3.2.

3.1 Ranking-Based Discrete Optimization Algorithm

Ranking-based Discrete Optimization Algorithm (RDOA) is based on a single agent random search algorithm for single-objective optimization. Since RDOA is a based on the random search, it does not guarantee convergence to the optimal solution, but goes to its neighborhood. On the other hand RDOA can be applied to a large variety of optimization problems including black-box-type problems for which mathematical expression of objective function is unknown.

The RDOA begins with an initial solution

$$X = \{x_1, x_2, \ldots, x_s\}, \tag{11}$$

which is a subset of the set L of candidate locations and is considered as the best solution found so far.

A new solution

$$X' = \{x'_1, x'_2, \ldots, x'_s\} \tag{12}$$

is derived from X by changing some locations for the new facilities. Each location x_i has probability $1/s$ to be changed and inverse probability $(1 - 1/s)$ – to be copied without change. In case of change, a new location is randomly sampled from the set L of all candidate locations excluding those which already forms X or X':

$$x'_i = \begin{cases} l \in L \setminus (X \cup X'), & \text{if } \xi_i < 1/s, \\ x_i, & \text{otherwise,} \end{cases} \tag{13}$$

where ξ_i is a random number uniformly generated over the interval $[0, 1]$, and $i = 1, 2, \ldots, s$.

Fig. 1 Illustration of the Pareto front of the DCFLP/FE

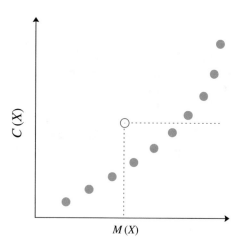

Each candidate location $l_i \in L$ has a rank value r_i which is expressed by positive integer value and defines fitness of l_i to form a new solution. At the beginning of the algorithm all candidate solutions have unit ranks, which are updated during the run-time of the algorithm by setting larger ranks to the more promising candidate location and lower ranks to the less promising ones. If the new solution X' has better fitness than the best known solution X, i.e. $M(X') > M(X)$, where $M(\cdot)$ is the objective function of the problem, then

(1) the ranks of candidate locations forming X' are increased by one, and
(2) the ranks of candidate locations forming X, but do not belong to X' are reduced by one.

Mathematically,

$$r_i = \begin{cases} r_i + 1, & \text{if } l_i \in X', \\ r_i - 1, & \text{if } l_i \in X \setminus X', \\ r_i, & \text{otherwise.} \end{cases} \tag{14}$$

If the fitness of the new solution is not better that the fitness of the best solution found so far, i.e. $M(X') \leq M(X)$, then the ranks of candidate locations which belong to X', but do not belong to X, are reduced by one:

$$r_i = \begin{cases} r_i - 1, & \text{if } l_i \in X' \setminus X, \\ r_i, & \text{otherwise.} \end{cases} \tag{15}$$

After update all ranks are adjusted so that the minimal rank would be equal to one.

The ranks are used to evaluate the probability π_i to sample l_i when forming a new solution. In the first strategy the sampling probability of l_i is proportional to the rank r_i:

$$\pi_i^r = \frac{r_i}{\sum_{j=1}^{|L|} r_j}. \tag{16}$$

The value of π_i^r is always between 0 and 1 and proportional to the rank value of the i-th candidate location – the larger rank value, the larger sampling probability will be assigned.

The second strategy includes CFLP feature – the geographical distance between locations. Besides the rank value r_i the sampling probability of l_i to change x_z' depends on the geographical distance between l_i and x_z' – the larger distance the lower sampling probability. Mathematically the probability can be expressed as

$$\pi_i^{rd} = \frac{r_i}{d(l_i, x_z') \sum_{j=1}^{|L|} \frac{r_i}{d(l_j, x_z')}}, \tag{17}$$

where $d(\cdot, \cdot)$ is the geographical distance between two given locations.

The algorithm which uses only rank values to evaluate the sampling probability is denoted by RDOA and the algorithm which additionally includes distance is denoted by RDOA/D.

The new solution improves the best solution found so far, then iteration is assumed to be successful and the new solution is used as the best known solution in further iterations. This iterative process continues till the maximum number of functions evaluations is exceeded.

3.2 Performance of RDOA

Both above algorithms have been applied to solve different instances of the CFLP/EF with binary and partially binary rules for customer choice (see Sect. 2.1). The database of geographical coordinates and population of 6960 municipalities in Spain has been used as demand points in the experimental investigation. The distance has been measured using Haversine formula which is based on the Great Circle principle (Sinnott 1984).

Three firms were considered as the market participants owning the same number of facilities per firm: 5 or 10. The preexisting facilities were located in the largest demand points measuring by the population. All preexisting facilities are located in the most populated demands points and are distributed among the firms as given in Table 1.

It was expected to locate $s = 5$ and $s = 10$ new facilities, selecting their locations from the set L of 500, 1000, and 5000 candidate locations.

The quality of the obtained solution is expressed as the ratio of the market share of the obtained solution to the market share of the optimal solution:

$$Q(X) = \frac{M(X)}{M(X_O)}, \tag{18}$$

where X_O stands for the optimal solution and X – for its approximation obtained by the algorithm under investigation. The quality value is restricted in the interval $[0, 1]$, where larger value means better quality of the solution and the maximal possible value means that $X \sim X_O$.

Table 1 Indices of pre-existing facilities for each firm

Firm	Indices of demand points	
	5 preexisting facilities	10 preexisting facilities
J_1	1, 4, 7, 10, 13	1, 4, 7, 10, 13, 16, 19, 22, 25, 28
J_2	2, 5, 8, 11, 14	2, 5, 8, 11, 14, 17, 20, 23, 26, 29
J_3	3, 6, 9, 12, 15	3, 6, 9, 12, 15, 18, 21, 24, 27, 30

An algorithm has been run for 10,000 function evaluations for each instance. Since both algorithms are based on the random search, statistical results of 100 independent runs has been analyzed.

Performance of RDOA and RDOA/D have been compared with the performance of the Genetic Algorithm (GA) (Davis 1991) on the same problem instances. The population of 100 individuals, uniform crossover with the rate of 0.8 and mutation rate of $1/s$ have been in the GA. The algorithm has been run for 100 generations thus performing 10000 function evaluations in total. Each experiment has been performed 100 times.

For the instances of the facility location problem with the binary rule of customer choice, the optimal solution X_O has been determined using deterministic algorithm Xpress (FICO Xpress Mosel 2014), whereas the best solution X, found in all experiments has been considered as X_O for the problem with the partially binary rule.

The performance of the algorithms have been compared by performance profiles when locating 10 new facilities among 5000 candidate locations – the most complicated DCFLP/EF instance. The performance profiles of RDOA, RDOA/D and GA are presented in Fig. 2, where the horizontal axis stand for the quality of the approximation after 10,000 function evaluations and the vertical – for the probability to achieve an approximation with particular quality. See Hendrix and Lančinskas (2015) for more details on comparison of algorithm by probability to achieve an approximation with a particular accuracy.

The figure shows that the RDOA/D after 1000 function evaluations gives notably better solution than the RDOA and GA. The RDOA/D demonstrates the probability close to one to find the solution with $Q > 0.9$. The probability to achieve similar approximation with RDOA is around 0.9 and with GA – around 0.6.

The same algorithms as in previous experiments have been applied to solve the same instances of the location problem, but with the different customer choice rule

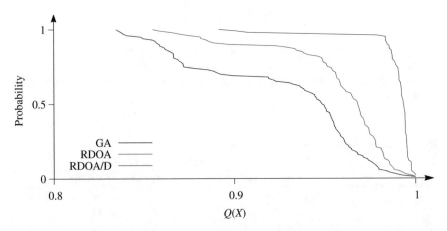

Fig. 2 Dependency of probability to achieve an approximation of the optimal solution on its accuracy when solving the facility location problem with the binary customer choice rule

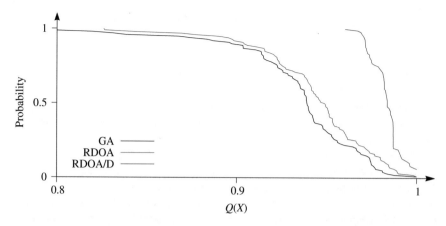

Fig. 3 Dependency of probability to achieve an approximation of the optimal solution on its accuracy when solving the facility location problem with the partially binary customer choice rule

– the partially binary rule. Since the problem with partially binary customer choice rule cannot be solved by the deterministic solver Xpress, the best solution found by a heuristic algorithm in all experiments has been considered as the optimal solution X_O. It is worth to note that RDOA/D gives the optimal solution in most runs on the problem with binary customer choice rule and arrives to the best known solution in most of runs on the problem with partially binary customer choice rule.

The results obtained for the problem with the partially binary customer choice rule is similar to those obtained for the problem with the binary rule, though solutions referring to the best market share are different. It was also realized that for the location problem with the partially binary rule, the proposed ranking based algorithms outperforms GA and the inclusion of the geographical distance factor has notable advantage.

The results presented in Fig. 3 show that partially binary customer choice rule makes the problem more complicated to all the algorithms – the curves are lower comparing with those in Fig. 2. Despite that RDOA/D still demonstrates better performance than RDOA and GA and its advantage is much more notable, comparing to the results presented in Fig. 2.

4 Solution of Multi-objective DCFLP/FE

This section aims to the description of the procedure of solution of the multi-objective DCFLP/FE. The heuristic algorithm for approximation of the Pareto front, based on the ranking of candidate locations, is described in the next section and performance of the algorithm is discussed in Sect. 4.2.

4.1   Multi-objective Random Search with Ranking

The ranking strategy has been applied to solve the bi-objective DCFLP/FE described in Sect. 2.2. The derived algorithm for multi-objective discrete optimization with ranking of the search space elements has been proposed in Lančinskas et al. (2016) and named by Multi-Objective Random Search with Ranking (MORSR). The MORSR is based on generation of new solutions by modifying the non-dominated solutions found so far.

The MORSR algorithm begins with a set P of non-dominated solutions. A new solution X' is derived from the reference solution X randomly selected from P using the same expression (13) as in RDOA (see Sect. 3). Similarly like in RDOA, each candidate location l_i has its rank r_i, which initially is equal to 1. Further the rank r_i is adjusted depending on the successes to form a new solution using l_i. If the newly generated solution X' is non-dominated in P and $l_i \in X'$ then

$$r_i = \begin{cases} r_i + 1, & \text{if } l_i \in X', \\ r_i, & \text{otherwise.} \end{cases} \tag{19}$$

If the non-dominated X' dominates a solution $X'' \in P$, then the ranks of candidate locations which form a X'', but do not form X' are reduced by one.

$$r_i = \begin{cases} r_i - 1, & \text{if } l_i \in X'' \setminus X', \\ r_i, & \text{otherwise.} \end{cases} \tag{20}$$

This rule is applied to all solutions $X'' \in P$, which are dominated by X'.

In order to avoid negative or non-proportionally large ranks, after every update all ranks are arranged so that the minimum rank value would be equal to one.

The ranks are used to identify the promising candidate locations and assign to them the corresponding probabilities to be sampled in (13): larger rank value r_i means larger sampling probability π_i of the candidate location l_i. The probability π_i is expressed by

$$\pi_i = \frac{r_i}{\sum_{j=1}^{|L|} r_j}. \tag{21}$$

If X' is a non-dominated in P, then P is updated by including X' and removing all the solutions dominated by X', if any:

$$P = (P \setminus P'') \cup \{X'\}, \tag{22}$$

where P'' is a set of all solutions from P, which are dominated by X'. The updated set P is used to sample a new reference solution X as a base for a new solution X' in the next iteration. Such an iterative process is continued till the number of function evaluations is exceeded. The the set P of non-dominated solutions is considered as an approximation of the true Pareto set.

4.2 Performance of MORSR

The performance of the MORSR algorithm has been investigated by solving different instance DCFLP/FE. The performance of the MORSRS algorithm has been compared with the performance of well-known algorithm for multi-objective optimization – the NSGA-II.

Consider two firms owning sets two J_1 and J_2 of facilities, respectively. Both firms compete for the market share in the market described by the same set of 6090 demand points in Spain, which was used in experimental investigation of RDOA in Sect. 3.2. Both firms have 10 preexisting facilities per each, randomly located in 20 largest demand points. See Table 2 for the indices of demand points, where the preexisting facilities of both firms are located.

The firm owning facilities in J_1 wants to extend its market share by establishing a number of new facilities thus facing the DCFLP/FE. The following three instances of the DCFLP/FE have been considered:

- locate 5 new facilities, when the number of candidate locations is 500;
- locate 5 new facilities, when the number of candidate locations is 1000;
- locate 10 new facilities, when the number of candidate locations is 1000.

In the first two cases the optimization problem has 5 variables but different search space; the third instance contains 10 variables and the larger search space.

The limit of 10,000 function evaluations has been set for each approximation of the Pareto front. Due to stochastic nature of the algorithms, each experiment has been run for 100 times and statistical estimators have been analyzed.

The NSGA-II has been implemented and adapted for CFLP/FE using the strategy to apply Genetic Algorithm for DCFLP described in Lančinskas et al. (2017). After a series of experiments with different parameters of NSGA-II, the following set of parameters has been chosen:

- population size: 100;
- probability for crossover: 0.6;
- probability for mutation: $1/s$, where s is the number new facilities.

Table 2 Indices of demand points, where preexisting facilities of the firms are located

Firm	Indices of the demand points									
J_1	9	15	17	1	4	14	2	16	18	20
J_2	6	11	13	5	12	8	10	7	3	19

The Rational Hyper-Volume (RHV) metric based on Hyper-Volume (HV) (Zhou et al. 2006) has been used to compare performance of the algorithms. The HV evaluates quality of the obtained approximation of the Pareto set by measuring the area made by the members of the approximation and the given reference point. The RHV additionally includes the results obtained by Pure Random Search (PRS) as the worst case results and measures how an algorithm under investigation performs better than PRS in the sense of HV and is expressed by

$$RHV_A = \frac{HV_A - HV_{PRS}}{HV_{PRS}}, \tag{23}$$

where HV_A and HV_{PRS} is the HV values of Pareto front approximations, obtained by the algorithm under investigation and PRS, respectively.

In order to compare performance of solution of different problem instances, the Pareto front is scaled to the unity square $[0, 1]^2$ with respect to extreme points in the true Pareto front and the point $(0, 1)$ is chosen as the reference point for evaluation of HV. See Fig. 4 for the illustration of RHV, where area between two Pareto fronts indicates quality of the approximation of the true Pareto front: the larger area, the higher quality of the approximation.

All ranks of candidate locations are equal to one at the beginning of the MORSR algorithm and is updated during the run-time of the algorithm. Figure 5 illustrates the rank values assigned to the candidate locations after 1000, 5000, and 10,000 function evaluations. The horizontal axis of the figure shows indices of elements in the set of 100 candidate locations with the largest ranks after 1000 function evaluations. The set is sorted by final rank values (after 10,000 function evaluations) in descending order: the candidate location with the highest ranks is in the first position and the candidate location with the lowest rank – in the last position.

Fig. 4 Illustration of the relative hyper-volume metric

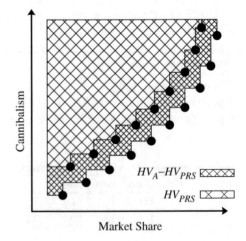

Market Share

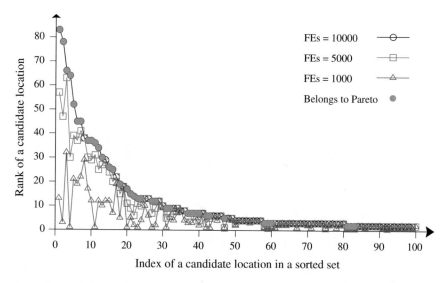

Fig. 5 Ranks of the candidate locations after 1000, 5000, and 10,000 function evaluations with candidate locations from Pareto set indicated

The final rank values are illustrated by circle-marked graph, where filled circles indicates candidate locations which form a non-dominated solution (optimal locations).

All optimal candidate locations has rank value larger than 1, and only 6 of them have rank value lower than 5. The triangle-marked graph shows that some optimal locations can be noticed after 1000 function evaluations – they have notably larger ranks. After 5000 function evaluations most of optimal locations can be highlighted by their rank values. It is worth to note that a candidate location, which rank is increased in the early stage of the algorithm, likely will be considered as an optimal locations in the final stage of the algorithm. Therefore, it is worth to consider them as more promising candidate locations.

There also exist a correlation between rank of an optimal location and its appearance in a non-dominated solution: the candidate locations with larger rank values in the final stage of the algorithm statistically forms more non-dominated solutions in the same approximation.

The performance of MORSR and NSGA-II is compared by RHV metric. The results are presented in Fig. 6 which shows dependency of the average RHV value on the number of function evaluations performed. Different colors correspond to different instances of the problem: choose locations for 5 new facilities from 500 candidate locations, choose 5 locations from 1000 candidates, and choose 10 locations from 10,000 candidates; different marks correspond to different algorithms: NSGA-II and MORSR. It is clearly seen from the figure, that MORSR notably outperforms NSGA-II independent on the problem instance and the number of function evaluations – NSGA-II always gave better average RHV value.

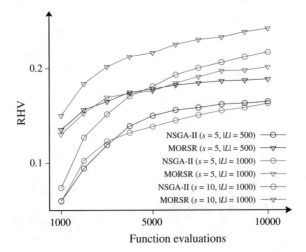

Fig. 6 Average results obtained by MORSR and NSGA-II for CFL/FE when selecting locations for 5 and 10 new facilities from 500 and 1000 candidate locations

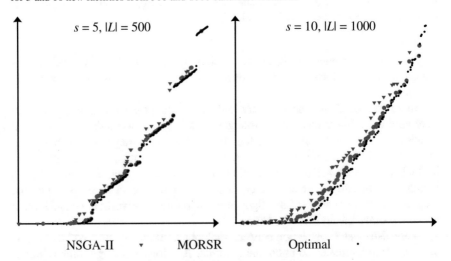

Fig. 7 Selected approximations of the true Pareto fronts, obtained by different algorithms, in the context of the true Pareto front

The correlation between accuracy of the approximation and the ranks of candidate locations has been also experimentally investigated. The results presented in Figs. 5 and 6 show that the ranking of candidate locations leads to the extraction of the most promising candidate locations what is useful when determining the probability to select a certain candidate location to form a new solution in MORSR (whereas all candidate locations have the same probability to be selected in NSGA-II) herewith leading to a better accuracy of approximation of Pareto-front within the same number of function evaluations.

The selected approximations of the Pareto front, obtained NSGA-II and MORSR are illustrated in Fig. 7, where the horizontal axis stand for the market share and the vertical axis – for the effect of cannibalism. The Pareto fronts are compared the true Pareto front which was determined by the complete enumeration algorithm. One can see from the figure, MORSR is able to determine most of Pareto-optimal solutions for the simplest instance of DCFLP/FE with 5 new facilities and 500 candidate locations. However only few Pareto-optimal solutions are determined by MORSR for the DCFLP/FE 10 new facilities and 1000 candidate locations. On the other hand, there is notable advantage of MORSR against NSGA-II in the sense of accuracy of the approximation – the approximation derived by MORSR is notably closer to the true Pareto front, comparing with the approximation derived by NSGA-II.

5   Conclusions

Two facility location problems have been described in the paper: the single-objective discrete competitive facility location problem for an entering firm and the multi-objective problem for firm expansion. The corresponding random search algorithms based on ranking of the candidate locations for the new facilities have been proposed to tackle the problems at hand.

The results of the computational experiments show that the presented algorithm RDOA/D for the single-objective problem notably outperforms the well-known Genetic Algorithm (GA) which is considered as suitable to solve similar problems.

The results of the computational experiments show that the presented algorithm MORSR for the multi-objective problem notably outperforms the well-known Non-Dominated Sorting Genetic Algorithm (NSGA-II) independently to the instance of the problem investigated and the number of function evaluations. Analysis of selected approximations of the Pareto front showed that MORSR is able to determine most Pareto-optimal solutions for smaller instances of the problem. Although it is complicated for MORSR to find locations for 10 new facilities having a set of 1000 candidate locations, the algorithm still shows notably better performance in the sense of accuracy of the approximation than NSGA-II, which is considered as a good algorithm for such a kind of optimization problems.

In general, it can be stated that the strategies for ranking of candidate locations can notably increase performance of a random search algorithm for single- or multi-objective discrete competitive facility location problems.

Acknowledgements This research has been supported by Fundación Séneca (The Agency of Science and Technology of the Region of Murcia, Spain) under the research project 20817/PI/18. This article is based upon work from COST Action CA15140 "Improving Applicability of Nature-Inspired Optimisation by Joining Theory and Practice (ImAppNIO)" supported by COST (European Cooperation in Science and Technology).

References

Aboolian R, Berman O, Krass D (2008) Optimizing pricing and location decisions for competitive service facilities charging uniform price. J Oper Res Soc 59(11):1506–1519

Ashtiani M (2016) Competitive location: a state-of-art review. Int J Ind Eng Comput 7(1):1–18

Balinski M (1965) Integer programming: methods, uses and computation. Manage Sci 24:253–313

Chinchuluun A, Pardalos PM (2007) A survey of recent developments in multiobjective optimization. Ann Oper Res 154(1):29–50

Chinchuluun A, Pardalos PM, Migdalas A, Pitsoulis L (eds) (2008) Pareto optimality, game theory and equilibria. In: Springer optimization and its applications, vol 17. Springer, New York

Christaller W (1950) Das grundgerust der raumlichen ordnung in europa : Die systeme der europaischen zentralen orte. Frankfurter Geographische Hefte 24:96S

Davis L (1991) Handbook of genetic algorithms. Van Nostrand Reinhold, VNR computer library

Deb K, Pratap A, Agarwal S, Meyarivan T (2002) A fast and elitist multiobjective genetic algorithm: NSGA-II. IEEE Trans Evol Comput 6:182–197

Doerner KF, Gutjahr WJ, Nolz PC (2009) Multi-criteria location planning for public facilities in tsunami-prone coastal areas. OR Spectrum 31(3):651–678

Drezner T (2014) A review of competitive facility location in the plane. Logistics Res 7(1):114

Drezner T, Drezner Z (2004) Finding the optimal solution to the huff based competitive location model. Comput Manage Sci 1(2):193–208

Eiselt HA, Marianov V, Drezner T (2015) Competitive location models. In: Laporte G, Nickel S, Saldanha da Gama F (eds) Location science. Springer International Publishing, Cham, pp 365–398

Farahani RZ, SteadieSeifi M, Asgari N (2010) Multiple criteria facility location problems: a survey. Appl Math Model 34(7):1689–1709

Fernández P, Pelegrín B, Lančinskas A, Žilinskas J (2017) New heuristic algorithms for discrete competitive location problems with binary and partially binary customer behavior. Comput Oper Res 79:12–18

FICO Xpress Mosel (2014) Fair Isaac Corporation

Fischer K (2011) Central places: the theories of von Thünen, Christaller, and Lösch. In: Eiselt HA, Marianov V (eds) Found Location Anal. Springer, US, pp 471–505

Francis R, Lowe T, Tamir A (2002) Demand point aggregation for location models. In: Drezner Z, Hamacher H (eds) Facility Location Appl Theor. Springer, Berlin Heidelberg, pp 207–232

Ghosh A, Craig C (1991) FRANSYS: a franchise distribution system location model. J Retail 64(4):466–495

Goldberg DE (1989) Genetic algorithms in search, optimization and machine learning, 1st edn. Addison-Wesley Longman Publishing Co., Inc, Boston, MA, USA

Hakimi L (1995) Location with spatial interactions: competitive locations and games. In: Drezner Z (ed) Facility location: a survey of applications and methods. Springer, New York, pp 367–386

Hakimi SL (1964) Optimum locations of switching centers and the absolute centers and medians of a graph. Oper Res 12(3):450–459

Hakimi SL (1965) Optimal distribution of switching centers in a communication network and some related theoretic graph problems. Oper Res 13:462–475

Hendrix E, Lančinskas A (2015) On benchmarking stochastic global optimization algorithms. Informatica 26(4):649–662

Holland JH (1975) Adaptation in natural and artificial systems. The University of Michigan Press, Michigan

Huff D (1964) Defining and estimating a trade area. J Mark 28:34–38

Jaramillo JH, Bhadury J, Batta R (2002) On the use of genetic algorithms to solve location problems. Comput Oper Res 29(6):761–779

Lančinskas A, Fernández P, Pelegrín B, Žilinskas J (2016) Solution of discrete competitive facility location problem for firm expansion. Informatica 27(2):451–462

Lančinskas A, Fernández P, Pelegrín B, Žilinskas J (2017) Improving solution of discrete competitive facility location problems. Optim Lett 11(2):259–270

Liao SH, Hsieh CL (2009) A capacitated inventory-location model: formulation, solution approach and preliminary computational results. In: Chien BC, Hong TP, Chen SM, Ali M (eds) Next-generation applied intelligence. Springer, Berlin Heidelberg, pp 323–332

Lösch A (1940) Die räumliche Ordnung der Wirtschaft: eine Untersuchung über Standort. Fischer, Wirtschaftsgebiete und internationalen Handel. G

Medaglia AL, Villegas JG, Rodríguez-Coca DM (2009) Hybrid biobjective evolutionary algorithms for the design of a hospital waste management network. J Heuristics 15(2):153

Montibeller G, Franco A (2010) Multi-criteria decision analysis for strategic decision making. In: Applied optimization, vol 103. Springer, Berlin Heidelberg

Peeters PH, Plastria F (1998) Discretization results for the Huff and Pareto-Huff competitive location models on networks. TOP 6:247–260

ReVelle CS, Swain RW (1970) Central facilities location. Geogr Anal 2(1):30–42

Serra D, Colomé R (2001) Consumer choice and optimal locations models: formulations and heuristics. Pap Reg Sci 80(4):439–464

Sinnott RW (1984) Virtues of the haversine. Sky Telescope 68:159

Srinivas N, Deb K (1994) Multiobjective optimization using nondominated sorting in genetic algorithms. Evol Comput 2:221–248

Suárez-Vega R, Santos-Penate DR, Dorta-Gonzalez P (2004) Discretization and resolution of the $(r|X_p)$-medianoid problem involving quality criteria. TOP 12(1):111–133

Suárez-Vega R, Santos-Penate DR, Dorta-González P (2007) The follower location problem with attraction thresholds. Pap Reg Sci 86(1):123–137

von Thunen JH (1910) Der isolierte Staat in Beziehung auf Landwirtschaft und Nationalokonomie. Verlag von Gustav Fischer, Jena

Villegas JG, Palacios F, Medaglia AL (2006) Solution methods for the bi-objective (cost-coverage) unconstrained facility location problem with an illustrative example. Ann Oper Res 147(1):109–141

Zhou A, Jin Y, Zhang Q, Sendhoff B, Tsang E (2006) Combining model-based and genetics-based offspring generation for multi-objective optimization using a convergence criterion. In: 2006 IEEE international conference on evolutionary computation, pp 892–899

Investigating Feature Spaces for Isolated Word Recognition

**Povilas Treigys, Gražina Korvel, Gintautas Tamulevičius,
Jolita Bernatavičienė and Bożena Kostek**

Abstract The study addresses the issues related to the appropriateness of a two-dimensional representation of speech signal for speech recognition tasks based on deep learning techniques. The approach combines Convolutional Neural Networks (CNNs) and time-frequency signal representation converted to the investigated feature spaces. In particular, waveforms and fractal dimension features of the signal were chosen for the time domain, and three feature spaces were investigated for the frequency domain, namely: Linear Prediction Coefficient (LPC) spectrum, Hartley spectrum, and cochleagram. Due to the fact that deep learning requires an adequate training set size of the corpus and its content may significantly influence the outcome, thus for the data augmentation purpose, the created dataset was extended with mixes of the speech signal with noise with various SNRs (Signal-to-Noise Ratio). In order to evaluate the applicability of the implemented feature spaces for isolated word recognition task, three experiments were conducted, i.e., 10-, 70-, and 111-word cases were analyzed.

Keywords Speech recognition · 2D feature spaces · Convolutional Neural Networks

1 Introduction

Much attention was given by researchers to the speech processing task in automatic speech recognition (ASR) over the past decades. The state-of-the-art methods applied to ASR are still often based on the extraction of features and the use of machine learning algorithms to recognize these features. A simplified scheme of a conventional

P. Treigys · G. Korvel (✉) · G. Tamulevičius · J. Bernatavičienė
Institute of Data Science and Digital Technologies, Vilnius University, Vilnius, Lithuania
e-mail: grazina.korvel@mif.vu.lt

B. Kostek
Audio Acoustics Laboratory, Faculty of Electronics, Telecommunications and Informatics,
Gdańsk University of Technology, Gdańsk, Poland

© Springer Nature Switzerland AG 2020
G. Dzemyda et al. (eds.), *Data Science: New Issues, Challenges
and Applications*, Studies in Computational Intelligence 869,
https://doi.org/10.1007/978-3-030-39250-5_9

speech recognition system may consist of the following components: speech signal input and processor, acoustical analysis and modeling component, and finally decision making, i.e., decoder. All additional components (like language models, semantic knowledge) are also included in a sequential manner. A strict separation between speech analysis and decoding processes led to the independent development of these processes.

The speech analysis part of the speech recognition system was always considered as crucial in speech recognition. The purpose of the analysis step was to extract the so-called feature vectors that had to be characterized by low order and high discriminative power (an ability to discriminate between similar sounds). Constant development of acoustical modeling techniques included various human speech production and perception models, and it led to a great variety of acoustical features: linear predictive coding parameters (Makhoul 1975) and their warped modifications, human perception motivated mel frequency cepstral coefficients (Mermelstein 1976), perceptual linear prediction coefficients (Hermansky 1990), wavelets (Hlawatsch and Boudreaux-Bartels 1992), power normalized cepstral coefficients (Kim and Stern 2016), and others based on feature vectors and machine learning algorithms (Noroozi et al. 2017; Vryzas et al. 2018). It would also be interesting to check whether such a mixture of techniques may be used in automatic labeling of speech-related structures (Vuegen et al. 2018).

As already mentioned, in the classical approach, the speech analysis process was organized in a pipeline manner: the signal was framed into overlapping frames, and these were analyzed consistently. The result was a sequence of feature vectors that were processed in subsequent stages. This feature organization ensured the temporal variation of features to be saved. A typical speech analysis configuration was 20–30 ms frames with the overlap of 10–20 ms. The main drawback of this scheme was the limited "scope of view", i.e., when analyzing the end of the utterance, the beginning of the utterance was no longer seen.

In recent years, there has been growing interest in the use of Deep Neural Network (DNN) in which the feature extraction process is discarded. A new paradigm of signal analysis has emerged due to huge power of parallel computing, i.e., deep learning structures can take entire speech utterance, and this guarantees a "full view" analysis and preservation of temporal information. DNNs used for ASR show satisfactory results (Li et al. 2018; He and Cao 2018). Our previous investigations also show that deep learning-based speech recognition returns high word classification rate (Korvel et al. 2018). The goal of the mentioned research was to evaluate the suitability of 2D audio signal feature maps for speech recognition based on deep learning. The current paper is an extension of our previous research by employing additional two-dimensional representations which are widely used in literature.

Cheng and Tsoi (2016) show that a method based on fractal dimension features can provide comparative performance for audio signal emotion recognition (Cheng and Tsoi 2016). Features referring to fractal dimensions are successfully used for spontaneous speech analysis in Alzheimer's disease diagnosis (López-de-Ipina et al.

2015). Taking into account these successful fractal dimension features-based applications in the area of audio signal analysis, we have decided to apply them to our current research using DNN.

According to the results obtained in the previous work (Korvel et al. 2018), the spectrum representation returns the best classification accuracy. Therefore, we decided to include 2D spectrum representation based on other transformation techniques, namely: Linear Prediction Coefficient spectrum and Hartley spectrum. Linear predictive coding (LPC) is well-established in the speech recognition area (Soong and Rosenberg 1988). Sankar and his colleagues show that LPC using Neural Network returns good results in speech sound classification and estimation (Sankar et al. 2018). The authors of that paper classified speech into five classes: voiced, unvoiced, silence, music, background noise. On the other hand, a comparison of Hartley and Fourier transforms is given by Paraskevas and Rangoussi (2008). The results presented in their paper show that Hartley phase spectrum outperforms Fourier phase spectrum in the phoneme classification task.

In addition, the Hartley phase spectrum, the cochleagram can be regarded as a modification of spectrogram according to the human auditory system (McDermott 2017; Lyon 2017). Sharan and Moir (2019) show that the proposed cochleagram time-frequency image representation gives the best classification performance when used with CNN for acoustic event recognition.

In recent years, end-to-end acoustic models for ASR using DNNs, which are taken directly from the windowed speech waveforms (WSW), are also investigated (Bhargava and Rose 2015). A preliminary analysis of redundancies in the WSW-based DNN acoustic models was performed, and a structured way to initialize the WSW-based DNN was provided by Bhargava and Rose (2015). The automatic recognition of spontaneous emotions from speech using a combination of CNNs with Long Short-Term Memory (LSTM) networks is given in the work of Trigeorgis et al. (2016). The input for this network was also WSW. Therefore, we choose to utilize the WSW-based input in our current research study.

Summarizing, according to the reviewed literature sources, we decided to include the following speech signal representations in our study: waveform, fractal dimension, Linear Prediction Coefficient spectrum, Hartley spectrum, and cochleagram.

2 Feature Spaces

In this study, we focused on two-dimensional (2D) feature maps (Korvel et al. 2018). The 2D representation of one-dimensional speech signal was obtained by analyzing particular characteristics in the time domain. We believe that the time-based analysis of speech signal can reveal all the static and dynamic discriminative information needed for recognition purposes.

The following speech signal feature spaces were chosen to be investigated:

1. Time-domain representations. Considering the abstracting power of the convolutional network, the first choice was a time-domain representation of the speech signals. We have employed the waveform-based and the fractal dimension-based representations of the signal.
2. Frequency-domain (spectral) representation. Previous studies have shown the effectiveness of spectrum-based representation. Therefore, in this study, we have chosen a few alternative calculation schemes for spectral analysis: Linear Prediction Coding-based analysis, Hartley Transform-based analysis, and cochleagrams—human perception-based representation. The first one gives a detailed spectral view as an alternative to Fourier transform-based analysis. Hartley transform combines magnitude and phase spectral data, which is also different from Fourier-based analysis. The cochleagram imitates the signal processing in the human auditory system and thus gives us an opportunity to estimate additional spectral representation of the speech signal.

Two-dimensional feature maps were obtained as follows. First of all, the signal was framed in such a way to obtain a time sequence of overlapping signal frames. Specific analysis for each signal frame was applied, resulting in a feature vector. Consecutive analysis of the entire utterance has formed an array of features, which was converted into a grey scale image. For this purpose, the feature values were normalized, and the array was exported to image format file.

The detailed description of each feature space is given in section below.

2.1   Fractal Dimension Features

Fractal-based analysis of speech signal has emerged from non-linear analysis techniques: energy operators, nonlinear predictors, various chaotic models. These techniques should be capable of describing natural and non-linear speech phenomena such as elision, co-articulation, assimilation, variations of pitch and speech rate (Teager and Teager 1993; Titze et al. 1993).

Fractal geometry analyzes issues of self-similarity, complexity, and irregularities of the objects. A fractal can be defined as an abstract mathematical object, exhibiting self-similarity and self-affinity. Consequently, the fractal objects may be used to describe the irregular or fragmented shape of natural features, time sequences, as well as other complex objects.

Fractal properties of the analyzed object are evaluated using a value of fractal dimension (FD). It is a numerical value and characterizes the fragmentation level of the data analyzed (Karbauskaite and Dzemyda 2016). Generally, the fractal dimension is estimated as follows (Eq. 1):

$$D \sim \lim_{s \to \infty} \frac{\log N}{\log 1/s}, \tag{1}$$

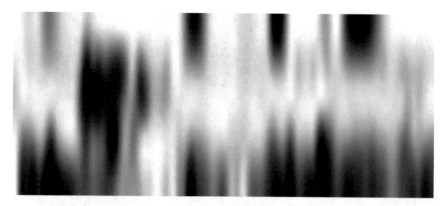

Fig. 1 The fractal dimension features of the Lithuanian word "respublika" (in English: "republic")

here, N is the length of the analyzed object, s is the scaling factor.

FD employment for speech analysis purposes is not a new idea. Various FD-based techniques were employed for voice pathology detection (López-de-Ipiña et al. 2015; Ali et al. 2016), speech recognition (Pitsikalis and Maragos 2006; Ezeiza et al. 2013), speech emotion and intonation (Tamulevičius et al. 2017; Phothisonothai et al. 2013; Zheng et al. 2015), speaker identification (Nelwamondo et al. 2006), speech segmentation (Fantinato et al. 2008).

In this study various FD algorithms were employed for isolated word recognition: Katz (1988), Castiglioni (2010), Higuchi (1988), Hurst exponent-based approaches (Barbulescu et al. 2010; Molino-Minero-Re et al. 2015).

In total, eight different FD values were extracted: Katz, Castiglioni, Higuchi and five different Hurst exponent-based (approximated by a polynomial, detrended fluctuation analysis, average wavelet coefficient, GewekePorter-Hudak estimator, and generalized version of the Hurst exponent). Considering the specificity of the speech signal, all utterances were analyzed in the classical frame-by-frame manner: FD values were extracted for each frame. A graphical representation of fractal dimension features is given in Fig. 1.

2.2 Waveforms

In comparison to other analysis techniques, the waveform of the speech signal is the primary source to be examined in classification tasks. Still, because of the redundancy of speech, the direct employment of the speech signal was limited in the pipeline manner-based analysis applications. Emerging deep learning techniques outgrew these limitations and waveforms appeared as an alternative to the conventional representation techniques (Hannun et al. 2014; Sainath et al. 2015; Trigeorgis et al. 2016). The signal waveform is sent to the network for abstraction and feature formation, and this enables to implement the end-to-end principle of the speech recognition.

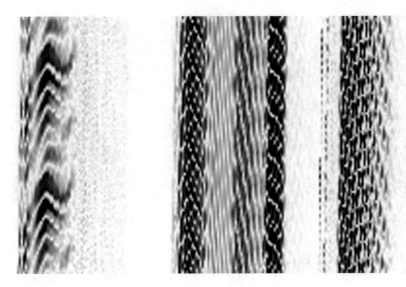

Fig. 2 Waveforms of the word "respublika" (in English: "republic")

Nevertheless, the optimization question of the waveform analysis is still unanswered. If, with regard to spectrograms, we can talk about spectral trends and generalizations (e.g., properties of the vocalized and non-vocalized phonemes are similar for all speakers), in the case of waveforms we have individual speaker properties (pitch, speaking rate, loudness, accent, etc.), abstractio, which can obscur speech content. Therefore, the effectiveness of the waveform based analysis is still theoretically ungrounded and is estimated using experiment-based speech recognition score only.

In this study, we have decided to employ the waveform analysis and to compare it with other 2D features. We have not made any presumption on waveform pre-processing for the mapping and analyzing in networks. Basically, a 2D form of the waveform was formed by framing the signal itself. The consecutive arrangement of signal frames gave us a map of waveform values (Fig. 2). The framing parameters (512-point frame length, the 400-point overlap of frames) were the same as in other feature map formation algorithms we have implemented in this study. This enabled us to obtain feature maps with the same and therefore comparable sizes. An example of waveforms is depicted in Fig. 2.

2.3 Hartley Spectrum

In digital signal processing, spectral analysis is one of the most dominant techniques, often considered as standard procedure. The main idea of spectral analysis is the Fourier transform-based representation of the signal frequency content. The complex value-based nature of the Fourier transform is estimated by distinguishing the magnitude and phase spectra. In some cases (speech signal processing, for example), the phase spectrum is ignored thus, only a real-value spectrum is obtained (Rabiner and Schafer 2007).

Hartley transform was proposed as a real-valued alternative to Fourier transform and is considered as Fourier-related transform (Hartley 1942; Bracewell 1983). Two main advantages of the Hartley transform can be distinguished. First of all, real-value-based results are obtained for the real-valued sequences, which means no additional processing of spectral data (such as rejection of imaginary part). Thus, the Hartley transform-based result contains full spectral information of the sequence. Secondly, the direct Hartley transform is also inverse transform, which simplifies the algorithmic implementation of Hartley transform-based cepstral analysis. Besides, the fast version of the Hartley transform exists and is considered as faster than Fast Fourier Transform (Bracewell 1984).

The discrete Hartley transform of the signal is defined as:

$$
\begin{aligned}
H(k) &= \frac{1}{N} \sum_{n=0}^{N-1} x(n) cas\left(\frac{2\pi nk}{N}\right) \\
&= \frac{1}{N} \sum_{n=0}^{N-1} x(n) \left[\cos\left(\frac{2\pi nk}{N}\right) + \sin\left(\frac{2\pi nk}{N}\right)\right], \quad k = 0, 1, \ldots, N - 1. \quad (2)
\end{aligned}
$$

Here, the function $cas(\cdot)$ is the cosine-and-sine or Hartley kernel function.

Is it important to emphasize, that Fourier and Hartley transforms represent the same spectral information but in different forms. Therefore, they are interrelated, and the Hartley transform can be expressed in terms of Fourier transform:

$$
H(k) = R[F(k)] - I[F(k)], \tag{3}
$$

where, $R[F(k)]$, $I[F(k)]$ are the real and imaginary parts of the Fourier transform respectively.

When applied to signal frames, consequently, Hartley transform returns the spectrogram of the signal (Fig. 3).

2.4 Cochlegram

In the spectrogram, we have two axes: time and frequency. According to the literature (Lyon 2017) a spectrogram image has too few dimensions to be analogous to the image that the eye transfers as information to the cortex. In order to have a full auditory image one more dimension should be added. Chochleagram has the special axis orthogonal to the frequency axis, which gives a movie-like representation of sound.

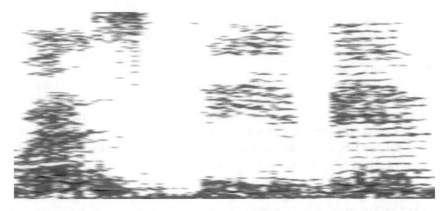

Fig. 3 The Hartley transform-based spectrogram of word „vyriausybė" (in English: "government")

In principle, cochleagram is the speech representation method which tries to emulate the human hearing system (McDermott 2017; Lyon 2017). The speech signal is converted into a multidimensional vector which shows how our brain process information received from the ear. This conversion is performed by deriving time-frequency representation of the speech signal.

The creation of cochleagram begins with performing auditory filtering. For this purpose, Gammatone Auditory Filterbank is used. Gammatone frequency cepstral coefficient (GFCC) is generated from applying a Discrete Cosine Transformation (DCT) on each column of a cochleagram because the dimensions of the cochleagram are highly correlated (Ma et al. 2015). In our work, the impulse response of the Grammatone filter is calculated according to the following Equation given in the work of Patterson and Holdsworth (1996):

$$g(t) = At^{n-1}e^{-2\pi Bt}\cos(2\pi f_0 t + \varphi) \tag{4}$$

where n is the filter order, φ is the phase, f_0 is the center frequency, A is the filter amplitude, B is the filter bandwidth, and t refers to the time.

In this paper, the fourth order filter is used ($n = 4$). The phase φ is equal to minimum phase. The center frequencies are equally spaced on Equivalent Rectangular Bandwidth (ERB) scale (Glasberg and Moore 1990). The number of filters used in this research is equal to 128. The filter bandwidth B is calculated by the following formula:

$$B = 1.019 \cdot 24.7 \cdot \left(\frac{4.37 f_0}{1000} + 1\right) \tag{5}$$

Though the bandwidths of filters with higher center frequencies are wider as those with lower center frequencies, they are narrow. Due to the narrow bandwidth, filter normalization should be used. For this purpose, gains, given as filter amplitude A [see Eq. (1)] are used. According to Wang and Brown (2006), gains are set in

Fig. 4 The cochleagram of word „vyriausybė" (in English: "government")

order to simulate the resonances due to the outer and middle ear parts, which boosts sound energy in the 2–4 kHz range. The AGC (Automatic Gain Control) is used for amplitude description in this paper (Slaney 1993).

Filtered responses of Grammatone filter bands are obtained by Fast Fourier Transform based on Finite Impulse Response (FIR) filtering using overlap-add method (Smith 2002). The output of the filter channel [denoted as $y(t)$] can be expressed as:

$$y(t) = x(t) * g(t) \qquad (6)$$

where $x(t)$ is a speech signal, the symbol * denotes the convolution.

In result, we obtain 128 filter channel outputs. The output is time–frequency decomposition of the speech signal. Each of them is divided into time frames, length of with 256 samples. Calculation of energy for each frame in each channel is performed. Central frequencies of the filterbank are equally distributed over the Bark scale to simulate the human auditory system.

The cochleagram of word "vyriausybė" with 128-channel gammatone filterbank is shown in Fig. 4. An overlap between consecutive frames is 50%.

2.4.1 Linear Predictive Coding Spectrum

The linear predictive coding technique is widely used by speech researchers due to the fact that the spectrum based on this technique retains human vocal tract parameters.

The basic idea of linear predictive technic is filter construction using the Liner Prediction (LP) technique. According to this technique, the predicted speech signal $\hat{x}(n)$ (where $n = 1, \ldots, N$, N is the number of samples of signal) can be expressed as a linear combination of its p previous samples:

$$\hat{x}(n) = \sum_{i=1}^{p} a_i x(n - i) + \text{Gu(n)} \qquad (7)$$

here $u(n)$ is the system excitation with the gain factor G.

The system function of all-pole linear filter in z-domain is:

$$H(z) = \frac{G}{1 + a_1 z^{-1} + a_2 z^{-2} + \cdots + a_p z^{-p}} \tag{8}$$

where $a = [1, a_1, \ldots, a_p]$ is a set of prediction coefficients. These coefficients are estimated by minimizing the squared prediction error (the difference between predicted and real signal value) (Makhoul 1975).

The frequency response of the filter H(z) reflects the spectrum of the speech signal and can be estimated through system polynomial:

$$A(z) = 1 + a_1 z^{-1} + a_2 z^{-2} + \cdots + a_p z^{-p} = 1 + \sum_{k=1}^{p} a_k z^{-k} = \prod_{i=1}^{p} \left(1 - z_i z^{-1}\right). \tag{9}$$

The complex-valued roots of the polynomial $A(z)$ give the values of the signal spectrum. The higher model order p we apply, the more detail spectrum of the signal we will obtained. The model order 10–20 is considered as describing spectral envelope of the speech signal (formant frequencies), the order 50–200 will return a detailed spectrum, i.e., spectral harmonics, comparable to Fourier transform result.

In our study, we decided to extract 10 formants for signal spectrogram. For this purpose, we employed the autocorrelation method for estimation of the LPC parameters and applied 20th order analysis (the order of the filter is two times the expected number of formants). The graphical representation of the signal spectrogram is shown in Fig. 5.

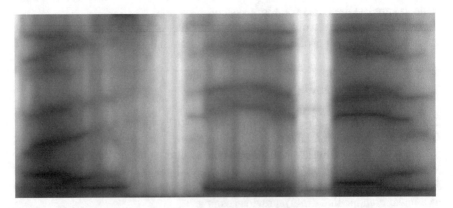

Fig. 5 The LPC-based spectrogram of the word „vyriausybė" (in English: "government")

Table 1 Neural network architecture

Layers	No. of filters	Stride size	Max pooling and kernel size	Is batch normalization used?
1st convolutional	32	3 × 3	3 × 3	Yes, before max pooling operation for all convolutional layers
2nd convolutional	64	2 × 2	2 × 2	
3rd convolutional	64	2 × 2	2 × 2	
1st dense	64	–	–	Yes, after activation
2nd dense	Depends on the output class count	–	–	–

3 Convolutional Neural Network Architecture

The neural network model consists of three convolutional and two dense layers. Each layer is followed by non-linear down-sampling with the window size 3 by 3. After each of the layers, except the last dense, the ReLu activation function is applied, and batch normalization is introduced after each convolutional layer with the view to speed-up model convergence. The model loss is estimated as categorical cross entropy, and the first-order gradient-based Adam optimization algorithm is applied. As the input for the neural network, several signal representations described in the previous Section were used. Characteristics of the neural network layers are presented in Table 1.

Full description of the neural network architecture can be found in the paper of Korvel et al. (2018). The network architecture has not been intentionally modified with the goal to investigate additional feature spaces and to compare the results with those obtained in the previous study.

4 Experimental Results

In order to evaluate the applicability of the implemented feature spaces for isolated word recognition task, three experiments were conducted: 10-, 70-, and 111-word cases were analyzed. These words were recorded to the .wav file of the audio format with the following parameters: 11 kHz; 32 bit; mono. The experiment was performed on the Lithuanian speech records obtained from 62 speakers. Each speaker pronounced 111 words 10 times. For the data augmentation purpose, the created dataset was extended by adding noise levels at various SNR (Signal-to-Noise) ratio levels, i.e., 15, 20, 25, and 30 dB and mixed with the speech signal. Since the purpose of adding noise to speech was data augmentation, thus white noise, mixed at proportions given above, constituted the background noise. This is because other types of background noise may have greater impact on speech signal intelligibility performance. 60% of all data samples were used for network training, and the rest for validation and testing. The network performance is shown in Tables 2, 3 and 4. As seen from Tables 2, 3 and 4 the applied data augmentation type did not influence the results. It is interesting that cochleagrams representation brought the highest scores of all indicators, i.e., Precision, Recall and f1-score. It showed a consistent effectiveness over all tasks performed.

The results shown in Tables 2, 3 and 4 also include Spectrum, Cepstrum, Chroma, Mel-scale cepstrum effectiveness collected from our previous experiments (Korvel et al. 2018). It was done in order to show a full picture of the research.

When comparing the results obtained in the current study with those from our previous research, it may be seen that mel-scale cepstrum retained high scores throughout the task of recognizing 10-, 170- and 111-words. In this context, it would be valuable to analyze whether cochleagrams and mel-scale cepstrum representations

Table 2 The network performance (10-word task)

Parameter space	Precision	Recall	f1-score	Validation accuracy	Validation loss
Waveforms	0.99	0.99	0.99	0.9924	0.0348
Fractals	0.99	0.99	0.99	0.9906	0.0332
Hartley spectrum	1.00	1.00	1.00	0.9989	0.0080
LPC spectrum	1	1	1	0.9973	0.0102
Cochleagram	1	1	1	0.9998	0.0008
Spectrum	1	1	1	0.9995	0.0022
Cepstrum	0.89	0.89	0.89	0.9000	0.3171
Chroma	0.99	0.99	0.99	0.9891	0.0375
Mel-scale cepstrum	0.99	0.99	0.99	0.9936	0.0270

Table 3 The network performance (70-word task)

Parameter space	Precision	Recall	f1-score	Validation accuracy	Validation loss
Waveforms	0.91	0.91	0.91	0.9096	0.3969
Fractals	0.92	0.92	0.92	0.9192	0.3274
Hartley spectrum	0.94	0.94	0.94	0.9420	0.2739
LPC spectrum	0.97	0.97	0.97	0.9666	0.1598
Cochleagram	0.99	0.99	0.99	0.9921	0.0529
Spectrum	0.99	0.99	0.99	0.9942	0.0374
Cepstrum	0.71	0.71	0.71	0.7140	1.1275
Chroma	0.84	0.84	0.84	0.8457	0.6286
Mel-scale cepstrum	0.94	0.94	0.94	0.9371	0.2616

Table 4 The network performance (111-word task)

Parameter space	Precision	Recall	f1-score	Validation accuracy	Validation loss
Waveforms	0.88	0.88	0.88	0.8855	0.5023
Fractals	0.88	0.88	0.88	0.8836	0.4943
Hartley spectrum	0.91	0.91	0.91	0.9089	0.4059
LPC spectrum	0.96	0.96	0.96	0.9593	0.1912
Cochleagram	0.98	0.98	0.98	0.9850	0.0936
Spectrum	0.99	0.99	0.99	0.9855	0.0951
Cepstrum	0.66	0.64	0.64	0.6390	1.4278
Chroma	0.76	0.76	0.76	0.7660	0.9845
Mel-scale cepstrum	0.91	0.91	0.91	0.9073	0.4019

share some common features that allow for effective speech recognition. The first answer that occurs is related to the Bark scale that is used in creating these spectral representations. The human auditory system, along with auditory cortex, can overcome distortions and fragmentations of speech. Moreover, this is also with the good agreement of other studies showing that filterbanks generated by mel-scaled filters on the spectrogram return a better performance than other primary feature representations (Hinton et al. 2012).

5 Conclusions

Obtained results from the experiments let us conclude that:

- Frequency-domain representations seem to be more effective than time-domain representations. The highest rate was obtained in the case of Fourier-based spectrograms. A little lower results were obtained in the case of cochleagrams, LPC-based spectrum, and Hartley spectrum.
- Mel-scale cepstrum has shown sufficient effectiveness and therefore, should be explored further. Analysis of the feature order, additional processing, and modeling techniques (human perception, e.g.) could be a potential direction of further studies.
- Time-domain representation did not achieve the effectiveness of frequency-domain. Despite the full signal information contained in the waveform, its discriminative power was not exclusive. Therefore, we may claim that in these particular experiments, the main speech characteristics are represented in the frequency domain.
- The chroma-based representation has shown the lowest effectiveness. This representation comes from music domain and does not gives an effective description of the speech signal specificity.

There are also several conclusions with regard to data augmentation:

- First of all, the notion known from the literature that deep learning technique requires big data seems not fully justified. It should certainly be a case of a sufficient amount of data, depending on the task performed by a neural network. Moreover, the Expert System group says: "machine learning is a good solution for big data analytics, but we prefer to add our semantic experience and expertise to it." (Machine Learning for Big Data Analytics 2019), thus our approach conforms to that observation.

- When looking at data augmentation which consisted in mixing noise at various SNR levels, it would be valuable to test the primary source of the data (i.e., recorded words) with other types of noise, e.g., babble (cocktail-party-effect noise) speech or environmental background noise. However, then, the task performance should be evaluated utilizing WER (Word Error Rate) along with typical recognition effectiveness measures. Moreover, neural networks may be employed for separation of speech and noise (Messaoud and Bouzid 2016) as this subject is of great importance in many application areas. This will constitute the basis for the continuation of our study.

Further experiments will be carried out on other languages, i.e., Polish and English, to test whether the experimental outcomes of this study with regard to the representation used are the same across several languages. Also, we would like to apply a similar approach to speech that carries on some emotional characteristics as in the current study, only neutrally spoken utterances were used. The applied experimental set up may be compared to other research outcomes (Vryzas et al. 2018).

Acknowledgements This research was partly supported by the statutory funds of Gdańsk University of Technology, Faculty of Electronics, Telecommunications and Informatics.

References

Ali Z, Elamvazuthi I, Alsulaiman M, Muhammad G (2016) Detection of voice pathology using fractal dimension in a multiresolution analysis of normal and disordered speech signals. J Med Syst 40(1):20

Bărbulescu A, Serban C, Maftei C (2010) Evaluation of Hurst exponent for precipitation time series. In: Proceedings of the 14th WSEAS international conference on Computers, vol 2, pp 590–595

Bhargava M, Rose R (2015) Architectures for deep neural network based acoustic models defined over windowed speech waveforms. In: Sixteenth annual conference of the international speech communication association

Bracewell RN (1983) Discrete Hartley transform. JOSA 73(12):1832–1835

Bracewell RN (1984) The fast Hartley transform. Proc IEEE 72(8):1010–1018

Castiglioni P (2010) Letter to the Editor: What is wrong in Katz's method? Comments on: a note on fractal dimensions of biomedical waveforms. Comput Biol Med 40(11–12):950–952

Cheng M, Tsoi AC (2016) Fractal dimension pattern based multiresolution analysis for rough estimator of person-dependent audio emotion recognition. arXiv preprint arXiv:1607.00087

Ezeiza A, de Ipiña KL, Hernández C, Barroso N (2013) Enhancing the feature extraction process for automatic speech recognition with fractal dimensions. Cogn Comput 5(4):545–550

Fantinato PC, Guido RC, Chen SH, Santos BLS, Vieira LS, Júnior SB, Rodrigues LC, Sanchez FL, Escola JPL, Souza LM, Maciel CD, Scalassara PR (2008). A fractal-based approach for speech segmentation. In: 2008 Tenth IEEE international symposium on multimedia. IEEE, pp 551–555

Glasberg BR, Moore BC (1990) Derivation of auditory filter shapes from notched-noise data. Hear Res 47(1–2):103–138

Hannun A, Case C, Casper J, Catanzaro B, Diamos G, Elsen E, Prenger R, Satheesh S, Sengupta S, Coates A, Ng AY (2014) Deep speech: scaling up end-to-end speech recognition. arXiv preprint arXiv:1412.5567

Hartley RV (1942) A more symmetrical Fourier analysis applied to transmission problems. Proc IRE 30(3):144–150

He L, Cao C (2018) Automated depression analysis using convolutional neural networks from speech. J Biomed Inform 83:103–111

Hermansky H (1990) Perceptual linear predictive (PLP) analysis of speech. J Acoust Soc Am 87(4):1738–1752

Higuchi T (1988) Approach to an irregular time series on the basis of the fractal theory. Physica D 31(2):277–283

Hinton G, Deng L, Yu D, Dahl G, Mohamed AR, Jaitly N, Senior A, Vanhoucke V, Nguyen P, Kingsbury B, Sainath T (2012) Deep neural networks for acoustic modeling in speech recognition. IEEE Sig Process Mag 28(6):82–97

Hlawatsch F, Boudreaux-Bartels GF (1992) Linear and quadratic time-frequency signal representations. IEEE Sig Process Mag 9(2):21–67

Karbauskaite R, Dzemyda G (2016) Fractal-based methods as a technique for estimating the intrinsic dimensionality of high-dimensional data: a survey. Informatica 27(2):257–281

Katz MJ (1988) Fractals and the analysis of waveforms. Comput Biol Med 18(3):145–156

Kim C, Stern RM (2016) Power-normalized cepstral coefficients (PNCC) for robust speech recognition. IEEE/ACM Trans Audio Speech Lang Process 24(7):1315–1329

Korvel G, Treigys P, Tamulevičius G, Bernatavičienė J, Kostek B (2018) Analysis of 2D feature spaces for deep learning-based speech recognition. J Audio Eng Soc 66(12):1072–1081. https://doi.org/10.17743/jaes.2018.0066

Li K, Xu H, Wang Y, Povey D, Khudanpur S (2018) Recurrent neural network language model adaptation for conversational speech recognition. In: Interspeech, Hyderabad, pp 1–5

López-de-Ipina K, Solé-Casals J, Eguiraun H, Alonso JB, Travieso CM, Ezeiza A, Barroso N, Ecay-Torres M, Martinez-Lage P, Beitia B (2015) Feature selection for spontaneous speech analysis to aid in Alzheimer's disease diagnosis: a fractal dimension approach. Comput Speech Lang 30(1):43–60

Lyon RF (2017) Human and machine hearing. Cambridge University Press

Ma C, Qi J, Li D, Liu R (2015) Improving bottleneck features for automatic speech recognition using gammatone-based cochleagram and sparsity regularization. In 2015 Asia-Pacific Signal and Information Processing Association Annual Summit and Conference (APSIPA). IEEE, pp 63–67

Machine Learning for Big Data Analytics. https://www.expertsystem.com/machine-learning-big-data-analytics/. Accessed Mar 2019

Makhoul J (1975) Linear prediction: a tutorial review. Proc IEEE 63(4):561–580

McDermott JH (2017) Stevens' handbook of experimental psychology and cognitive neuroscience, vol 2, Sensation, perception, and attention, 4th edn

Mermelstein P (1976) Distance measures for speech recognition, psychological and instrumental. In: Chen RCH (ed) Pattern recognition and artificial intelligence. Academic, New York, NY, USA, pp 374–388

Messaoud MAB, Bouzid A (2016) Speech enhancement based on wavelet transform and improved subspace decomposition. J Audio Eng Soc 63(12):990–1000. https://doi.org/10.17743/jaes.2015.0083

Molino-Minero-Re E, García-Nocetti F, Benítez-Pérez H (2015) Application of a time-scale local Hurst exponent analysis to time series. Digit Sig Proc 37:92–99

Nelwamondo FV, Mahola U, Marwala T (2006) Improving speaker identification rate using fractals. In: The 2006 IEEE international joint conference on neural network proceedings. IEEE, pp 3231–3236

Noroozi F, Kaminska D, Sapinski T, Anbarjafari G (2017) Supervised vocal-based emotion recognition using multiclass support vector machine, random forests, and adaboost. J Audio Eng Soc 65(78):562–572. https://doi.org/10.17743/jaes.2017.0022

Paraskevas I, Rangoussi M (2008) Phoneme classification using the Hartley phase spectrum. In: Second ISCA workshop on experimental linguistics

Patterson RD, Holdsworth J (1996) A functional model of neural activity patterns and auditory images. Adv Speech Hearing Lang Process 3(Part B):547–563

Phothisonothai M, Arita Y, Watanabe K (2013) Extraction of expression from Japanese speech based on time-frequency and fractal features. In: 2013 10th international conference on electrical engineering/electronics, computer, telecommunications and information technology. IEEE, pp 1–5

Pitsikalis V, Maragos P (2006) Filtered dynamics and fractal dimensions for noisy speech recognition. IEEE Sig Process Lett 13(11):711–714

Rabiner LR, Schafer RW (2007) Introduction to digital speech processing. Now Publishers Inc

Sainath TN, Weiss RJ, Senior A, Wilson KW, Vinyals O (2015) Learning the speech front-end with raw waveform CLDNNs. In: Sixteenth annual conference of the international speech communication association, pp 1–5

Sankar A, Aiswariya M, Rose DA, Anushree B, Shree DB, Lakshmipriya PM, Sathidevi PS (2018) Speech sound classification and estimation of optimal order of LPC using neural network. In: Proceedings of the 2nd international conference on vision, image and signal processing. ACM, p 35

Sharan RV, Moir TJ (2019) Acoustic event recognition using cochleagram image and convolutional neural networks. Appl Acoust 148:62–66

Slaney M (1993) An efficient implementation of the Patterson-Holdsworth auditory filter bank. Apple Computer, Perception Group, Tech Report, 35(8)

Smith S (2002) Digital signal processing: a practical guide for engineers and scientists. Elsevier

Soong FK, Rosenberg AE (1988) On the use of instantaneous and transitional spectral information in speaker recognition. IEEE Trans Acoust Speech Signal Process 36(6):871–879

Tamulevičius G, Karbauskaitė R, Dzemyda G (2017) Selection of fractal dimension features for speech emotion classification. In: 2017 Open Conference of Electrical, Electronic and Information Sciences (eStream). IEEE, pp 1–4

Teager HM, Teager SM (1993) Evidence for nonlinear sound production mechanisms in the vocal tract. In: Hardcastle WJ, Marchal A (eds) Speech production and speech modelling. NATO ASI series (series D: behavioural and social sciences), vol 55. Springer, Dordrecht

Titze IR, Baken R, Herzel H (1993) Evidence of chaos in vocal fold vibration. In: Titze IR (ed) Vocal fold physiology: new frontier in basic science. Singular, San Diego, pp 143–188

Trigeorgis G, Ringeval F, Brueckner R, Marchi E, Nicolaou MA, Schuller B, Zafeiriou S (2016) Adieu features? end-to-end speech emotion recognition using a deep convolutional recurrent network. In: 2016 IEEE international conference on acoustics, speech and signal processing (ICASSP). IEEE, pp 5200–5204

Vryzas N, Kotsakis R, Liatsou A, Dimoulas C, Kalliris G (2018) Speech emotion recognition for performance interaction. J Audio Eng Soc 66(6):457–467. https://doi.org/10.17743/jaes.2018.0036

Vuegen L, Karsmakers P, Vanrumste B, Van Hamme H (2018) Acoustic event classification using low-resolution multi-label non-negative matrix deconvolution. J Audio Eng Soc 66(5):369–384. https://doi.org/10.17743/jaes.2018.0018

Wang D, Brown GJ (2006) Computational auditory scene analysis: principles, algorithms, and applications. Wiley-IEEE Press

Zheng WQ, Yu JS, Zou YX (2015) An experimental study of speech emotion recognition based on deep convolutional neural networks. In: 2015 International conference on affective computing and intelligent interaction (ACII). IEEE, pp 827–831

Developing Algorithmic Thinking Through Computational Making

Anita Juškevičienė

Abstract Programming needs algorithmic thinking—the main component of computational thinking. Computational thinking is overlapping with many digital age skills necessary for digital learners. However, it is still a challenge for educators to teach CT in an attractive way for learners, also to find support to CT teaching content design and assessment. To address this problem, the literature review on CT in education was conducted and the main ideas of CT implementation and assessment were identified. The results show that modern technologies are widely used for learning enhancement and algorithmic thinking improvement. The implications of these results is that modern technologies can facilitate effective learning, CT skills gaining and learning motivation.

Keywords Computational thinking · Algorithmic thinking · Problem-based learning

1 Introduction

The demand for programmers around the world is growing. However, the carrier of the programmer is chosen not by many learners. Therefore, the understanding of the IT and knowledge in this field in general education needs to be encouraged. However, it is important not only to increase the number of IT learners, but also the number of successfully graduated. Moreover, digital age learners need to have a broader set of computational skills, thus there is the need of educational approaches changes. In order to motivate and attract learners there is the need to identify the causes of difficulties in developing computing education competences (including programming) and propose appropriate teaching/learning strategies, theories and tools for education process encouragement.

Computer programming is the craft of analysing problems and designing, writing, testing, and maintaining programs (Massachusetts DLCS 2016). Programming needs

A. Juškevičienė (✉)
Institute of Data Science and Digital Technologies, Vilnius University, Vilnius, Lithuania
e-mail: anita.juskeviciene@mif.vu.lt

© Springer Nature Switzerland AG 2020
G. Dzemyda et al. (eds.), *Data Science: New Issues, Challenges and Applications*, Studies in Computational Intelligence 869,
https://doi.org/10.1007/978-3-030-39250-5_10

algorithmic thinking (AT). And our daily life is surrounded with the algorithms thus AT is deemed as one of the key elements to be able to be an individual in line with the age of informatics. Algorithmic thinking is the main component of computational thinking (CT). And most of the CT definitions have its roots in AT—the ability of formulating problems that transform an input into the desired output using algorithms. CT could be defined as the skills of being able to develop creative solutions for the problem with an algorithmic approach by handling a problem by the individuals that could establish healthy communication in a cooperative environment (Korkmaz et al. 2017).

Furthermore, programming is a complex intellectual process that must be easily and engagingly presented to the learner, thus attracting learners. In order to manage causes of difficulty to program, it is proposed to use one of the methods of constructivist teaching theory—problem-based learning and computational thinking theory. Computational thinking should be compulsory in all schools, as it helps to get acquainted with programming and stimulate interest in computer science and modern technology use. Because, the ways of computing teaching differ in countries or even in schools: computing subject could be mandatory, elective, integrated and there are some countries where learners have had no experience with CS related subjects at all. Thus, the learning process is enhanced by improving the learning environment conditions where the learners can construct. The design of the world's objects as well as solving the problem that is related to real world can be implemented by modern and intelligent technologies, (for example, the Smart House, the Internet of Things). So, in order to make programming attractive, the content of learning for basic school needs to be drawn up in accordance with the principles of the aforementioned theories.

The focus of this paper was to identify the causes of difficulties in developing computing education competences (including programming) and propose appropriate teaching/learning strategies for education process encouragement. Toward this direction the literature analysis on that topics was conducted. The main discussed aspects fall into categories of problem-based learning approach, teaching materials, assessment, and technologies for developing computational thinking skills through making.

The paper is structured as follows: Sect. 2 introduces the concept of programming and the problem based learning theory as a solution for the causes of difficulties to program management. Computational thinking and making, main components of CT are discussed in Sect. 3. Section 4 covers the literature on computational thinking implementation strategies (for teaching content, CT assessment and CT development technologies) needed for educators' support. The paper ends up with conclusions.

2 Programming and Problem-Based Learning

Programming is a complex intellectual process. It must be simple and appealing to the learner, thus attracting learners. In the study of programming learning problems, the following causes of difficulties groups are identified (Table 1): (1) teaching methods, (2) the use of learning techniques, (3) learner skills and attitudes, (4) the nature of programming, (5) psychological reasons (Jadzgeviciene and Urboniene 2013).

In order to solve such problems, it is proposed to use one of the methods of constructivist teaching theory—problem-based learning (PBL). PBL is an instructional (and curricular) learner-centered approach that empowers learners to conduct research, integrate theory and practice, and apply knowledge and skills to develop a viable solution to a defined problem (ill-structured, real world) (Savery 2015, p. 5). Thus it can be used to avoid 'learners' skills and attitudes' difficulties causes. S. Papert stated that learning is the best when the learner actively develops objects of the real world (for example, a sand castle), and not just ideas or knowledge. In such way the pioneer of the constructivist theory, has expanded the theory of constructivism. The learning process itself is enriched by improving the opportunities for the learner to

Table 1 Causes of programming training difficulties

Component	Cause of difficulties
Teaching methods	Programming training is still not personalized
	Teacher used training methods are not consistent with learning styles of the students
	Dynamic concepts are often taught through static content
	A teacher is more focused on teaching a programming language and its syntax rather than dealing with task solving through a programming language and environment
The use of learning techniques	Learners use irrelevant learning techniques or methodology
	Learners work not enough independently to acquire programming expertise
Skills and attitudes of learners	Learners must have acquired or wish to acquire a wide range of skills related to program development: understanding of problems, knowledge linking to a problem, reflection of a task and its solution, persistence in task solving, application of basic mathematical and logical knowledge, specific knowledge of programming
	It has been observed that the main difficulty for learners is not to get the result itself, i.e. to write a program, but to go through the development process
	A lot of beginners improperly use their skills of writing a stepwise specification in a natural language, i.e. they incorrectly transform natural language semantics into a programming language
The nature of programming	Programming requires a high level of abstraction
	Programming language syntax is very complex
Psychological reasons	Learners are not motivated
	Generally they begin programming learning in a complicated period of their life, e.g. adolescence

construct. Modern and smart devices are very suitable for the construction of world objects. The smart device is an electronic device that usually communicates with other devices and can interact with each other independently. The main component of this unit is a controller that controls connected sensors (temperature, humidity, sound, motion and etc.) and other devices (e.g. motors, RFID tags, bulbs). Often it consists of a board integrating a microcontroller with a microprocessor, sensors and other microschemas. For example, Arduino, Codebug, Micro:bit, Makey Makey, and a minicomputer—Raspberry Pi. Their use in schools from the primary education is rapidly increasing because of the successful use in teaching programming (Davis et al. 2013; Vasudevan et al. 2015). These controllers are programmed in a variety of environments. Some of them are similar or even are the same to programming environments familiar to Lithuanian students. For example, Scratch can be used to program the Raspberry Pi microcomputer or Arduino and Makey Makey controllers, or Ardublock (suitable for Arduino), Blocky (for CodeBug) and JavaScript Blocks Editor (PXT) for Micro:bit controller programming. These controls are usually coded with blocks, similar to Scratch.

Problem-based learning (PBL) is widely and successfully used in primary, basic, secondary, vocational and higher education (Torp and Sage 2002), and in different subjects, in different contexts, learners of different ages are trained (Maudsley 1999). Because problem-based learning is learner-centered it means that the learner ir therefore responsible for his/her learning and solving problems on his/her own (Savery 2015). This means that programming learning difficulties group "learning techniques" is addressed, such as, learners do not have enough work independently to acquire programming skills. Students also choose their own suitable learning methods (i.e. studying their own learning paths: on their own or in a group, actively or passively, etc.), the speed and content of learning (e.g., text material, video, audio recording, etc. Independence also encourages learner motivation (Savery and Duffy 1995), and it is associated with the fifth group of programming learning problems— "psychological motives", i.e. when the learners have no motivation. Motivation grows when the problem is ill-structured and complicate, i.e. the learner is looking for more detailed problems (or formulates the problem) and solves it by their solution through their experience. The problem is to be real world. Henceforth, the use of intelligent devices for the real world problem solving (e.g. smart home, Internet of things) is appropriate.

3   Computational Thinking and Making

The theory of computational thinking is closely related to the PBL method and programming. Jeanette Wing introduced the term 'computational thinking' published in Wing (2006, p. 33), which states that CT involves problem solving, system design, and understanding of human behaviour through the use of fundamental computer science concepts. According to (Serafini 2011), CT must be compulsory in all schools, as it helps to become familiar with programming and stimulate interest in computer science. One of the CT category is the programming or algorithmization—planning and organizing step-by-step sequences for solving a problem. Therewithal, CT consists of three basic levels: concepts (used by programmers, such as variables), practices (problem-solving practices, such as iteration) and perspectives (students' understandings of themselves, their relationships to others, and the technological world around them, for example, questioning about the technology world) proposed by Brennan and Resnick (2012) that may be considered in CT assessment process. Besides, Lye and Koh (2014) review results showed that there is the need of assessment of learning outcomes, taking into account not only the concepts level of CT (the majority of the studies researched only this level), but also the study of computational practices and perspectives levels. Moreover, such data could be collected by using on-screen recording and students' thinking aloud techniques. Additionally, Lye and Koh identified the lack of the study of computational thinking in a naturalistic classroom settings (mostly studies were carried out as after-school activities) and the study of students' thinking using the method of natural communication such as think-aloud protocol, because current surveys are most often done by using questionnaires, field observations and surveys. Combining previously mentioned techniques allows better understanding of the computational practices and computational perspectives of CT.

Traditional teaching methods are criticized thus educational approaches are changing. Digital age learners need to have a broader set of computational skills, such as creativity, critical thinking, and problem-solving. Those are often referred to as CT. There is no unified definition and skills set of CT. Thus developing CT curricula and assessing CT learning is a persistent challenge. CT is mostly defined as a problem solving process that includes a number of particular characteristics (such us, skills, competencies, abilities and etc.). Additionally, in the attempt to define CT, researchers often focus on the core components of CT. Literature review made on CT allows to identify eight main groups of such components: Data analysis and representation (data collection, analysis, representation; generalisation; patterns finding; drawing conclusions), Computing Artefacts (artefact development), Decomposition (breaking into parts), Abstraction (details suppression; modelling), Algorithms (Sequence of steps; set of instructions; automation), Communication and collaboration (Communication; collaboration), Computing and Society (Computing influence, implication and concepts), Evaluation (Evaluation; correction) (Juškeviciene and Dagiene 2018) in order to better understand the CT theory. The concepts involved in each CT component are given in Table 2.

One of CT component is abstraction that can be defined as the solution for a more general problem by ignoring certain details (Krauss and Prottsman 2017). And the processes of programming are impossible without such a skill. Thus CT deals with the programming difficulties group 'nature of programming' because programming requires a high level of abstraction. Although, the CT is mostly criticized due ambiguity and vagueness there are a lot of studies and successful examples of the potential benefits of CT integration into curriculum. And the most effective way to foster CT from early ages is by including CT activities in the school curriculum. However, before integration into curriculum, the teachers should be provided with support on CT development techniques including teaching material, implementation examples,

Table 2 CT components groups

Components	Concepts involved
Data analysis and representation	Data collection
	Data analysis
	Data representation
	Generalisation
	Patterns finding
	Drawing conclusions
Computing Artefacts	Artefacts development
	Artefacts designing
Decomposition	Breaking into parts
Abstraction	Details suppression
	Modelling
	Information filtering
Algorithms	Sequence of steps
	Procedures
	Set of instructions
	Automation
Communication and collaboration	Communication
	Collaboration
	Computational analysis
Computing and Society	Computing influence
	Computing implication
	Computing concepts
Evaluation	Evaluation
	Correction

easy to use devices and clear assessment methodology. The support for teachers is more discussed in next chapter.

Educators and researchers widely agree that hands on projects motivate and attract learners (Martin-Ramos et al. 2017; El-Abd 2017). Such hands on projects and practical application often is named as computational making (Rode et al. 2015) or Physical computing paradigm that takes the computational concepts "out of the screen" and into the real world so that the student can interact with them (Rubio et al. 2013). Physical computing involves creative arts and design processes and brings together hard- (such, as sensors, LEDs, servos) and software components (Przybylla and Romeike 2014). It covers the design and realization of interactive objects and installations, that require learners to use their imagination for tangible products of the real world development. If learners failed to solve a problem at an abstract level, often they succeeded using tangible objects, because of the use of multiple representations of the same knowledge. Learners are motivated by their own creativity that is achieved by using modern prototyping boards that allows them to create significant projects (Jamieson and Herdtner 2015). New teaching methodologies might help the student to overcome programming learning difficulties, such as learning programming concepts, no interest or motivation, feel uncomfortable. For this purpose, it was proposed the use of the physical computing paradigm (Rubio et al. 2013). Thus computational making can deal with 'psychological reasons' and 'teaching methods' programming difficulties groups.

4 How to Support Teachers in CT Implementation?

Studies on teacher's self-efficiency show that teacher's self-efficiency correlates with students' motivation. Moreover, students are affected indirectly by teacher's instructional strategies, approaches and competences planned to build (Nordén et al. 2017). Thus teachers need support in CT implementation process by providing appropriate content, methods, tools and curriculum for teaching and assessment. There are a lot of different strategies proposed for CT teaching, such as, PLB, peer coaching methodology, prototyping, gamification approach. Using constructionism-based problem solving learning makes it possible to reach the positive learning outcomes (Lye and Koh 2014). Peer coaching approach improves presentation and communication skills, self-confidence, and ability to apply abstract ideas (Martin-Ramos 2017). Additionally, it facilitated the communication and relaxed the classroom atmosphere. Gamification approach helps to maintain learners' attention while exploring CT concepts (Rode et al. 2015). Moreover, educational games are often utilized for teaching programming and developing CT, also, games promote problem-solving abilities, engagement, motivation and result in the development of knowledge in various subjects (Nouri and Mozelius 2018). Prototyping is a common method in physical computing. It is suitable for CT teaching as it includes critical thinking development trough testing and debugging (Przybylla and Romeike 2014).

4.1 Computational Thinking Teaching Material

Studies show that CT teaching content should involve introductory theoretical part (on CS, CT and engineering), tools, examples and instructions of CT implementation in class activities, interconnections between CT components and teaching content. Introductory theoretical part should explain the CS and CT and programming concepts, physical computing—sensing technologies, microcontrollers, actuators (Przybylla and Romeike 2014; Martin-Ramos et al. 2017; Rubio et al. 2013; Martin-Ramos et al. 2017). The aim of class demonstration should focus on physical examples of computational concepts in order to enhance the student understanding due the use of diverse perceptive paths (Rubio et al. 2013). Moreover, it is important to underline that interactive objects cannot act without programming them by humans, even learners. This is the powerful idea of programming that makes CS visible and understandable in many aspects of our lives (Przybylla and Romeike 2014). Also, traditional methods are widely critiqued and focus on the one right result thus the educators and researchers provide new strategies for learning motivation. To keep learners engaged and motivated the step-by-step complexity of tasks should be employed. For example, learners do several mini-projects before implementing the final project that consequence as a high satisfaction in Arduino course (Martin-Ramos et al. 2017). Learners are more interested to learn when presented with hands on projects and practical applications (El-Abd 2017; Rode et al. 2015; Rubio et al. 2013). Projects are widely and successfully implemented by using prototyping boards. Learners like the ease of use of prototyping boards that, do not need much time for learning how to configure and allow them to create significant projects by solving real life problems and focusing only on tasks (Jamieson and Herdtner 2015; Candelas et al. 2015; Martin-Ramos et al. 2017). Also, current generation of learners prefers to receive immediate feedback that is possible due prototyping, such us, moving, sounding or blinking. Prototyping is a dominant learning method in physical computing. Such boards are often adopted for gamification approach implementation in classes (Nouri and Mozelius 2018; Rode et al. 2015). Laboratory classes link demonstrations to class activities. It is recommended to use various responsive, reactive and control systems in laboratory activities in order to enchase learning. Another important aspect of successful learning by using modern devices is a lot of available information and examples for learners ranging from basic documentation to specific courses (Candelas et al. 2015). Of course, in order to assess learning outcomes in CT development the educators need easy to use assessments. Research in this field is in its infancy and require huge efforts in order to support educators in CT implementation (Moreno-León et al. 2018).

Based on literature review the list of main tips for teaching material content was developed based on above mentioned references:

- Theoretical introduction and class demonstrations (devices, programming, skills, algorithms, concepts, tasks and components interconnection, examples of implementation and etc.)
- Laboratory modules and hands on activities with real problems solving tasks and prototyping methods
- Step-by-step complexity of tasks
- Easy to use, not time consuming devices
- A lot of available data of examples
- Easy assessments and implementations.

4.2 Computational Thinking Assessments

Easy to use assessments must be developed in order to support educators in the inclusion of CT in classroom and curriculum. However, the scholars are only starting to research this is a field. To promote CT implementation in schools requires a huge effort.

Usually educators assess learners' competencies, skills and abilities that are listed in the curriculum, as well as learners' interest or satisfaction. So called standard assessment paradigm where knowledge can be transmitted from teacher to student and subsequently measured through standardized forms of assessment, for example, tests that objectively measure the amount of knowledge (Mislevy 2012). Assessment as procedure (e.g., portfolios) still objectively measure students' abilities. Differ only methods for data collection.

More flexible approach that describes a wide range of activities and goals associated with assessment is Evidence-Centered Design (ECD). ECD could be used as an approach for CT practices assessment. ECD is a comprehensive framework for describing the conceptual, computational and inferential elements of educational assessment (Mislevy 2012). It could include standard assessment tools extended by the various informal assessment activities, such interacting with learners, simulation tasks and situations or role playing with the main aim to structure situations to evoke particular kinds of evidence. ECD, also, formalizes framework for designing and implementing assessments. CT in the context of modelling and simulations was presented in project GUTS (Lee 2011; Lee et al. 2011). CT assessment could be implemented trough modelling and simulations as an exploratory method to assess transference of CT—an approach to problem investigation and problem-solving when faced with a new scenario. Moreover, such transference provides descriptions of CT concepts application in solving problems rather than just recall CT definitions or use programming constructs in the form such as ability to abstract problems, validate models or conceptualize automation usage. The pair interviews with a joint tasks as collaborative endeavor was used for examining learners' programming processes focusing on a key aspect of CT—interpreting and navigating multiple representations in order to support students' learning and assessment of CT (Barth-Cohen et al. 2018). FACT assessment system for algorithmic thinking aspects of CT was developed combining several data measures, such as, prior experience survey, formative quizzes (captures learners progress and targets of difficulty), Scratch projects (directed and open-ended programming projects), preparation of future learning assessment design (assesses transfer of learning from block-based to a text-based programming language), pre-post CS perceptions survey (includes free response question), pre-post CT assessments (measures pre- and post- CT skills) in order to capture the cognitive, affective and transfer aspects of deeper learning, as well as comprehensive views of learners' CT understanding (Grover 2017, p. 279). Such FACT assessment methodology was designed based on literature review (gaps of CT assessment in formal K-12 CS education), and research made in middle school classrooms, findings. Thus it provided the opportunities for learners to demonstrate their understanding in multiple way, because different assessments do not work equally well for all learners (Grover 2017, p. 285). It is advisable to combine different assessments in order to assess different CT aspects and levels. For this purpose, artifact-based interviews, and design scenarios approaches can be used (Brennan and Resnick 2012). Interviews can be in discussion-based format that allow to assess how learners were employing CT practices while developing projects. Design scenarios allow to systematically explore different ways of knowing, to highlight a developmental or formative approach and, to observe learner's design practices and debugging strategies.

Moreover, three main tools that are commonly used by educators worldwide for CT assessment are identified: CT-test, Bebras and Dr. Scratch. CT-test used for diagnostic-summative assessment; Bebras used for skills onto real and significant problems transference assessment; and Dr. Scratch used for formative assessment (that fosters student's CT) (Moreno-Leon 2018).

However, such assessment requires time and efforts from educators in order to identify learners' skills, knowledge and abilities improvement level. Thus computerized methods have been proposed in the literature for teachers' support.

One of such computerized method is Learning Analytics (LA). LA is defined as 'the measurement, collection, analysis and reporting of data about learners and their contexts, for purposes of understanding and optimizing learning and environments in which it occurs' (Papamitsiou and Economides 2014). LA allows to analyse learner behaviour patterns. By using LA systems learners' interaction in CT environments is analysed and conclusions are made in order to enhance the evaluation or improve the learning process (Amo et al. 2018). Thus LA approach offers great possibilities in terms of learning improvement. Such as, to ensure in time and quality feedback through the analysis of students' behaviour in programming practice (Amo et al. 2018). This can be reached, for example, by using Clickstream—a technique that consists in the collection and analysis of data generated by users. Amo et al., developed the modified Scratch tool with LA that allowed to capture students made clicks in order to analyse learners' behaviours and obtain a score and a taxonomy for each assessed project in terms of rubric and behaviour. Project portfolio analysis can be used as an CT assessment approach. Scrape tool (the "User Analysis" tool) was used in order to analyse the portfolio of Scratch projects that generates a visual representation of the blocks used (or not used) in every project (Brennan and Resnick 2012). It allows to identify computational concepts that were being encountered by user as CT concepts could be mapped to particular Scratch blocks.

Next approach, Educational data mining (EDM) differ from ECD as it discovering meaningful patterns in available data. EDM—'developing, researching, and applying computerized methods to detect patterns in large collections of educational data that would otherwise be hard or impossible to analyze due to the enormous volume of data within which they exist' (Papamitsiou and Economides 2014). EDM is often combined with others methods in order to assess CT. Data mining and qualitative analytics methods were used by Tissenbaum, Sherman, Sheldon in Tissenbaum et al. (2018) in order to assess learners' CT practices. Two data mining approaches were applied using MIT App Inventor—blocks-based programing environment for building mobile apps: block selection mining and moment of failing based on Sherman (2017) study. Furthermore, in order to detect what is missed "above the screen" the qualitative approaches could be applied. Rutstein et al. (2014) develop the framework to analyse the discourse among learners that was used in combination to data mining approaches for collaborative CT nuances analyse.

LA and EDM differ in their origins, techniques, fields of emphasis and types of discovery (Papamitsiou and Economides 2014). ECD and educational data mining

(EDM) might seem in conflict: structuring situations to evoke particular kinds of evidence, versus discovering.

Based on literature review, four group of assessments can be identified:

- Standard assessment paradigm (as knowledge measurement). Methods: Test, exam testing, portfolios
- Evidence-Centered Design (ECD)—structuring situations to evoke particular kinds of evidence (knowledge application). Methods: Role playing, simulated situations, game-based assessment, portfolio assessment, modelling, surveys, interviews, blogs, think aloud protocols, live presentations, reports
- Learning analytics (LA)—allows to analyse student behaviour patterns when they interact with tools and online learning environments, set them in context with their learning outcomes and draw conclusions to enhance the evaluation or improve the learning process. Methods: Click stream, project portfolio analysis (Scrape tool)
- Educational Data mining (EDM)—discovering meaningful patterns in available data. It develops and adapts statistical, machine-learning and data-mining methods to study educational data generated by interacting with environment. Methods: Block selection mining, moments of failing.

The use of just one type of assessment methods could provide a limited or inaccurate sense of learners' computational competencies. Combining methods lead to deeper understanding and improvements for assessment design and analysis. Such data could be collected by using on-screen recording and students' thinking aloud techniques that allows better understanding of the computational practices and computational perspectives of algorithmic thinking.

4.3 Computational Thinking Development Technologies

CT skills could be developed by an unplugged and computerized activities (programing environments and physical devices) (Moreno-Leon 2018). Many tools are used for this purpose. Lockwood and Mooney (2017) reviewed over 50 different tools used for CT teaching, such as, Scratch—visual programming language, MIT App Inventor, RaBit EscApe wooden magnetised pieces game with the aim to place pieces together in a predefined path and help the rabbit escape from the apes, Pandemic board game adoption for CT concepts incorporating and learning whilst playing collaboratively, Program your robot—serious game that is based on a framework to map CT concepts to the game structure, Lego Mindstorm—series of kits contain software and hardware to create customizable, programmable robots; CS Unplugged—collection of free learning activities that teach Computer Science through engaging games and puzzles that use cards, string, crayons and lots of running around; Arduino and etc. meaningful patterns in available data [Mislevy].

Above mentioned tools can be grouped into four categories (Moreno-León et al. 2018; Bocconi et al. 2018):

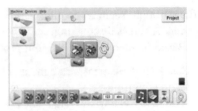

(a) Bebras card's task example   (b) Lego WeDo environment

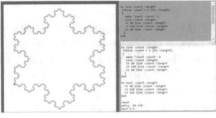

(c) LOGO online   (d) Arduino traffic light prototype

Fig. 1 CT skills development tools

- Unplugged activities;
- Arrow or Block-based visual environments;
- Textual programming languages (LOGO);
- Connected with the physical world.

Examples of such tools are presented in Fig. 1 (a) unplugged activity example from Bebras card task; (b) block-based visual programming screenshot of Lego WeDo environment; (c) textual programming example of LOGO online tool; (d) physical world object prototype of traffic light for car and pedestrian.

Unplugged activities avoid the use of digital devices, as they are focused on ideas and concepts, and not on its implementation (e.g. Bebras cards). Such activities are experience-based and students may use Lego blocks, pencil and paper or other material found in their environment (Weigend et al. 2018). Arrow or block based visual environments allow learners visually program a sequence of commands making use of arrows, icons or blocks (e.g. LEGO WeDo programming environment, Micro:bit in Blocs, Ardublock). Learners have to type the commands in the specific syntax of the language in textual programming (e.g. LOGO). In the group 'Connected with the physical world' learners write programs that control and have effects on objects of the physical world. Such as, programmable toys, robots, boards that are controlled by the source code written in a tablet, smartphone or computer, and then is transferred to the device itself (e.g. Lego, Arduino, Bee Bot, Blue Bot) (Moreno-León et al. 2018; Bocconi et al. 2018).

Computational making is implemented by using physical devices such as prototyping boards. Educators widely use Arduino, Micro:bit, Raspberry Pi, BeagleBone Black and etc. in order to benefit students and develop CT skills. Students like ease of use of these devices. Studies justifies adopting these boards in a curriculum (Jamieson and Herdtner 2015). For example, Arduino board is named as an attractive device that do not need much time for learning how to configure, suitable both for programmers and engineers (Candelas et al. 2015). Arduino became more and more popular among academics (El-Abd 2017). Of course despite the benefits, educators face challenges of using modern prototyping boards, such as it is hard to assess what students have done because of a lot of available content online, but there are some solutions provided, such as make tasks that requires the low-level interfacing without libraries, allow learners to use any codebase and assess the system based on the final product, or allow learners to use any code but provide the report how their code is differentiated from the existing code (Jamieson and Herdtner 2015; Candelas et al. 2015).

5 Discussion and Conclusions

Programming difficulties could be managed by using problem-based learning and computational thinking approaches. In such a way, computational thinking as well as algorithmic thinking skills are also improved.

Modern technologies are widely used for learning enhancement and algorithmic thinking development. Hands on projects motivate and attract learners. Educators could be supported in such activities design by providing with necessary instructional material, appropriate teaching methods, tools and clear assessment methodology.

Prototyping boards (e.g. Arduino) can facilitate effective learning and algorithmic thinking skills gaining trough computational making approach and combination of methods for assessment. Combining assessment methods allows better identification of learner's computational thinking skills. Moreover, it is recommended to implement assessment in the naturalistic classroom settings.

Acknowledgements This project has received funding from European Social Fund (project No 09.3.3-LMT-K-712-02-0066) under grant agreement with the Research Council of Lithuania (LMTLT).

References

Amo Filvá D, Alier Forment M, García Peñalvo FJ, Fonseca Escudero D, Casany Guerrero MJ (2018) Learning analytics to assess students' behavior with scratch through clickstream. In: Proceedings of the learning analytics summer institute Spain 2018: León, Spain, June 18-19, 2018, pp. 74–82. CEUR-WS.org

Barth-Cohen LA, Jiang S, Shen J, Chen G, Eltoukhy M (2018) Interpreting and navigating multiple representations for computational thinking in a robotics programming environment. J STEM Educ Res 1(1–2):119–147

Bocconi S, Chioccariello A, Earp J (2018) The Nordic approach to introducing computational thinking and programming in compulsory education. Report prepared for the Nordic@ BETT2018 Steering Group

Brennan K, Resnick M (2012) New frameworks for studying and assessing the development of computational thinking. In: Proceedings of the 2012 annual meeting of the American Educational Research Association, Vancouver, Canada, 1–25

Candelas FA, Garcia GJ, Puente S, Pomares J, Jara CA, Pérez J, Torres F (2015) Experiences on using Arduino for laboratory experiments of automatic control and robotics. IFAC-PapersOnLine 48(29):105–110

Davis R, Kafai Y, Vasudevan V, Lee E (2013, June) The education arcade: crafting, remixing, and playing with controllers for Scratch games. In: Proceedings of the 12th international conference on interaction design and children. ACM, pp. 439–442

El-Abd M (2017) A review of embedded systems education in the Arduino age: lessons learned and future directions. Int J Eng Pedagogy (iJEP) 7(2):79–93

Grover S, Basu S (2017, March) Measuring student learning in introductory block-based programming: examining misconceptions of loops, variables, and boolean logic. In: Proceedings of the 2017 ACM SIGCSE technical symposium on computer science education. ACM, pp. 267–272

Jadzgevičienė V, Urbonienė J (2013) The possibilities of virtual learning environment tool usability for programming training. In: Proceedings of the 6th international conference innovative information technologies for science, business and education, pp 14–16

Jamieson P, Herdtner J (2015) More missing the Boat—Arduino, Raspberry Pi, and small prototyping boards and engineering education needs them. In: 2015 IEEE frontiers in education conference (FIE). IEEE, pp 1–6

Juškevičienė A, Dagienė V (2018) Computational thinking relationship with digital competence. Inform Educ 17(2):265–284

Korkmaz Ö, Çakir R, Özden MY (2017) A validity and reliability study of the Computational Thinking Scales (CTS). Comput Hum Behav 72:558–569

Krauss J, Prottsman K (2017) Computational thinking and coding for every student. In: The teacher's getting-started guide. Corwin Press Inc

Lee I (2011) Assessing Youth's Computational Thinking in the context of modeling & simulation. In: AERA conference proceedings

Lee I, Martin F, Denner J, Coulter B, Allan W, Erickson J, Werner L (2011) Computational thinking for youth in practice. Acm Inroads 2(1):32–37

Lockwood J, Mooney A (2017) Computational thinking in education: where does it fit? A systematic literary review. arXivpreprint arXiv:1703.07659

Lye SY, Koh JHL (2014) Review on teaching and learning of computational thinking through programming: what is next for K-12? Comput Hum Behav 41:51–61

Martin-Ramos P, Lopes MJ, da Silva MML, Gomes PE, da Silva PSP, Domingues JP, Silva MR (2017) First exposure to Arduino through peer-coaching: impact on students' attitudes towards programming. Comput Hum Behav 76:51–58

Massachusetts Department of Elementary and Secondary Education (2016) 2016 Massachusetts digital literacy and computer science (DLCS) curriculum framework, Malden, MA. http://www.doe.mass.edu/frameworks/dlcs.pdf. Accessed 25 Sept 2018

Maudsley G (1999) Do we all mean the same thing by "problem-based learning"? A review of the concepts and a formulation of the ground rules. Acad med: J Assoc Am Med Coll 74(2):178–185

Mislevy RJ, Behrens JT, Dicerbo KE, Levy R (2012) Design and discovery in educational assessment: evidence-centered design, psychometrics, and educational data mining. JEDM| J Educ Data Min 4(1):11–48

Moreno-León J, Román-González M, Robles G (2018, April) On computational thinking as a universal skill: a review of the latest research on this ability. In: 2018 IEEE global engineering education conference (EDUCON). IEEE, pp 1684–1689

Nordén LÅ, Mannila L, Pears A (2017) Development of a self-efficacy scale for digital competences in schools. In: 2017 IEEE frontiers in education conference (FIE). IEEE, pp 1–7

Nouri J, Mozelius P (2018) A framework for evaluating and orchestrating game based learning that fosters computational thinking. In: EduLearn 2018, vol 10

Papamitsiou Z, Economides AA (2014) Learning analytics and educational data mining in practice: a systematic literature review of empirical evidence. J Educ Technol Soc 17(4):49–64

Przybylla M, Romeike R (2014) Physical computing and its scope-towards a constructionist computer science curriculum with physical computing. Inform Educ 13(2):241–254

Rode JA, Weibert A, Marshall A, Aal K, von Rekowski T, El Mimouni H, Booker J (2015) From computational thinking to computational making. In: Proceedings of the 2015 ACM international joint conference on pervasive and ubiquitous computing. ACM, pp 239–250

Rubio MA, Hierro CM, Pablo APDY (2013) Using Arduino to enhance computer programming courses in science and engineering. In: Proceedings of EDULEARN13 conference, pp 1–3

Rutstein DW, Snow E, Bienkowski M (2014, April) Computational thinking practices: analyzing and modeling a critical domain in computer science education. In: Annual meeting of the American Educational Research Association (AERA), Philadelphia, PA

Savery JR (2015) Overview of problem-based learning: definitions and distinctions. In: Essential readings in problem-based learning: exploring and extending the legacy of Howard S. Barrows, pp 5–15

Savery JR, Duffy TM (1995) Problem based learning: an instructional model and its constructivist framework. Educ technol 35(5):31–38

Serafini G (2011, October) Teaching programming at primary schools: visions, experiences, and long-term research prospects. In: International conference on informatics in schools: situation, evolution, and perspectives. Springer, Berlin, Heidelberg, pp 143–154

Sherman MA (2017) Detecting student progress during programmingactivities by analyzing edit operations on theirblocks-based programs (Doctoral dissertation, University of Massachusetts Lowell)

Tissenbaum M, Sheldon J, Sherman MA, Abelson H, Weintrop D, Jona K, Snow E (2018) The state of the field in computational thinking assessment. International Society of the Learning Sciences, Inc. [ISLS]

Torp L, Sage S (2002) Problem-based learning for K-16 education. In: Association for supervision and curriculum development. Alexandria, VA

Vasudevan V, Kafai Y, Yang L (2015, June) Make, wear, play: remix designs of wearable controllers for scratch games by middle school youth. In: Proceedings of the 14th international conference on interaction design and children. ACM, pp. 339–342

Weigend M, Pluhár Z, Juškevičienė A, Vaníček J, Ito K, Pesek I (2018) WG5: constructionism in the classroom: creative learning activities on computational thinking. In: Dagienė V, Jasutė E (eds) Constructionism 2018, Vilnius, Lithuania, pp 884–900

Wing JM (2006) Computational thinking. Commun ACM 49(3):33–35

Improving Objective Speech Quality Indicators in Noise Conditions

Krzysztof Kąkol, Grażina Korvel and Bożena Kostek

Abstract This work aims at modifying speech signal samples and test them with objective speech quality indicators after mixing the original signals with noise or with an interfering signal. Modifications that are applied to the signal are related to the Lombard speech characteristics, i.e., pitch shifting, utterance duration changes, vocal tract scaling, manipulation of formants. A set of words and sentences in Polish, recorded in silence, as well as in the presence of interfering signals, i.e., pink noise and the so-called babble speech, also referred to as the "cocktail-party" effect is utilized. Speech samples were then processed and measured utilizing objective indicators to check whether modifications applied to the signal in the presence of noise increased values of the speech quality index, i.e., PESQ (Perceptual Evaluation of Speech Quality) standard.

Keywords Lombard speech · Speech quality indicators · Speech modification

1 Introduction

The Lombard speech is an effect discovered in 1909 by Etienne Lombard, a French otolaryngologist (Lombard 1911). This effect occurs when the speaker unconsciously changes certain acoustic features of uttered speech in noise. Lombard observed that for example, speakers in the presence of the crowd speak in a slightly different way than when they have the opportunity to speak in a more intimate situation.

Numerous studies on the Lombard language (Boril et al. 2007b; Egan 1972; Kleczkowski et al. 2017; Lu and Cooke 2008; Stowe and Golob 2013; Therrien

K. Kąkol
PGS Software S.A., Gdańsk, Poland

G. Korvel (✉)
Institute of Data Science and Digital Technologies, Vilnius University, Vilnius, Lithuania
e-mail: grazina.korvel@mif.vu.lt

B. Kostek
Audio Acoustics Laboratory, Faculty of Electronics, Telecommunications and Informatics,
Gdańsk University of Technology, Gdańsk, Poland

© Springer Nature Switzerland AG 2020
G. Dzemyda et al. (eds.), *Data Science: New Issues, Challenges and Applications*, Studies in Computational Intelligence 869,
https://doi.org/10.1007/978-3-030-39250-5_11

et al. 2012; Vlaj and Kacic 2011; Zollinger and Brumm 2011) have identified many features characteristic of this type of expression (Bapineedu 2010; Boril et al. 2007a; Junqua et al. 1999; Kleczkowski et al. 2017; Lau 2008), among others increasing the fundamental frequency or shifting energy from lower frequency bands to medium and higher frequencies. This work focuses primarily on analyzing speech signal when uttered in a simulated noisy environment, then modifying clean speech and artificially mixing it with some noise types and finally applying some objective measures such as PESQ (Perceptual Evaluation of Speech Quality) or P.563 and comparing quality of speech samples.

Work on the Lombard speech often is related to subjective studies of speech intelligibility. There are, however, objective indicators such as PESQ (Perceptual Evaluation of Speech Quality) or P.563, which are used in quality studies of telecommunications channels (Beerends et al. 2009, 2013; ITU-T Recommendation P.563 2004; ITU-T Recommendation P.800 1996; Recommendation P.862:Whitepaper PESQ 2001). That is why we have decided to employ such measures in our study.

The research carried out as a part of this work included several stages: (1) recording of sound samples (words and sentences) without and in the presence of pink noise and the interference signal (i.e., babble speech). The reference (original source) signal was recorded first in silent conditions, and then the same sentences were recorded in the presence of additional disturbances enforcing the Lombard effect appearance in speech recordings; (2) the source files (reference files) processed by increasing the f_0 value, utterance duration changes, vocal tract scaling as well as formants modification; (3) the same files were mixed with pink noise and babble speech interfering signal (4) a comparison of PESQ measures applied to steps (1) and (3) was performed.

2 Characteristics of the Lombard Speech

Features of the Lombard speech are diverse and include the following phenomena (Bapineedu 2010; Junqua et al. 1999; Kleczkowski et al. 2017; Lau 2008; Lu and Cooke 2008; Nishiura 2013):

– increasing the level of sound intensity,
– raising the fundamental frequency of the signal,
– shifting energy from lower frequency bands to higher frequency bands,
– increase in the value of formants, mainly F1 and F2,
– increasing the duration of vowels,
– increasing the spectral tilt (spectral tilt), etc.

Most of these features can easily be determined, but observing changes in these features in the context of Lombard speech is not so simple. The measurement of the instantaneous value of the frequency of the formants is not reliable in this case, because it may indicate, for example, a temporary change associated with the emotions contained in the utterance. In the case of Lombard speech detection, it is easier to use long-term measures, e.g., median values of the fundamental frequency in a longer time.

3 Objective Quality Level Indicators

From the point of view of speech quality, the best measure of quality is the subjective measurement of intelligibility. ITU defines standards for subjective measurements—using listeners' experts. The most important standards have been formulated in the form of the standard P.800/P.830 (ITU-T Recommendation P.800 1996; ITU-T Recommendation P.800.1 2006). The results of this type of measurements are presented as MOS (Mean Opinion Score) (ITU-T Recommendation P.800.1 2006). Such measurement should be performed in the form of listening tests in a group of listeners. Despite the standardization of this type of tests (requirements regarding the acoustics of the listening room, the admissible interference value, the monitoring system, the way of testing, the reliability of testers, etc., included in the standards, among others: ITU-R BS.1116 (2016) and ITU-R BS.1284 (2003), they are usually subject to errors due to the fact that each listener may have different auditory experiences. However, it should be remembered that speech intelligibility is by definition, a subjective indicator, which is why this type of research is conducted as the final verification of indicators.

However, quality of the speech signal can also be measured objectively (ITU-T Recommendation P.563 2004). Objective indicators regarding the speech signal (e.g., clarity) are most often used in the measurement of quality of telecommunications channels. It is important primarily because of the need to ensure the proper quality of services.

There are many factors that influence speech signal quality, including:

– a narrow transmission band or coding using a low transmission rate,
– compression and coding algorithms,
– background noise,
– delay in packets in digital transmission,
– quality of transmission devices (e.g., mobile phones).

Quality of speech in telecommunications channels can be measured in two ways:

– double-ended measurement—performed by comparing the signal before and after the channel (reference signal and tested),
– single-ended measurement—estimation of speech quality based on perceptual factors, without knowledge of the reference signal.

The first standard for measuring speech signal quality was P.861, known as PSQM (Perceptual Quality Measurement System). Unfortunately, this standard did not take into account many factors present in modern digital transmission channels, e.g., loss of packets in VoIP transmission, background noise, or variable delay. All these factors were included in the P.862 standard (PESQ) (Recommendation P.862 2001). In comparative measurements of objective and subjective tests, the correlation coefficient of the PSQM test with subjective measurements reached the value of 0.26, while in the case of PESQ analysis—0.93. In PESQ measurement, the original and degraded signals are mapped onto an internal representation using a perceptual model. The difference in this representation is used by a cognitive model to predict the perceived speech quality of the degraded signal (Beerends et al. 2009, 2013).

The scheme of the PESQ algorithm includes the following stages:

- leveling the signal level; when comparing the original and degraded signal, both should be aligned to the same power level,
- input filtering; in PESQ, signal filtering is modeled, which takes place in telephone devices and the telecommunications network,
- time alignment; transmission systems (e.g., VoIP) introduce a significant delay to the signal, which must be compensated before the measurement,
- auditory (perceptual) transformation; the reference and degraded signal is processed by the auditory transformation system to simulate the characteristics of human hearing—for example, this system removes those parts of the signal that are not heard by the listener,
- interference calculation. The calculated disturbance parameters are converted into PESQ scores in the range from -1 to 4.5. These results are converted on the MOS-LQO (Listening Quality Objective) scale (ITU-T Recommendation P.862.1 2003), i.e., values from 1 to 5 (where 1 means bad speech quality and 5—excellent quality).

4 Single-Ended Measurements

In single-sided measurements, the MOS value is estimated exclusively on the basis of the interference signal. In the case of the P.563 standard, the use of a real expert listening to the conversation on the test device should be simulated. This device can be any receiver, e.g., a mobile phone. Since, in this case, the degraded signal is not compared to the original signal, the speech quality indicator depends on the listening device. It is, therefore, an important element of the P.563 standard (ITU-T Recommendation P.563 2004).

Each signal subjected to MOS measurement using P.563 must be pre-processed by using the model of the listening device. Next, a speech detector (VAD—Voice Activity Detector) is used to mark the speech-related signal fragments. In the next stage, the speech signal is subjected to a series of analyses and assigned to a given class of disturbances. Parameterization of the signal in P.563 can be divided into three basic function blocks (ITU-T Recommendation P.563 2004):

- analysis of the vocal tract and speech abnormality; in this case, it is possible to distinguish the indication of unnaturalness separately for female and male voices and the so-called the "robot" effect,

– additional noise analysis; in this case, it is important to detect static background noise and noise associated with the signal envelope,
– interruptions, mute, and time cuts.

The test signal must also meet the requirements specified in the standard, so that there is the possibility of detecting speech quality using the P.563 algorithm, including:

– the sampling frequency must be greater than or equal to 8 kHz,
– the digital signal resolution must be 16-bit,
– the signal cannot be longer than 20 s, and the speech in the signal cannot be shorter than 3 s.

5   PESQ MOS Measurement Method

As mentioned earlier, PESQ (Perceptual Evaluation of Speech Quality) is a measure of signal quality in the telecommunications channel (ITU-T Recommendation P.862 2001). It should be noted that factors for which PESQ can be used for objective speech quality measurement include testing environmental noise.

In this work, the implementation of the PESQ algorithm available on the ITU websites is used. This implementation enables to carry out PESQ measurements by comparing two signals, one of which is the original (reference) signal and the other a test signal—with interference (in the case of the tests conducted—a signal with a mixed pink noise or a babble speech signal imitating the buzz of human voices).

6   Recordings

Recordings of all speech samples were performed in a room acoustically prepared for that purpose. Recordings were performed first without any additional noise and then with the presence of pink noise and babble speech. To provide the possibility of processing the speech signal itself without the noise path, the distortion signal was played through the person's headphones. This way, the microphone was only recording the uttered speech and not noise. The noise was generated using the Noise Generator Soft application. The thresholds were set up on the levels of 70% and 90% of the generator sound level, which means, respectively, 72.5 dB and 83.8 dB.

17 sentences with different prosody (indicative, imperative and questioning utterances) and additionally 10 separate words uttered by both women and men (three men and three women) were recorded. In the current analysis, only men utterances with neutral prosody were used. Moreover, the files recorded in the presence of noise were used to analyze Lombard speech effects, only.

6.1   Illustration of Lombard Effect in Speech Utterances

In Fig. 1 spectrogram examples of a sentence uttered in silent conditions and in noise (babble speech) are shown. The increase in level and duration of the utterance are visible; other effects are discerned only after quantitative analysis.

7   Recording Modifications

In this work, recordings with the presence of noise were used. However, the sound path does not contain noise itself—according to the description of the recording

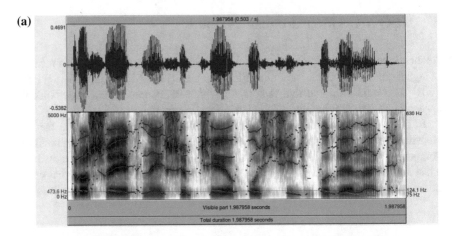

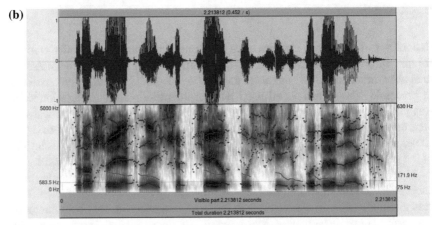

Fig. 1 Spectrogram of a sentence in Polish: original sample, silent recording conditions (**a**), Lombard speech occurrence (an increase of the level and duration change are visible) (**b**)

method. Therefore, the recordings were mixed with the pink noise of a different signal to noise ratio (SNR). For the consecutive samples the SNR was set respectively to −10 dB, −5 dB, 0 dB, 5 dB and 10 dB.

In order to improve objective quality indicators of speech utterances in noise conditions, various signal modifications are considered in this research study. For this purpose, Praat software scripts created by Corretge (2012), their modification introduced by the authors as well as new scripts designed within this study were used. All measurements and manipulations were made with Praat software version 6.0.39 (Boersma and Weenink 2018). The detailed description of signal modifications is given in this Section. The authors decided to focus on frequency-domain filtering, manipulation of duration, modifying pitch, vocal tract, and manipulating formants.

7.1 Frequency-Domain Filtering

The first modification considered in this paper is related to focusing on the specific frequency regions of the speech signal. These regions are obtained by applying a filter in the frequency domain. Two signal filtering options are considered:

(1) Application of a low-pass filter to the speech signal, which cuts high frequencies.
(2) Application of a high-pass filter to the speech signal, which cuts low frequencies.

In this research study, the following cutoff frequency values: 60, 120, 180, 240 Hz are used in the case of a high-pass filter. For a low-pass filter, these values are 6000, 7000, 8000 Hz. The cutoff frequencies were chosen because we wanted to keep all relevant spectrum characteristics related to the speech and at the same time, eliminate irrelevant frequencies as well as noise.

Filtering was applied with the Praat built-in filtering function (Ubul et al. 2009). This function multiplies a complex spectrum in the frequency domain by real-valued filter function, which has the symmetric Hann-like band shape. The width of the region between pass and stop equals to 100 Hz.

7.2 Manipulation of Duration

Changing of speech duration results in speeding up or slowing down the analyzed speech signals. In order to change the signal duration, the Pitch Synchronous-Overlap-and-Add (PSOLA) approach, described by Moulines and Charpentier (1990), is used. This approach is a time domain technique, therefore the modification of signal duration is performed directly in the time domain. The algorithm modifies the time position of speech frames according to some desired pitch contour. The steps of the PSOLA algorithm given as a pseudo-code are given below:

Step 1. Extracting pitch contour of speech sound

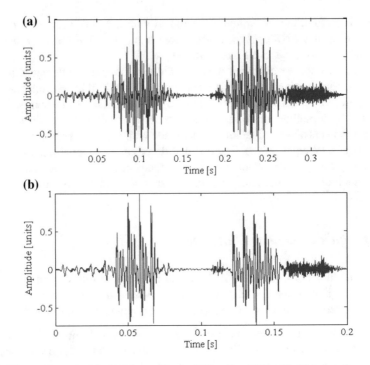

Fig. 2 The oscillogram of the Polish word "zakaz" (Eng. "prohibition"): (**a**) before the vocal tract change, (**b**) after decreasing the signal length by 40%

Step 2. Separating the frames

Step 3. Adding waves back.

The applied frames are centered around the pitch marks and extended to the previous point and the next one. Modifications of duration are made in the following way:

– To increase the length of the signal, the frames are replicated.
– To decrease the length of the signal, the frames are discarded.

An example of a duration change is given in Fig. 2.

7.3 Applying Modification to Pitch

Modifications of pitch are achieved by varying fundamental frequency (f_0) contour. The speech signal contains both periodic and non-periodic parts. Fundamental frequency was extracted only for periodic parts of the signal. The autocorrelation method was used for that purpose. This method was chosen because it gives precise results for periodic signals. The signal was divided into frames of the length of 0.05 s.

Two manipulations of f_0 contours are considered here. The first one is f_0 increasing and decreasing. In this study, the new pitch values are calculated by the following formulas:

$$f_0^{(new)} = f_0 \frac{M f_0^{(new)}}{M f_0} \tag{1}$$

where symbol M denotes the median (the midpoint of the pitch array).

The pitch modification, as well as the process of the duration modification, is also implemented by employing the PSOLA approach. An example of f_0 increasing is given in Fig. 3.

As we see from Fig. 3, the increase of the fundamental frequency moves only frequencies to the right. The values of other frequencies remain the same.

The second manipulation, which is carried out, is called smoothing. This operation transforms f_0 contour into a continuous curve. We want to check if smoothing of sharp peaks improves speech recognition in noise conditions. The literature shows that spectral smoothing is a useful technique for speech recognition (Ghai and Sinha 2009). The authors of that study used a smoothing procedure in the context of children's speech recognition on the adults' speech trained models. In this research study,

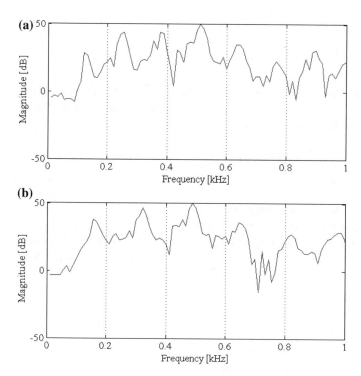

Fig. 3 Fast Fourier transform (FFT) spectrum of the Polish word "zakaz" (Eng. "prohibition"): (**a**) before the f_0 change, (**b**) after increasing f_0 by 40%

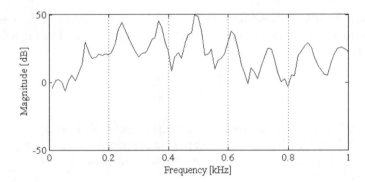

Fig. 4 The Fast Fourier transform (FFT) spectrum of the Polish word "zakaz" (Eng. "prohibition")
after pitch smoothing

we applied the Praat built-in smoothing function. Smoothing is achieved through
changes in the underlying spectrum of the raw f_0 contour that try to eliminate some
of its higher frequency components (Arantes 2015). An example of pitch smoothing
is given in Fig. 4. It should be noted, that the modified signal retains the same duration
as the original one.

7.4   Modification of the Vocal Tract

The goal of this modification is to scale the spectral envelope. The process of manipulation first begins with extracting pitch contour of speech sound. Then the scaling
coefficient is calculated by the following formula:

$$c = \begin{cases} \frac{1}{1+\frac{x}{100}} \ in \ the \ case \ of \ the \ increase \ of \ the \ vocal \ tract \ length \\ 1+\frac{x}{100} \ in \ the \ case \ of \ the \ decrese \ of \ the \ vocal \ tract \ length \end{cases} \tag{2}$$

where x is an decrease or increase given in percentage.

For changing the vocal tract length, the four-step algorithm was applied. The
algorithm was created based on the method given in the work by Darwin et al. (2003).
This method was used by the authors for separating talkers of different gender. The
steps of the algorithm are given below:

Step 1. Multiplying pitch by sc
Step 2. Multiplying duration by $1/sc$
Step 3. Resampling at the original sampling frequency multiplied by sc
Step 4. Playing the samples at the original sampling frequency.

To save the results of the vocal tract changing, the PSOLA approach was used.
The graphical presentation of the vocal tract length changing is given in Fig. 5.

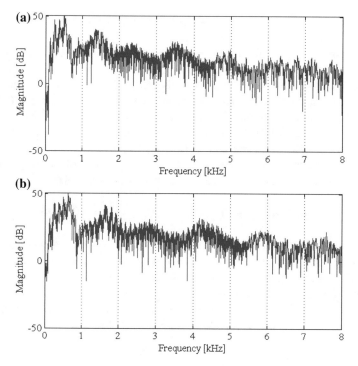

Fig. 5 The Fast Fourier transform (FFT) spectrum of the Polish word "zakaz" (Eng."prohibition"):
(**a**) before the vocal tract change, (**b**) after increasing the vocal tract length by 20%

Graphical representation of the vocal tract modification, depicted in Fig. 5, shows
that the increase of the vocal tract length scales the spectral envelope by moving all
frequencies to the right. It should be mentioned that the modified signal retains the
same duration and fundamental frequency as the original one.

7.5 Manipulation of Formants

The speech signal can be considered as a combination of excitation signal (voice
source) and vocal tract filter. The model describing this theory of speech production
is the so-called source-filter model. The vocal tract filter is modeled based on formants
(the peaks in the frequency spectrum). The excitation signal is constructed from the
noise residual. In order to change formants, the LPC analysis, which decomposes
speech sounds into the two above mentioned parts, is performed. Manipulation of
formants is performed only for periodic parts of the signal. The non-voiced parts of
the speech signal are ignored.

First of all, the spectrum of the speech signal is filtered. Information below the
lower edge f_1 and above the upper edge f_2 is removed. (In this research study

$f_1 = 0$ and $f_2 = 5000$). Then resampling of the speech signal from the real sampling frequency to frequency $2f_2$ is performed.

In the next step, the voiced parts of the speech signal are selected, and the LPC analysis is performed. The LPC analysis represents the spectrum with the chosen number of formants. The number of formants used in this research was equal to five. The LPC filter can be considered as an approximation to the vocal tract. The noise residual is obtained by filtering the speech signal through this LPC filter.

Then the formant frequencies of the speech signal are calculated. The process of calculation of formant frequencies is performed on the spectrum divided into short segments. Based on the Liner Prediction (LP) technique, each of these segments can be approximated as a linear combination of its p previous samples:

$$\hat{x}(n) \approx a_1 x(n-1) - a_1 x(n-2) - \cdots - a_p x(n-p) \tag{3}$$

where $\hat{x}(n)$ is the predicted signal, $a = \left[1, a_1, \ldots, a_p\right]$ are a set of coefficients.

The number of previous samples is called the order of the filter. In this research, we use the general rule that the order of the filter is twice as long as the expected number of formants plus 2 (i.e., $p = 12$ in our experiment).

The formant frequencies are obtained as the roots of the polynomial given by Eq. (3). New formants are calculated by the following formula:

$$F_1 = F_1 + \frac{x \cdot F_1}{100} \tag{4}$$

$$F_2 = F_2 + \frac{x \cdot F_2}{100} \tag{5}$$

where x is a decrease in percentage.

The signal with modified formants is obtained by filtering the residual signal with formants given in Eqs. (4)–(5) through the filter with original LPC coefficients. In the last step of the modification procedure, reassembling of the speech signal to its initial frequency is performed.

8 Estimation of PESQ MOS Values for the Analyzed Set of Recordings

8.1 Mixing the Recording with Pink Noise and Babble Speech

As already mentioned, processing of the recorded utterances encompasses two types of distortions that were mixed with the recording:

– pink noise,
– babble speech, which is a typical noise for places with a lot of people.

Both types of distortions were mixed with an original signal with the same signal-to-noise ratios (SNR) as when comparing recordings without and with Lombard speech (see Recording Modifications Section).

8.2 PESQ Estimation

The following charts present the results of MOS estimation using PESQ algorithm. Every bar represents one proposed modification of the input signal. The resulting estimated MOS values should be compared with the ones obtained by measuring the speech quality of the clean, non-Lombard speech signal. Therefore every chart presents the obtained PESQ MOS values for different distortion type and different level of SNR. The modification types have been described using the following denotations contained in the charts:

- FR1—Formants F1 and F2 raised respectively by 10% and 8%
- FR2—Formants F1 and F2 raised respectively by 3% and 2%
- FR3—Formants F1 and F2 raised respectively by 4% and 2%
- FR4—Formants F1 and F2 raised respectively by 5% and 4%
- FR5—Formants F1 and F2 raised respectively by 8% and 6%
- FR6—Formants F1 and F2 raised by 10%
- FR7—Formants F1 and F2 raised by 20%
- FR8—Formants F1 and F2 raised by 30%
- PS1—Pitch smoothing (100%)
- PS2—Pitch smoothing (25%)
- PS3—Pitch smoothing (50%)
- PS4—Pitch smoothing (75%)
- CD1—Increased duration by 20%
- CD2—Increased duration by 40%
- CD3—Decreased duration by 20%
- CD4—Decreased duration by 40%
- CVT1—Vocal tract moved 20% to the left
- CVT2—Vocal tract moved 10% to the left
- CVT3—Vocal tract moved 10% to the right
- CVT4—Vocal tract moved 20% to the right
- LPF1—Low-pass filter with cutoff frequency 6 kHz
- LPF2—Low-pass filter with cutoff frequency 7 kHz
- LPF3—Low-pass filter with cutoff frequency 8 kHz
- HPF1—High-pass filter with cutoff frequency 120 Hz
- HPF2—High-pass filter with cutoff frequency 180 Hz
- HPF3—High-pass filter with cutoff frequency 240 Hz
- HPF4—High-pass filter with cutoff frequency 60 Hz
- REF—Reference signal (original speech recorded in silent conditions).

Examples of these analyses are shown below. They contain the comparison of speech signal modifications with the reference signal (original speech recorded in silent conditions). Figure 6 shows an example of estimated PESQ MOS values for SNR = 0 pink noise (a) and babble speech distortions (b). Some modifications seem to be more effective than others. For instance, raising the formants and changing duration have a positive impact on the estimated MOS. Other such analyses (SNR: −10 dB up to +10 dB; pink noise/babble speech) show that regardless of the noise applied and SNR this tendency stays true for all other cases. Below, charts from Fig. 7 show only four modifications—CD1, CD2, FR7, and FR8 for additional comparison. In all cases of analyses, PESQ MOS values are higher for pink noise disturbance than for babble speech. However, the differences between them are rather small. That

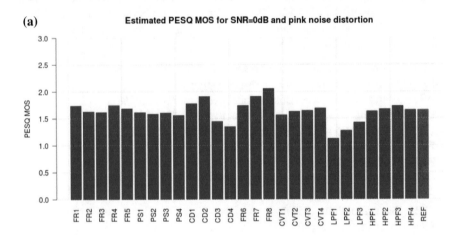

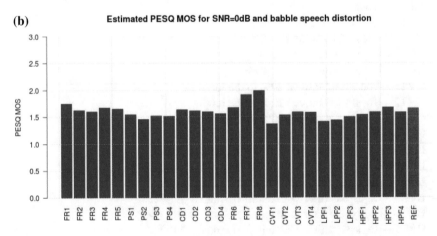

Fig. 6 Estimated PESQ MOS values for SNR = 0, pink noise (**a**) and babble speech (**b**) distortions

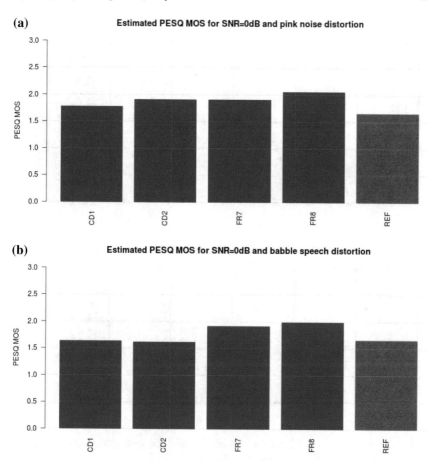

Fig. 7 Estimated PESQ MOS values for SNR = 0, pink noise (**a**) and babble speech (**b**), denotations are as follows: CD1—increased duration by 20%, CD2—increased duration by 40%, formants raised: 20% (FR7) and 30% (FR8)

is why Fig. 8 contains a comparison of the results of processing the most effective changes (raising formants) as a function of SNR. In all cases, an improvement of SNRs occurs. Overall, changes in MOS are present for both positive and negative SNRs.

In principle, modifications performed on the original speech set to increase PESQ MOS values worked with differentiated effectiveness depending on the type of the modification applied. However, some of the results obtained are encouraging as PESQ MOS values differ by one class between the reference and the modified signals. Contrarily, there are some cases when modification decreases PESQ MOS values.

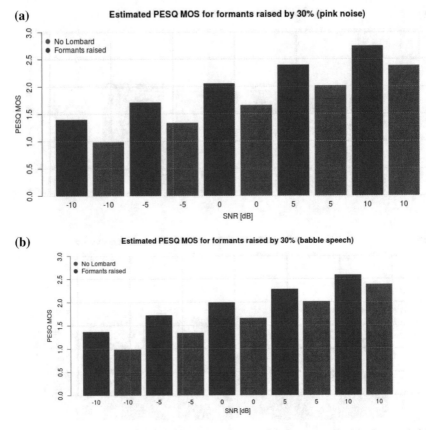

Fig. 8 Comparison of the PESQ MOS values of processing the most effective changes (raising formants) as a function of SNR, pink noise (**a**), babble speech (**b**)

9 Classification

In the second part of the experiment classification of unprocessed and processed files in noise conditions is performed. Babble speech is used as a damping factor. The aim of this analysis is to check which of signal modifications, given in section "Recording modifications", improve recognition of speech utterances in the presence of the babble speech. Speech samples were recorded first without any additional noise, and then—as mentioned above—babble speech was used as a disturbance. For each of the speech samples, the acoustic parameters were extracted. In the experiment, 18 Mel-Frequency Cepstral Coefficients (MFCCs) coefficients (Mermelstein 1976) and their first-order derivatives were used. Before feature extraction, the signal preprocessing is carried out. The speech signal is divided into 512 sample frames, and an overlap constitutes 50% of the segment length is used . To compare classification rates, the k-Nearest Neighbors (kNN), a "classical" machine learning algorithm was employed. Based on obtained classification accuracies, the percent classification change was calculated:

$$change = Acc_1 - Acc_2 \qquad (6)$$

where Acc_1 is the classification accuracy given for signal after modification, and Acc_2—accuracy given for signal without Lombard effect.

The percent classification change is given in Table 1, where positive values indicate the percentage increase and negative ones—percentage decrease. As seen from Table 1 the results are non-conclusive; they are dependent on the type of the modification used, SNR and the "deepness" of the interference applied. Thus, this part of the experiment will further be pursued, especially as the training set was very small.

Table 1 The percentage change of classification

SNR (dB)	Changes applied to the vocal tract			
	20% to the left	10% to the left	10% to the right	20% to the right
−10	5.9	−11.8	−5.9	−5.9
−5	0	−17.7	−5.9	−5.9
0	5.88	−23.5	0	−23.6
5	−35.3	−35.3	−41.2	−52.9
10	−41.2	−11.8	−23.6	−41.2

SNR (dB)	Increasing the first five formants		
	10%	by 20%	by 30%
−10	−11.8	5.9	0
−5	11.8	0	0
0	17.7	−5	−11.8
5	−23.6	−35.3	−41.2
10	−11.8	−35.3	−23.6

SNR (dB)	Increasing the first two formants				
	F1 10%, F2 8%	F1 3%, F2 2%	F1 4%, F2 2%	F1 5%, F2 4%	F1 8%, F2 6%
−10	−5.9	−11.8	−11.8	−11.8	5.9
−5	−5.9	0	11.8	0	0
0	11.8	11.8	17.7	17.7	11.8
5	−17.7	−17.7	−23.6	−29.4	−17.7
10	−5.9	−5.9	−11.8	−5.9	−5.9

(continued)

Table 1 (continued)

	Low-pass filter		
SNR (dB)	Cutoff 6000 Hz	Cutoff 7000 Hz	Cutoff 8000 Hz
−10	−5.9	−5.9	−5.9
−5	−5.9	−5.9	−5.9
0	11.8	0	0
5	0	−11.8	−11.8
10	0	0	0

	High-pass filter			
SNR (dB)	Cutoff 60 Hz	Cutoff 120 Hz	Cutoff 180 Hz	Cutoff 240 Hz
−10	5.9	−5.9	−5.9	−5.9
−5	0	−5.9	−5.9	−5.9
0	5.9	−5.9	0	5.9
5	−35.3	−11.8	−11.8	−5.9
10	0	0	0	0

	Change of duration			
SNR (dB)	Increase by 20%	Increase by 40%	Decrease by 20%	Decrease by 40%
−10	−11.8	−5.9	0	0
−5	0	0	−5.9	0
0	5.9	35.3	5.9	0
5	−17.7	5.9	−23.5	−11.8
10	−5.9	−5.9	−17.7	−5.9

	Pitch smoothing			
SNR (dB)	By 100%	By 75%	By 50%	By 25%
−10	0	0	−11.8	−5.9
−5	−5.9	−5.9	−17.7	−11.8
0	−17.7	−17.7	11.8	−5.9
5	−47.1	−53.0	−17.7	−29.4
10	−35.4	−35.3	−29.4	−29.4

10 Conclusions

Performed experiments demonstrated that there is a possibility to improve speech quality by performing relatively simple operations on the input signal, judging by PESQ MOS measures. However, it was demonstrated only that the estimated Mean Opinion Score factor could be improved. The real value of MOS improvement was not verified, and it can only be examined in subjective tests. Nevertheless, PESQ MOS algorithm is used in telecommunications in parallel with listening tests or independently, due to the fact that PESQ results are correlated with subjective experiments in 93%. We can therefore assume, that the results of experiments performed with PESQ measurements may converge with the subjective examination.

Modifications applied are simple to obtain and may successfully be used in real-time systems working in noise, e.g., hearing aids or threat warning systems. It is worth noticing that the additional aspect, connected with Lombard speech, is the problem of automatic speech recognition in noise. Based on modifications introduced, some experiments were performed and show that classification results are non-conclusive. The obtained changes in accuracies are dependent on the type of the modification used, SNR and the "deepness" of the interference applied. Thus, this part of the experiment will be further pursued, especially as a set of the training set was very small.

References

Arantes P (2015) Time-normalization of fundamental frequency contours: a hands-on tutorial. In: Courses on speech prosody, p 98

Bapineedu G (2010) Analysis of Lombard effect speech and its application in speaker verification for imposter detection. M.Sc. thesis, Language Technologies Research Centre, International Institute of Information Technology

Beerends JG, Buuren RV, Vugt JV, Verhave J (2009) Objective speech intelligibility measurement on the basis of natural speech in combination with perceptual modeling. J Audio Eng Soc 57(5):299–308

Beerends JG, Schmidmer C, Berger J, Obermann M, Ullmann R, Pomy J, Keyhl M (2013) Perceptual objective listening quality assessment (POLQA), the third generation ITUT standard for end-to-end speech quality measurement part ii perceptual model. J Audio Eng Soc 61(6):385–402

Boersma P, Weenink D (2018) Praat: doing phonetics by computer [Computer Program]. Version 6.0.39. Retrieved May 2018

Boril H, Fousek P, Höge H (2007a) Two-stage system for robust neutral/Lombard speech recognition. InterSpeech

Boril H, Fousek P, Sündermann D, Cerva P, Zdansky J (2007b) Lombard speech recognition: a comparative study. InterSpeech

Corretge R (2012) Praat vocal toolkit. http://www.praatvocaltoolkit.com

Darwin CJ, Brungart DS, Simpson BD (2003) Effects of fundamental frequency and vocal-tract length changes on attention to one of two simultaneous talkers. J Acoust Soc Am 114(5):2913–2922

Egan JP (1972) Psychoacoustics of the Lombard voice response. J Auditory Res 12:318–324

Ghai S, Sinha R (2009) Exploring the role of spectral smoothing in context of children's speech recognition. In: 10th Annual conference of the international speech communication association

ITU-R BS.1116 (2016) Methods for the subjective assessment of small impairments in audio systems including multichannel sound systems

ITU-R BS.1284 (2003) General methods for the subjective assessment of sound quality

ITU-T (1996) Methods for subjective determination of transmission quality. Recommendation P.800, Aug

ITU-T (2003) Mapping function for transforming P.862 raw result scores to MOS-LQO. Recommendation P.862.1, Nov

ITU-T (2001) Perceptual evaluation of speech quality (PESQ), an objective method for end-to-end speech quality assessment of narrow band telephone networks and speech codecs. Recommendation P.862, Feb

ITU-T (2004) Single-ended method for objective speech quality assessment in narrow-band telephony applications. Recommendation P.563

ITU-T (2006) Mean opinion score (MOS) terminology. Recommendation P.800.1, July

Junqua J-C, Fincke S, Field K (1999) The Lombard effect: a reflex to better communicate with others in noise. In: 1999 IEEE international conference on acoustics, speech, and signal processing proceedings. ICASSP99 (Cat. No. 99CH36258), vol 4, pp 2083–2086

Kleczkowski P, Żak A, Król-Nowak A (2017) Lombard effect in Polish speech and its comparison in English speech. Arch Acoust 42(4):561–569. https://doi.org/10.1515/aoa-2017-0060

Lau P (2008) The Lombard effect as a communicative phenomenon. UC Berkeley Phonology Lab Annual Report

Lombard E (1911) Le signe de l'élévation de la voix (translated from French). Ann des Mal l'oreille du larynx 37(2):101–119

Lu Y, Cooke M (2008) Speech production modifications produced by competing talkers, babble, and stationary noise. J Acoust Soc Am 124:3261–3275

Mermelstein P (1976) Distance measures for speech recognition, psychological and instrumental. In: Chen RCH (ed) Pattern recognition and artificial intelligence. Academic, New York, NY, USA, pp 374–388

Moulines E, Charpentier F (1990) Pitch-synchronous waveform processing techniques for text-to-speech synthesis using diphones. Speech Commun 9(5–6):453–467

Nishiura T (2013) Detection for Lombard speech with second-order mel-frequency cepstral coefficient and spectral envelope in beginning of talking speech. J Acoust Soc Am

Stowe LM, Golob EJ (2013) Evidence that the Lombard effect is frequency-specific in humans. J Acoust Soc Am 134(1):640–647. https://doi.org/10.1121/1.4807645

Therrien AS, Lyons J, Balasubramaniam R (2012) Sensory attenuation of self-produced feedback: the Lombard effect revisited. PLoS One 7(11):e49370

Ubul K, Hamdulla A, Aysa A (2009) A digital signal processing teaching methodology using Praat. In: 2009 4th international conference on computer science & education. IEEE, pp 1804–1809

Vlaj D, Kacic Z (2011) The influence of Lombard effect on speech recognition. In: Speech technologies, Chap. 7, pp 151–168

Whitepaper PESQ (2001) An introduction. Psytechnics Limited

Zollinger SA, Brumm H (2011) The evolution of the Lombard effect: 100 years of psychoacoustic research. Behaviour 148:1173–1198

Investigation of User Vulnerability in Social Networking Site

Dalius Mažeika and Jevgenij Mikejan

Abstract The vulnerability of the social network users becomes a social networking problem. A single vulnerable user might place all friends at risk therefore, it is important to know how the security of the social network users can be improved. In this research, we aim to address issues related to user vulnerability to a phishing attack. Short text messages of the social network site users were gathered, cleaned and analyzed. Moreover, phishing messages were build using social engineering methods and sent to the users. K-means and Mini Batch K-means clustering algorithm were evaluated for the user clustering based on their text messages. A special tool was developed to automate the users clustering process and a phishing attack. Analysis of users responses to the phishing messages built using different datasets and social engineering methods was performed, and corresponding conclusions about user vulnerability were made.

Keywords User vulnerability · Phishing attack · Social network

1 Introduction

Number of social networking sites and virtual communities are growing constantly over the past years. Virtual communities have more than 2.7 billion active users worldwide, and it accounts for 65.8% of the total Internet user (Number of social media users worldwide from 2010). When people join social networking sites, they make personal profiles and connections to existing friends as well as those they meet through the site (Dwyer et al. 2007; Gundecha et al. 2014). A profile usually contains identifying information and sensitive private data as well. Disclosing private information or damaging it may force a user to be compromised, loosed confidence in the company and etc. Moreover, private information such as individual's place, date of birth, and short messages can be exploited to predict social security number

D. Mažeika (✉) · J. Mikejan
Vilnius Gediminas Technical University, Saulėtekio al. 11, 10223 Vilnius, Lithuania
e-mail: Dalius.Mazeika@vgtu.lt

© Springer Nature Switzerland AG 2020
G. Dzemyda et al. (eds.), *Data Science: New Issues, Challenges and Applications*, Studies in Computational Intelligence 869,
https://doi.org/10.1007/978-3-030-39250-5_12

or demographic attributes (Acquisti and Gross 2009; Burger et al. 2011; Vitak 2012; Burke and Kraut 2014).

Online content obtained from social network sites is also valuable for marketing, personalization, and legal investigation. Target groups of the users are defined using machine learning methods and advertisements are send to them. However, some advertisements may be harmful to consumers as for instant for the children or teenagers (Staksrud et al. 2013). Social network users also use different applications that are provided by social networking site, share them according to common interests in their groups, and invite friends to share their application as well. These applications can be malicious and harmful which can affect the users' personal information and affect their friends in the same community. It means that a single vulnerable user might place all friends at risk.

Different social engineering methods as for example social phishing can be used to affect virtual community user and to identify weaknesses of the user as well as disclose confidential information. Applying social engineering methods user's data can be obtained easier than trying to penetrate the software or hardware (Krombholz et al. 2015; Bullée et al. 2018). Two types of social engineering are identified: direct social engineering and indirect social engineering. Direct social engineering is used when person is suddenly compromised. Usually people who have been targeted in such a way do not understand the situation immediately therefore confidential information is disclosed. Indirect social engineering can be defined as a situation when the person is forced to apply for assistance that is provided by attacker and thus achieves the desired effect. Social networking sites is a good environment for attacker to hide his identity, to gain trust in a virtual community and to retrieve information (Ivaturi and Janczewski 2011). Therefore, it is important to determine vulnerability of the social network users and to know how security of the user can be improved.

2 Related Works

The vulnerability of social network site users is an important topic nowadays. There has been a considerable increase in the number of research papers that studied this topic. Different vulnerabilities arising from user's personal data as user names, demographic attributes, lists of friends, short text messages, photos, audios and videos were analyzed by researches. It is evidence that social networks and virtual communities have attained academic significance and various case studies are analyzed.

Kosinski et al. showed that highly sensitive personal attributes of the virtual community users can be automatically and accurately inferred using the variety of Facebook likes. Personal attributes include sexual orientation, ethnicity, religious and political views, personality traits, intelligence, happiness, use of addictive substances, parental separation, age and gender (Kosinski et al. 2013). Narayanan et al. showed that users are not protected on a social networking site by successfully de-anonymizing network data solely based on network topology. Authors concluded that privacy laws are inadequate, confusing, and inconsistent amongst nations making social networking sites more vulnerable (Narayanan and Shmatikov 2009). Krishnamurthy et al. investigated the problem of leakage of personally identifiable information and showed that it is possible for third-parties to link personally identifiable information through user actions both within social networks sites and elsewhere on non-social network sites. They showed the ability to link personal information and combine it with other information and identified multiple ways by which such leakage occurs (Krishnamurthy and Wills 2009). Liu et al. analyzed privacy awareness and found large number of social network user profiles in which people describe themselves using a rich vocabulary of their passions and interests (Liu and Maes 2005). It confirms the necessity to make vulnerability research on social networking site to make users aware of privacy risks. Wondracek et al. showed that information about the group memberships of a social network site user is sufficient to uniquely identify this person, or, at least, to significantly reduce the set of possible candidates (Wondracek et al. 2010). To determine the group membership of a user, they leverage well-known web browser history stealing attacks. Whenever a social network user visits a malicious website, this website can launch de-anonymization attack and learn the identity of its visitors. The implications of attack was manifold, since it required a low effort and had the potential to affect millions of social networking users.

Different data analysis methods are used to analyze the behavior of social network site users and to find dependencies and similarities between users (Pennacchiotti and Popescu 2011; Rao et al. 2011; Arnaboldi et al. 2013; Conti et al. 2012). Pennacchiotti et al. investigated user classification problem in social media with an application to Twitter. User profile, user tweeting behavior, linguistic content of user messages and user social network features were used for the classification framework (Pennacchiotti and Popescu 2011). Rao et al. used hierarchical Bayesian models for latent attribute detection in social media. They introduced new models that provide a natural way of combining information from user content and name morphophonemics (Rao et al. 2011). Arnaboldi et al. presented a detailed analysis of a real Facebook dataset aimed at characterizing the properties of human social relationships in online environments (Arnaboldi et al. 2013). Authors introduced linear models that predicted tie strength from a reduced set of observable Facebook variables and obtained good accuracy. Parsons et al. performed a scenario-based experiment of users' behavioral response to emails. Social engineering methods were used to build phishing emails (Parsons et al. 2013). Obtained results indicated that the participants who were informed that they were undertaking a phishing study were significantly better at correctly managing phishing emails and took longer to make decisions. Authors concluded that when people are primed about phishing risks, they adopt a more diligent screening approach to emails. Buglass et al. examine how the use of social network sites increases the potential of experiencing psychological, reputational and physical vulnerability (Buglass et al. 2016). Their findings indicated a positive association between Facebook network size and online vulnerability mediated by both social diversity and structural features of the network. In particular, network clustering and the number of non-person contacts were predictive of vulnerability. Authors support the notion that connecting to large networks of online friends can lead to increasingly complex online socializing that is no longer controllable at a desirable level. Mansour analyzed social networking vulnerability (Mansour 2016). Author concluded that even without soliciting information directly from a user of a social networking site, hackers, malware distributors, or other internet social engineers could quite easily infer a great deal of personal information from users, based simply on the users' profiles and networking behavior. Also malware and phishing programs are potential risk of social networking sites security.

Analysis of social network site users' responses to the phishing messages constructed based on available tweets is performed in this paper. Also, two different clustering methods are evaluated using different datasets. Social engineering was applied to make the content of phishing messages. Analysis of users vulnerability to phishing attack was performed.

3   Data Set Description

Short text messages or tweets of Twitter social network site users that follow one of the popular Lithuanian politician on the Twitter social network were collected and analyzed. This politician has 5890 followers. Twitter publicly shows profile information such as the user name, the location, and other information. Twitter API provides access to the additional user information, such as the number of a user's friends, followers and tweets. When the user is reading short text messages using web browser he also can see the following data as picture, text, tags, and links. However, a large part of the data is not available to the user, but can be accessed using REST API of Twitter Stream API. Example of user information obtained using the Twitter Stream API is shown in Fig. 1. Data are presented in JSON format.

Secure communication between Tweeter server and web browser is implemented using HTTPS. Authentication based on OAuth protocol is required to access user's data. Twitter Application Management tool was used to gain *access_token*, *access_token_secret*, *consumer_key* and *consumer_secret* keys. These keys were used for special tool that was developed for this project. It must be pointed that connection to the Twitter server may be terminated because of the high number of logins from the same account or low data reading rate.

The goal of the study was to find out the follower's hobby, interest, and the relevant topics by analyzing the content of the tweets. Such variables as user names and tweets were collected for the study. The length of the tweet is up to 140 characters and it is provided as JSON data object. Only English tweets were collected and stored in database. In order to prepare datasets for analysis, the following data cleaning operations were performed: tokenization, case converting, and removal of punctuation characters, links, usernames, stop words and short words. Lemmatizer and stemmer were applied to make final text preparation for the analysis. The last step was to build vectors of the keywords. Keywords were defined as the words whose recurrence frequency is significantly higher than the frequency of other words.

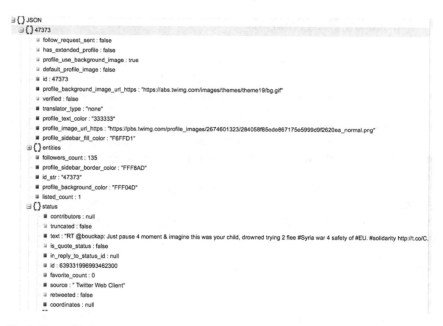

Fig. 1 Data of Twitter user in JSON format

Four different datasets were prepared for the study. The datasets contained tweets of 100, 500, 1000 and 2000 followers that follow Lithuanian politician. Characteristics of datasets are presented in Table 1 where the following parameters are shown: number of users in dataset, amount of words in dataset after removing special characters, punctuation marks, links, and user names. Also the number of unique words with different stem that were obtained after removing the insignificant words. The minimal and maximal word frequencies are shown in Table 1 as well.

When the number of users and tweets is increasing then the number of different unique words used in tweets is increasing as well. It can be explained because of the different vocabulary used by different users.

Table 1 Characteristics of datasets

Dataset no.	Number of Twitter users	Total number of words	Number of unique words	Word frequency (min)	Word frequency (max)
1	100	801	37	2	15
2	500	3946	94	3	15
3	1000	7838	121	4	15
4	2000	16,449	277	4	15

4 Models for Data Analysis

We divide data analysis into two parts. The first part was related with the user clustering based on the vector of the keywords obtained from tweets while the second part was related to constructing phishing text message. The goal of the first task was to find clusters of similar users and to understand their common interests and topics. Clusters allowed to distinguish user groups and to build particular phishing messages for the particular user groups. The phishing text message was build based on the most frequently used words in the tweets of the users belonging to the same cluster.

K-means and Mini Batch K-means unsupervised clustering algorithms were chosen for the text analysis of the tweets and for users grouping into the clusters (Conti et al. 2012; Sculley 2010; Duda et al. 1996; Celebi et al. 2013). We wanted to compare processing time of aforementioned clustering algorithms on different dataset in order to decide which algorithm is more suitable for tweet analysis. Four datasets showed in Table 1 were used. Four clustering models containing 3, 5, 10, 15 clusters were analyzed. Silhouette coefficient was used for model evaluation. Model validation workflow is shown in Fig. 2.

Results of Silhouette coefficient calculations are shown in Table 2. Processing time consumed for the data clustering applying K-means and Mini Batch K-means methods were measured as well, and results are presented in Fig. 3.

Analyzing obtained results of the model evaluation it can be noticed that in most cases K-means clustering algorithm performs more precise data clustering than Mini Batch K-means algorithm. However, the highest value of Silhouette coefficient of 0.375 was obtained when dataset No. 3 was tested using Mini Batch K-means algorithm. The highest Silhouette coefficient of 0.365 was obtained using K-means algorithms when dataset No. 4 was tested and number of clusters was 15. The difference between coefficient values is 2.66%. Also it must be mentioned that value of Silhouette coefficient is increasing in all datasets when the number of clusters is increasing as well.

Graphs presented in Fig. 3 show that K-means method is faster than Mini Batch K-means and the processing time difference is increasing when dataset size is increasing too. The average difference between processing time is 83.29%. Also, it must be

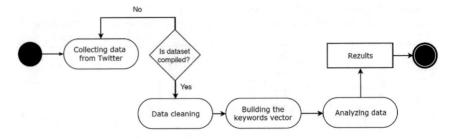

Fig. 2 Model validation activity diagram

Table 2 Results of model evaluation

Dataset no.	Number of clusters	Silhouette coefficient (K-means)	Silhouette coefficient (Mini Batch K-means)
1	3	0.258	0.328
	5	0.269	0.246
	10	0.296	0.290
	15	0.335	0.376
2	3	0.291	0.338
	5	0.320	0.267
	10	0.317	0.263
	15	0.321	0.370
3	3	0.310	0.265
	5	0.353	0.306
	10	0.326	0.347
	15	0.361	0.375
4	3	0.360	0.347
	5	0.362	0.320
	10	0.363	0.322
	15	0.365	0.332

pointed that processing time depends on the number of clusters. Moreover, when number of clusters is increasing, the processing time is increasing too. Comparing two clustering methods based on Silhouette coefficient, it can be concluded that the analysis of large datasets applying the K-means algorithm is more appropriate, therefore K-means was used for the further experiments with Twitter users' messages.

5 Experimental Study

A special tool was developed for automatic text message mining and determination of the keywords that can be used for phishing message. The following Python libs were used: NLTK—natural language processing toolkit, SCiPy—package of open-source software for mathematics, science, and engineering, NumPy—fundamental package for scientific computing that includes n-dimensional array objects, and Scikit-learn— machine learning library. The main functions of the developed tool was as follows: to connect to the Twitter server, get users data and tweets, to make text analysis, obtain keywords vector and to send phishing message to the users. UML class diagram of the tool is shown in Fig. 4.

The tool consists of four classes, i.e. Twitter User, Data, Program, and Interface. Class Twitter User is used to logging into the Twitter server and to get user metadata in the JSON format. Authorization method is included in this class as well. Class

Fig. 3 Dependence of
processing time versus
number of clusters: K-means
(**a**), Mini Batch K-means (**b**)

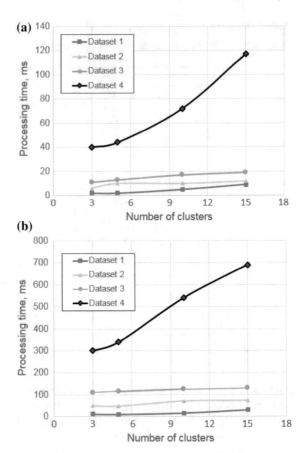

Data is used for data reading from JSON file, data processing, and clustering. Class Program is designed to build phishing message and to send it to the user while class Interface is used for graphical interface and for configuration of the tool. The user interface of the developed tool consists of two main blocks i.e. upper and lower (Fig. 5). The fields in the upper left side are used to enter initial data such as number of clusters, the minimum and maximum number of keywords and number of users. The clustering results and keywords are shown in the center part of the upper block. Lower block of the window is used to configure phishing messages and to send them to the specified cluster of the Tweeter users.

The model was evaluated under different conditions and settings: (1) three phishing messages were built for the dataset of 300 and 500 Twitter users and sent them; (2) five phishing messages were built for the dataset of 1000 and 2000 Twitter users and sent them.

Users were grouped into the 3 clusters in the first case while 5 clusters were built in the second case. K-means algorithm was used. Results of the clustering are shown in Fig. 6. Analyzing result it can be noticed that similar distribution of user numbers

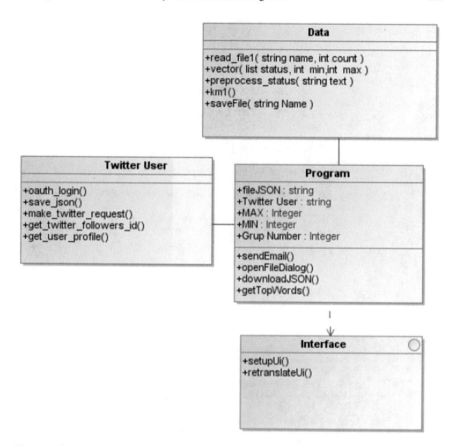

Fig. 4 UML class diagram of the developed tool

between clusters was obtained for datasets 1 and dataset 2. The largest number of users that is more than 63% were assigned to cluster 0 (Fig. 6a). The similar results were obtain when clustering of the dataset 2 and dataset 3 was performed. The largest number of users were assigned to cluster 0 and it includes 62.1% of users in dataset 2 and 55.5% of users in dataset 3 (Fig. 6b). It can be concluded that the majority of the followers use the similar words in their tweets and clustering of the users based on the tweets can be used for investigation.

The next step of the study was to form keywords vector for the each cluster defined in different datasets. Keywords vector consisted of 10 words. Frequency of the keywords that were included was limited from 3 to 15. Values of the keyword limits were set based on dataset properties shown in Table 1. Results of the experiment are presented in Table 3 where keyword obtained in different clusters are shown. Keywords are sorted by frequency of recurrence in the particular cluster. Also, the corresponding phishing messages and their type is provided in Table 3 where notations C, G, F mean curiosity, greed, and fear respectively.

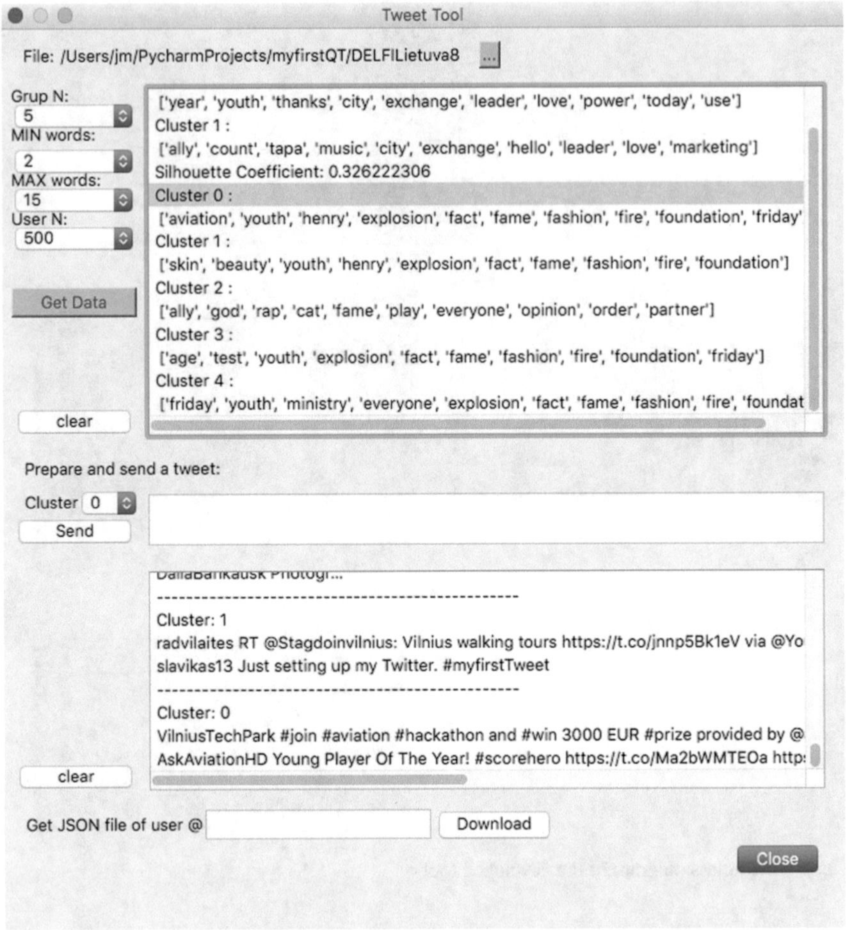

Fig. 5 User interface of the tool

Analyzing obtained results it can be seen that the context of the keywords in the datasets is related to music, business, job, and market activities and it correspond to the discussion topics of the users. Obtained keywords were used to search for the up to date topics on the web in order to know the possible contexts of the phishing messages. The information contained in first the website ranked by the search engine was selected as a context for the phishing message. Then the first three keywords were used to search for the information in the website content. The sentence with the smallest distance between all three keywords was selected. If such a sentence was not found, then a text contained all three keywords were selected. Finally, the content of the selected sentence or text was modified manually to make it interesting to the user. Each cluster has different keywords vector. Therefore particular phishing messages were created for each cluster. Phishing messages also contained a link to the

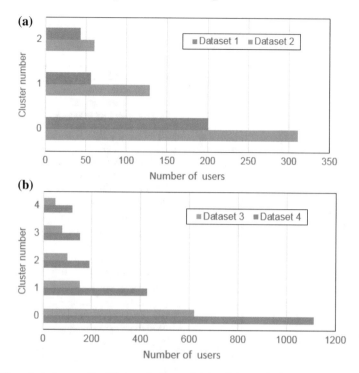

Fig. 6 User clustering results of dataset 1, dataset 2 (**a**) and dataset 3, dataset 4 (**b**)

corresponding websites on which basis the text was formed. A real website address was replaced by shortened one obtained from Google URL shortening service to acquire clicks of the links.

The text of the message was formulated to gain the trust of the user and to persuade him to click on the link. We build phishing messages applying psychological approaches such as curiosity, greed, and fear. Type of the phishing message is shown in Table 3. Distribution of message types was as follows: 9 curiosity type messages, 4 greed type messages, and 3 fear type messages. It is already known that applying social engineering approaches human emotions can be influenced so that he would click on the source of an unknown message, regardless of the awareness of the inherent risks involved (Staksrud et al. 2013; Krombholz et al. 2015). Phishing messages were sent from the impersonated user and each user received one message. Results of the experiment are presented in Fig. 7 where the number of clicks versus number of messages and clusters is shown.

Distribution of the ratio between the number of clicks within datasets and number of the sent phishing messages to the dataset users is as follows: 2.3% (dataset No. 1), 1.2% (dataset No. 2), 1% (dataset No. 3), and 1.15% (dataset No. 4). The largest number of clicks was 9, and it was obtained from phishing message sent to the users grouped in cluster 2 from dataset No. 4. It made up 39.1% of all clicks in this dataset. This phishing message had fear type psychological aspect and it strongly affected

Table 3 Results of datasets analysis

Dataset	Number of clusters	Cluster no.	Keywords and phishing message	Message type
1	3	0	Music, course, house, year, change, code, pun, message, crash, tweet (please visit Music House and get course for free)	G
		1	Remix, spring, year, welcome, place, course, crash, house, reason, join (only this Spring the best remix party of the year)	C
		2	Music, crash, punk, year, message, change, code, course, house, join (amazing music—the crash—punk revolution)	C
2	3	0	Win, basketball, bet, person, twitter, friend, chance, code, crash, win, head (how to win money betting on basketball)	G
		1	Workshop, leader, job, message, life idea, head, fun, hall, pun (the best workshop leader jobs)	G
		2	Fun, welcome, life, hall, pun, workshop, leader, job, idea, head (welcome to the fun side of life)	C
3	5	0	Dog, pun, club, easter, question, miss, brain, job, center, login (cat and dog puns)	C
		1	Football, market, club, pun, dog, easter, number, browse, experience, leader (globalization of the football transfer market)	C
		2	Job, market, energy, club, pun, dog, easter, number, chapter, job, globe (find the best global jobs of energy market)	C

(continued)

Table 3 (continued)

Dataset	Number of clusters	Cluster no.	Keywords and phishing message	Message type
		3	Plane, box, crash, workshop, food, enjoy, experience, face, fact, energy (black box of crashed plane found in Black Sea)	C
		4	Market, value, club, pun, dog, easter, number, state, value, friend, home (how to estimate the market value of your home)	G
4	5	0	James, house, music, pun, birthday, review, follow, creativity, issue, official, page (Jame's music house official web page)	C
		1	Town, pun, birthday, review, follow, creativity, James, issue, mouse, advice (the best remix party in the town)	C
		2	Australia, mouse, pun, birthday, review, follow, creativity, James, issue, answer (Australian conditions favorable for mouse plague)	F
		3	Pun, birthday, review, follow, creativity, James, issue, mouse, harry, route (routes migrants use and numbers arriving at Europe's)	F
		4	Information, is, issue, pun, birthday, review, follow, creativity, James, mouse (what is 'Islamic State'? More information about IS)	F

emotion of the users. The second largest number of click was 6 and it was obtained from cluster 0 in dataset 3. It made up 60% of all clicks in dataset 3. This phishing message affected users that were discussing about dogs. The distribution of clicks ratio between different types of phishing messages with a total number of the sent messages is as follows: 0.9% (curiosity type messages), 1.5% (greed type messages),

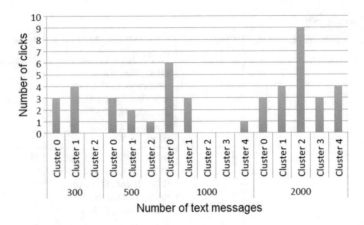

Fig. 7 Number of clicks versus number of phishing messages

and 3.4% (fear type messages). Also it must be mentioned that 3 types of phishing messages out of 16 different messages were not clicked at all. This can be explained that phishing messages were not interesting to the users although it had the main keywords from the tweets.

Based on the experimental results it can be concluded that social site user can be successfully attacked using phishing messages that were built based on their tweets. The fear type phishing messages had the highest click ratio.

6 Conclusions

In this paper, the Twitter user vulnerability to the phishing attach was analyzed. Utilizing a mixed methods to the tweet collections, the results provided a useful information about social networking characteristics. It was shown that user's personal information, as well as tweets, can be accessed by the attacker using Twitter Stream API. Users can be grouped into the clusters applying tweets clustering techniques, and corresponding phishing messages can be build based on the keywords. Evaluation of four different size datasets of the tweets showed that K-means algorithm is more precise compare to Mini Batch K-means when large amounts of data are processed. The experiment of the phishing attack showed that the largest number of clicks were attained from users that received fear type phishing message.

References

Acquisti A, Gross R (2009) Predicting social security numbers from public data. Proc Natl Acad Sci 106(27):10975–10980

Arnaboldi V, Guazzini A, Passarella A (2013) Egocentric online social networks: analysis of key features and prediction of tie strength in Facebook. Comput Commun 36:1130–1144

Buglass SL, Binder JF, Betts LR, Underwood JDM (2016) When 'friends' collide: social heterogeneity and user vulnerability on social network sites. Comput Hum Behav 54:62–72

Bullée JWH, Montoya L, Pieters W, Junger M, Hartel P (2018) On the anatomy of social engineering attacks—a literature-based dissection of successful attacks. J Investig Psychol Offender Profiling 15:20–45

Burger JD, Henderson J, Kim G, Zarrella G (2011) Discriminating gender on twitter. In: Proceedings of the conference on empirical methods in natural language processing. Association for Computational Linguistics, pp 1301–1309

Burke M, Kraut RE (2014) Growing closer on facebook: changes in tie strength through social network site use. In: Proceeding of the SIGCHI conference on human factors in computing systems, pp 4187–4196

Celebi ME, Kingravi HA, Vela PA (2013) A comparative study of efficient initialization methods for the k-means clustering algorithm. Expert Syst Appl 40(1):200–210

Conti M, Passarella A, Pezzoni F (2012) From ego network to social network models. In: Proceedings of the third ACM international workshop on mobile opportunistic networks, MobiOpp'12. ACM, New York, pp 91–92

Duda R, Hart PE, Stork DG (1996) kPattern classification and scene analysis. Wiley, New York

Dwyer C, Hiltz SR, Passerini K (2007) Trust and privacy concern within social networking sites: a comparison of Facebook and MySpace. In: AMCIS 2007 proceedings, p 339

Gundecha P, Barbier G, Tang J, Liu H (2014) User vulnerability and its reduction on a social networking site. ACM Trans Knowl Discov Data 9(2). Article 12

Ivaturi K, Janczewski L (2011) A taxonomy for social engineering attacks. In: Proceedings of international conference on information resources CONF-IRM 2011 proceedings, p 15

Kosinski M, Stillwell D, Graepel T (2013) Private traits and attributes are predictable from digital records of human behavior. Proc Natl Acad Sci

Krishnamurthy B, Wills C (2009) On the leakage of personally identifiable information via online social networks. ACM SIGCOMM Comput Commun Rev 40(1):112–117

Krombholz K, Hobel H, Huber M, Weippl E (2015) Advanced social engineering attacks. J Inf Secur Appl 22:113–122

Liu H, Maes P (2005) Interestmap: harvesting social network profiles for recommendations. Beyond Personalization IUI-2005, USA

Mansour RF (2016) Understanding how big data leads to social networking vulnerability. Comput Hum Behav 57:348–351

Narayanan A, Shmatikov V (2009) De-anonymizing social networks. In: 30th IEEE symposium on security and privacy

Number of social media users worldwide from 2010 to 2021. Available at: https://www.statista. com/statistics/278414/number-of-worldwide-social-network-users/. Last accessed 25.03.2019

Parsons K, McCormac A, Pattinson M, Butavicius M, Jerram C (2013) Phishing for the truth: a scenario-based experiment of users' behavioural response to emails. In: Security and privacy protection in information processing systems, IFIP advances in information and communication technology, vol 405. Springer, Berlin, pp 366–378

Pennacchiotti M, Popescu AM (2011) A machine learning approach to twitter user classification. In: Proceedings of the fifth international AAAI conference on weblogs and social media

Rao D, Paul MJ, Fink C, Yarowsky D, Oates T, Coppersmith G (2011) Hierarchical Bayesian models for latent attribute detection in social media. In: Proceedings of the fifth international AAAI conference on weblogs and social median

Sculley D (2010) Web-scale k-means clustering. In: Proceedings of the 19th international conference on World Wide Web. ACM, New York, pp 1177–1178

Staksrud E, Ólafsson K, Livingstone S (2013) Does the use of social networking sites increase children's risk of harm? Comput Hum Behav 29(1):40–50

Vitak J (2012) The impact of context collapse and privacy on social network site disclosures. J Broadcast Electron Media 56(4):451–470

Wondracek G, Holz T, Kirda E, Kruegel C (2010) A practical attack to de-anonymize social network users. In: 2010 IEEE symposium on security and privacy, Berkeley/Oakland, USA, pp 223–238

Zerocross Density Decomposition: A Novel Signal Decomposition Method

Tatjana Sidekerskienė, Robertas Damaševičius and Marcin Woźniak

Abstract We developed the Zerocross Density Decomposition (ZCD) method for decomposition of nonstationary signals into subcomponents (intrinsic modes). The method is based on the histogram of zero-crosses of a signal across different scales. We discuss the main properties of ZCD and parameters of ZCD modes (statistical characteristics, principal frequencies and energy distribution) and compare it with well-known Empirical Mode Decomposition (EMD). To analyze the efficiency of decomposition we use Partial Orthogonality Index, Total Index of Orthogonality, Percentage Error in Energy, Variance Ratio, Smoothness, Ruggedness, and Variability metrics, and propose a novel metrics of distance from perfect correlation matrix. An example of modal analysis of a nonstationary signal and a comparison of decomposition of randomly generated signals using stability and noise robustness analysis is provided. Our results show that the proposed method can provide more stable results than EMD as indicated by the odd-even reliability and noise robustness analysis.

Keywords Signal decomposition · Intrinsic modes · Modal analysis

1 Introduction

Decomposition of a composite signal into components is a popular approach in signal analysis with numerous applications such as signal compression, detection, denoising, reconstruction of missing data, classification, pattern recognition, and computer

T. Sidekerskienė (✉)
Department of Applied Mathematics, Kaunas University of Technology, Studentu 50, Kaunas, Lithuania
e-mail: tatjana.sidekerskiene@ktu.lt

R. Damaševičius
Department of Software Engineering, Kaunas University of Technology, Barsausko 69, Kaunas, Lithuania

R. Damaševičius · M. Woźniak
Faculty of Applied Mathematics, Silesian University of Technology, Kaszubska 23, Gliwice, Poland

© Springer Nature Switzerland AG 2020
G. Dzemyda et al. (eds.), *Data Science: New Issues, Challenges and Applications*, Studies in Computational Intelligence 869,
https://doi.org/10.1007/978-3-030-39250-5_13

235

vision (Rios and de Mello 2012). Fourier and wavelet transforms based on time-frequency analysis (Cohen 1995), time scales analysis methods (Boashash 2015), and Empirical Mode Decomposition (EMD) based on Hilbert-Huang Transform (Huang et al. 1998) are extensively used in signal and image processing. Other conceptually similar methods are Karhunen–Loève Decomposition (KLD), aka Proper Orthogonal Decomposition (POD) (Kerschen et al. 2005), finds the coherent structures in a collection of spatial data, which represents an optimum basis function in terms of energy. Smooth Decomposition (SD) (Bellizzi and Sampaio 2015) is a projection of spatial data such that the projected vector directions have the largest variance but the motions alongside these vector directions are as smooth in time as possible. Dynamic mode decomposition (DMD) (Schmid 2010) computes a set of signal components with a fixed oscillation frequency and decay/growth rate. Koopman mode decomposition is a method for data analysis that identifies fixed shapes (modes) which evolve by exponential growth/decay and oscillation (Hernandez-Ortega and Messina 2018). The Prony's method (Fernández Rodríguez et al. 2018) decomposes a signal into a linear aggregate of damped exponential functions. Hankel Total Least-Squares (HTLS) Decomposition (Van Huffel et al. 1994) represents a signal as a sum of delayed damped sinusoids (DDSs). Synchrosqueezing transform (Daubechies et al. 2011) performs a nonlinear time-frequency reassignment of the signal spectrum so that most prominent frequencies have more energy, resulting in a sparse and more focused time-frequency representation of a signal. Intrinsic chirp component decomposition (ICCD) (Chen et al. 2017) separate multi-component chirp signals in the time domain, whereas instantaneous frequencies and amplitudes of the intrinsic chirp components are modeled using Fourier series. Sparsity-enhanced signal decomposition (Cai et al. 2018) method which uses the generalized minimax-concave (GMC) penalty as a nonconvex regularizer to enhance sparsity in the sparse approximation and thus to improve the decomposition accuracy proposed Adaptive Variational Mode Decomposition (AVMD) (Lian et al. 2018) determines the amount of signal modes automatically based on the characteristic of signal component. Spectral intrinsic decomposition (SID) (Sidibe et al. 2019) decomposes any signal into a combination of eigenvectors of a partial differential equation (PDE) interpolation operator. Frequency-domain intrinsic component decomposition (FICD) (Liu et al. 2019) extracts multimodal signals with nonlinear group delays (GDs), while the estimation of GDs and signal amplitudes is performed in the frequency domain. Nonnegative matrix factorization decomposition (Sidekerskienė et al. 2017) uses factorization of the signal's Hankel matrix to revise signal component. Swam decomposition is based on swarm filtering (SwF) approach, where the processing is intuitively considered as a virtual swarm-prey hunting, while the iterative application of SwF yields an individual component of the input signal (Apostolidis and Hadjileontiadis 2017). Extreme-point weighted mode decomposition (EWMD) uses adjacent extreme-points for sifting process to construct a new mean curve by using the weights of adjacent extreme-points (Zheng et al. 2018).

In all cases, the objective is to understand the contents of the signal by analyzing its stand-alone components (modes). The oscillatory components that arise out of the decomposition are often called the intrinsic mode functions (IMF). IMFs were introduced in the context of Hilbert-Huang Transform (HHT) and Empirical Mode Decomposition (EMD). Using the EMD method, any complicated data set can be decomposed into a finite and often small number of components, described as intrinsic mode functions (IMF), which form a complete and nearly orthogonal basis for the original signal, and a trend. An IMF is defined as a function that satisfies the following requirements: the number of extrema and the number of zero-crossings must either be equal or differ at most by one, and the mean value of IMF is zero. Finding signal modes can be used for solving some very important practical goals such as removing noise by mode thresholding (Damasevicius et al. 2015). However, signal decomposition methods often have problems, namely, mode mixing (Damaševičius et al. 2017) and boundary problem (Jaber et al. 2014), which still remain unsolved. As a result, the researchers look for novel signal decomposition methods, which allow to mitigate the problems based by EMD and produce more meaningful IMFs.

We propose an adaptive, data-driven technique for processing and analyzing nonstationary signals, called Zerocross Density Decomposition (ZCD). The novelty is the application of Zero Crossing Based Spectral Analysis approach (Zierhofer 2017), which claims that rather than sampling the input signal, the spectrum of a signal can be calculated from the zero crossing positions only (Afzal 2000), while the number of zero-crossings and higher order crossings (HOC) (Kedem 1994) for a time series is directly related to correlation and statistical properties of a signal (Raz and Siegel 1992). We compare our results with those obtained by applying EMD, which is a classical method usually used as a baseline method by researchers proposing novel decomposition methods.

2 Method

We present the derivation and algorithm of ZCD method. We also describe and discuss measures (metrics) for evaluation of quality of signal decomposition.

2.1 Derivation

Let signal $x(t)$ be real valued, and its values are contained within the range of $-1 < x(t) < 1$. A nonlinear, finite-memory system with input $x_1(n)$ and output $y_1(n)$ is related as

$$y_1(n) = f(x_1(n), x_1(n-1), \ldots, x_1(n-N+1)) \tag{1}$$

where N is the size of system memory.

Such system can be approximated using the Stone–Weierstrass theorem that claims "Given $f: [a, b] \rightarrow R$ continuous and an arbitrary $\varepsilon > 0$ there exits

an algebraic polynomial $P(x)$ such that $|f(x) - P(x)| \leq \varepsilon$, $\forall x \in [a, b]$"
(Venkatappareddy and Lall 2017).

The factorization of $P(u)$ yields

$$P(u) = (u - u_1)(u - u_2)\ldots(u - u_{2N}) \tag{2}$$

Following the fundamental theorem of algebra, $P(u)$ has exactly $2N$ roots
$u_1, u_2, \ldots, u_{2N}$, which represent the zero crosses of $f(x)$. Let $x \to Z$ be a mapping
to extract the set representing the positions of real valued zero-crossings of $x(t)$. For
a sequence $f(n) \in R^N$, the zero-crossing representation (ZCR) is defined by the set
Z of zero-crossings location given by

$$Z(f) = \{n : f(n) \cdot f(n+1) \leq 0 \ \ \forall n = 0, 1, \ldots, N - 2\} \tag{3}$$

Let define a function $zrc(t)$ as it changes over time by dividing a signal into shorter
segments of equal length and then computing ZCR separately on each shorter segment. This reveals the short-time ZCR spectrum. Let x be a length N complex valued
signal, $(x_0, x_1, \ldots, x_{N-1})$, and let $n \leq N$ be the window size. This implies that the
zero-cross transform (ZCT) computes zero-crossings in $W = N - n + 1$ total windows. The short-time ZCT is computed by applying the ZCT to a continuous subset of
the data with W windows. First, we compute the ZCT of $(x_0, x_1, \ldots, x_{n-1})$. Second,
we move the window by one time index and the compute the ZCT of $(x_1, \ldots, x_n)$.
We repeat this procedure until the window covers the last n data points of the input
and compute the ZCT of $(x_{N-n}, \ldots, x_{N-1})$.

As demonstrated by Kumaresan and Wang (2001), the envelope of a periodic
signal $x(t)$ can be represented implicitly by zero-crossings of certain signals derived
from $x(t)$ using inverse signal analysis, i.e. by minimizing the energy in the error
signal. Although these results are valid for periodic signals, they can be applied to
non-periodic signals by appropriately windowing the signal, leading to the possibility
of representing signals by solely using zero-crossing information.

Following Okamura (2011), we can perform marginal distribution analysis of
the ZCT using histograms of short term signals over time window as follows. Let
$x(t)$ be a measurement which can have one of T values contained in the set $X =
\{x_1, \ldots x_T\}$. Consider a set of n elements whose measurements of the value of x are
$A = \{a_1, \ldots a_n\}$ where $a_t \in X$. The histogram of the set A along measurement x is
$H(A)$ which is an ordered list consisting of the number of occurrences of the discrete
values of X. Let $H_i(A), 1 \leq i \leq T$, denote the number of elements of A that have
value x_i, then:

$$H_i(A) = \sum_{t=1}^{n} C_{i,t}^A, \quad C_{i,t}^A = \begin{cases} 1 \ if \ a_t = x_i \\ 0 \ otherwise \end{cases} \tag{4}$$

Following Coifman et al. (2017), we define a histopolant, which is a periodic
continuous function constructed by interpolation, whose cell averages are the same

as the histograms with accuracy $\varepsilon > 0$ if $f(t) = A(t)e^{i\phi(t)}$ with $A \in C^1(T, R_+)$, while $\inf_{t \in R} \phi'(t) > 0$ and $\sup_{t \in R} \phi'(t) < \infty$, which satisfies

$$\left|A'(t)\right| \leq \varepsilon \phi'(t), \quad \left|\phi''(t)\right| \leq \varepsilon \phi'(t) \tag{5}$$

2.2 Algorithm

We summarize the proposed decomposition algorithm as follows.

1. Perform signal windowing.
2. For each signal window

    a. Calculate zero-crosses.
    b. Calculate histogram of zero-crosses by thresholding the signal at different evenly-spaced levels.
    c. Select a median histogram point as an interpolation point of histopolant.

3. Perform interpolation over histopolant points of all windows using cubic spline interpolation.
4. Save the resulting histopolant as the first mode.
5. Subtract the first mode from the signal.
6. Repeat 1–4 until the desired number if modes is extracted.
7. Save the remainder of the signal as a trend component.

2.3 Measures of Evaluation

The define the following evaluation measures for a time series $f(n)$ with sample duration T. Let $f(n)$ be decomposed into K modes and a residue, then the input sequence $f(n)$ can be written as

$$f(n) = \sum_{i=1}^{K+1} M_i(n) \tag{6}$$

The Partial Orthogonality Index (IO) (Huang et al. 1998) between IMFs is calculated using the following formula

$$IO_{j,k} = \frac{\sum_{n=0}^{T} I_j(n) I_k(n)}{\sum_{n=0}^{T} \left|I_j(n)\right|^2 + \sum_{n=0}^{T} \left|I_k(n)\right|^2} \tag{7}$$

The Total Index of Orthogonality IOT (Huang et al. 1998) is defined as

$$IO_T = \frac{\sum_{n=0}^{T} \sum_{j=1}^{K+1} \sum_{k=1}^{K+1} I_j(n) I_k(n)}{\sum_{n=0}^{T} |f(n)|^2} \quad j \neq k \tag{8}$$

The percentage error in energy (P_{EE}) (Venkatappareddy and Lall 2017) is defined as

$$P_{ee} = \frac{E_x - E_{emd}}{E_x} * 100 \tag{9}$$

here E_x is the energy of the signal, and E_{emd} is defined in Eq. 11.

$$E_x = \sum_{n=0}^{T} |f(n)|^2 \tag{10}$$

$$E_{emd} \triangleq \sum_{n=0}^{T} \sum_{i=1}^{K+1} |I_i(n)|^2 \tag{11}$$

We also use the following measures (characteristics) of decomposition:

Variance ratio defines how much of a total variance of a signal is explained by the variance of individual signal components as follows:

$$\frac{\sigma(M_i)}{\sum \sigma(M_i)}, \tag{12}$$

here σ is the standard deviation.

Smoothness of a mode function is defined as correlation between odd and even elements is defined as follows:

$$\left(\frac{\sum \dot{M}_i \ddot{M}_i}{\sqrt{\sum \dot{M}_i^2} \sqrt{\sum \ddot{M}_i^2}} \right)^2, \tag{13}$$

here $\dot{M}_i$ are the odd elements of $i-$th mode, and $\ddot{M}_i$ are the even elements of the $i-$th mode.

Sum of energy leakage (Singh et al. 2015) quantifies energy leaks between different signal modes, that is normalized by total signal energy as follows:

$$\sum \left\{ \left(\frac{MM^T}{\sum M^2} \right)^2 - 1 \right\}. \tag{14}$$

Ruggedness is defined as the number of times the derivative of the function crosses the zero value divided by the length of the time series:

$$\sum \left[M'(t) M'(t+1) < 0 \right] / |M_i|. \tag{15}$$

here $[\cdot]$ is the Iverson bracket. The Inverson bracket converts any logical proposition into a number that is 1 if the proposition is satisfied, and 0 otherwise.

Odd-even reliability (Cronbach 1946) is calculated as the correlation between odd and even values of the function:

$$R_{o/e} = diag \left(corr \left(\dot{M}_i, \ddot{M}_i \right) \right). \tag{16}$$

Variability measure evaluates the number of outliers in the derivative of the mode time series as follows:

$$\frac{1}{NP} \sum \sum \left[|M'_i(t)| > \left| \hat{M}'_i(t) \right| \right], \tag{17}$$

here $[\cdot]$ is the Iverson bracket operator, $\hat{M}_i$ is the random permutation of M_i and P is the number of permutations. Random permutation is used to obtain an artificially created time series for comparison, which has the same statistical characteristics as the original time series.

The distance from perfect correlation matrix is a novel measure proposed in this paper to characterize the relationship between the signal modes:

$$\sqrt{\sum \left(I - \frac{\sum M_i M_j}{\sqrt{\sum M_i^2} \sqrt{\sum M_j^2}} \right)^2}, \tag{18}$$

here I is the identity matrix, and M_i is the i−th mode.

3 Illustrative Example and Results

3.1 Modal Analysis

We use the Elecentro earthquake signal data set (http://www.vibrationdata.com/elcentro.htm) registered on May 18, 1940, north–south time component series (0–

Fig. 1 Elecentro earthquake
signal in USA (1940),
north–south time series data

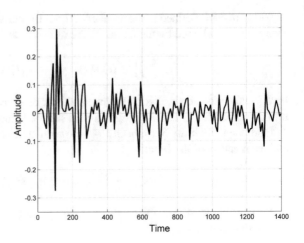

10 s) data (Fig. 1) as an illustrative example to demonstrate how signal decomposition using the proposed ZCD method is performed. We apply the proposed method as follows. First, we perform signal windowing (with 50% of overlap) as illustrated in Fig. 2. For each signal window, we calculate zero-crosses and histogram of zero-crosses by thresholding the signal at different evenly-spaced levels. Then, we select a median histogram point as an interpolation point of histopolant. The procedure is illustrated and the result is presented in Fig. 3.

We perform interpolation over histopolant points of all windows using cubic spline interpolation and save the resulting histopolant as the first mode (see Fig. 4.). Next, we subtract the first mode from the signal, and repeat steps 1–4 (see Sect. 2.2) until

Fig. 2 Signal windowing

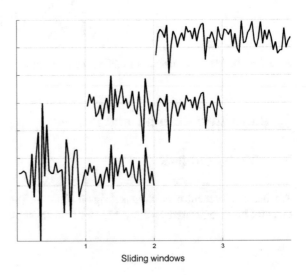

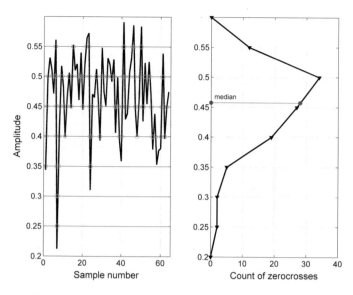

Fig. 3 Threshold-zerocrossing and identification of maximum zerocrossing probability

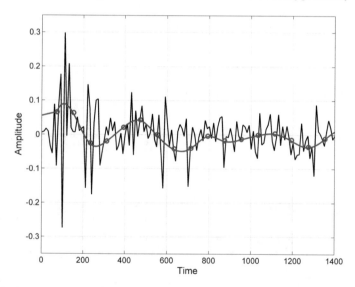

Fig. 4 Cubic spline interpolation over maximum zerocrossing density points: derivation of first signal mode

the desired number if modes is extracted. Finally, we save the remainder of the signal as a trend component.

Four signal modes obtained using the proposed method are given in Fig. 5. For comparison, we show the IMFs derived by a classical decomposition method EMD in Fig. 6.

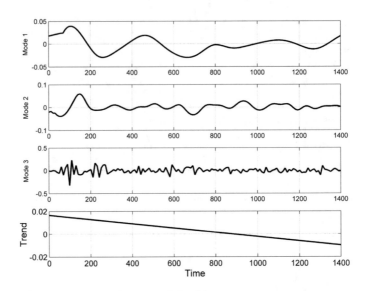

Fig. 5 Signal modes derived by the proposed algorithm

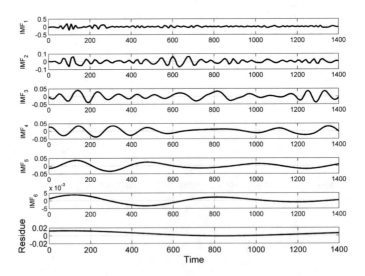

Fig. 6 IMFs derived by the EMD algorithm

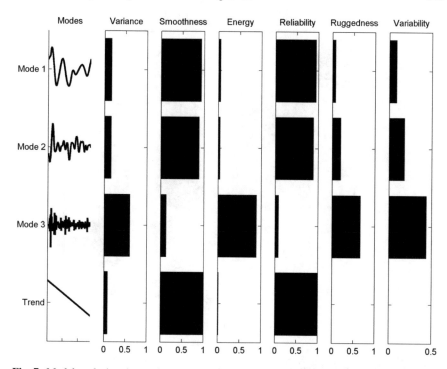

Fig. 7 Modal analysis using variance smoothness, energy, reliability, ruggedness and variability measures

We demonstrate the analysis of the stand-alone modes derived by the proposed method in Fig. 7. We used the variance, smoothness, energy, reliability, ruggedness and variability measures. From Fig. 7, we can observe that highly oscillatory modes have high variance and variability, while slowly changing modes have high smoothness and low variability.

An important measure of the quality of decomposition is the energy leakage measure, which shows interaction between different modes. We demonstrate the mode interaction for the analyzed signal as a pie matrix diagram in Fig. 8. Pie matrix diagram visualizes matrix values using colored area of a pie (fully black circle corresponds to maximum value while fully white circle corresponds to the minimum value). The diagram shows that there is some interaction between modes 3 and modes 1 and 2.

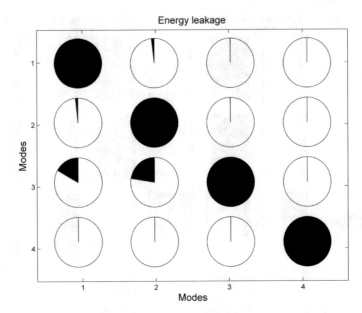

Fig. 8 Modal interaction analysis using energy leakage measure

3.2 Comparison with EMD

Here we compare the performance of our proposed decomposition method with classical EMD proposed by Huang et al. (1998). For a large scale comparison, we constructed random time series by generating 20 uniformly-distributed random numbers. Then we perform the cubic spline interpolation between random points and perform resampling to obtain smooth quasi-periodic times series. For comparison, we use 200 such time series, and perform decomposition using the proposed method and EMD. The result of cross-correlation between the signal modes obtained using the proposed method and the EMD modes are presented as pie matrix diagram in Fig. 9.

Note that the ZCD method presents the modes in reverse order as compared to EMD. Here we can see that, e.g., the modes with highest frequency (ZCD mode 4 and EMD mode 1), have some correlation.

For comparison, we use the distance from perfect correlation matrix measure. The result presented in Fig. 10 shows that the proposed method is better since the mixing between modes is lower as indicated by the lower value of the metric.

The comparison of smoothness of both methods is presented in Fig. 11 (lower value is better). Although the mean and median values characterizing both methods are quite similar, the ZCD method has lower variation in value, which means that the results are more stable.

The comparison of methods by sum of energy leakage is presented in Fig. 12 (lower value is better). In this case, the mean values for EMD are better, although

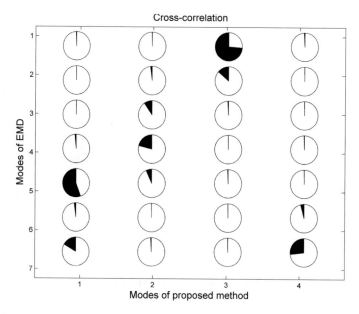

Fig. 9 Cross-correlation between the proposed method and EMD

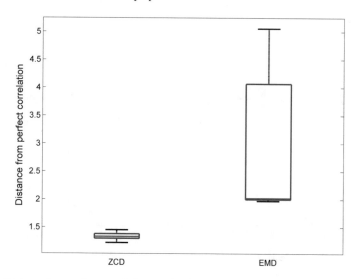

Fig. 10 Distance from perfect correlation matrix of signal modes derived by the proposed method and EMD

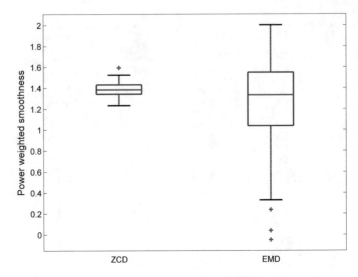

Fig. 11 Power weighted smoothness by the proposed method and EMD

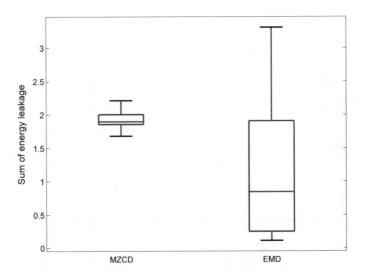

Fig. 12 Sum of energy leakage by the proposed method and EMD

in some cases the results for EMD are much worse, which is demonstrated by the outliers in the metric values.

Figures 13 and 14 show the results of stability analysis using odd-even reliability. We demonstrate the modes obtained by splitting the original signal in half and using only odd and only even parts of the signal. The correlation between the resulting modes is used to evaluate the stability of the decomposition methods. The results show that ZCD is much more stable than EMD for low frequency signal components.

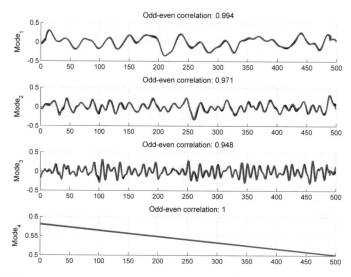

Fig. 13 Stability analysis (split-half (odd-even) reliability, ZCD)

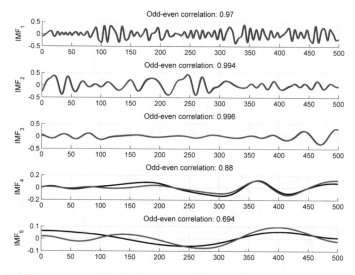

Fig. 14 Stability analysis (split-half (odd-even) reliability, EMD)

3.3 Noise Robustness Analysis

For analysis of noise robustness of the proposed decomposition method, we generated random signals as explained in Sect. 3.2 and added white Gaussian noise with different Signal-to-Noise Ratio (SNR) values. The robustness was measured as correlation of the modes extracted from the noise-contaminated signal with the modes extracted from the noise-free signal. The results are presented in Fig. 15.

Finally, we present the numerical results for mode orthogonality analysis to compare the modes obtained using EMD and the proposed method for the El Centro signal in Tables 1 and 2, respectively. Note that the EMD method generated 7 modes and residue, while our method has yielded 3 modes and residue. The results show

Fig. 15 Noise robustness of the proposed decomposition method for random signals contaminated with additive white Gaussian noise

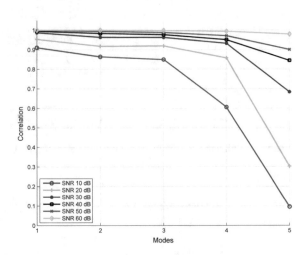

Table 1 Orthogonality index between IMFs obtained using EMD algorithm

IMF	1	2	3	4	5	6	7	Residue
1	0.500	0.042	−0.021	0.004	−0.021	−0.077	−0.010	−0.022
2	0.011	0.500	−0.032	−0.022	−0.033	0.052	0.078	0.041
3	−0.003	−0.020	0.500	−0.037	−0.000	0.000	−0.001	−0.019
4	0.000	−0.004	−0.012	0.500	−0.008	−0.034	−0.013	−0.035
5	−0.001	−0.010	−0.000	−0.012	0.500	−0.234	−0.622	−0.465
6	−0.001	0.000	0.000	−0.012	−0.055	0.500	0.061	0.093
7	−0.000	0.002	−0.000	−0.002	−0.063	0.026	0.500	0.117
Res	−0.000	0.000	−0.000	−0.002	−0.025	0.021	0.061	0.500

Table 2 Orthogonality index between modes obtained using the proposed ZCD algorithm

Mode number	1	2	3	Residue
1	0.5000	0.0628	−0.2043	0.0000
2	0.0677	0.5000	−0.2373	0.0000
3	−0.0145	−0.0156	0.5000	0.0000
Residue	0.0000	0.0000	0.0000	0.5000

that in some case EMD fails due to low orthogonality between modes, which is indication of the mode mixing effect (e.g., see orthogonality index of 5th and 7th IMFs in Table 1).

4 Conclusion

We presented the novel method of signal decomposition, based on the histogram of zero-crosses of a signal across different scales. Decomposition is performed in accordance with the local features (zero-crossings) of a signal. As a result, the method captures local signal oscillations across multiple scales well. The proposed method can provide more stable results than EMD as indicated by the odd-even reliability (average correlation for ZCD is 0.9783, while for EMD only 0.9068) and noise robustness analysis. The limitation of the proposed method is the boundary problem, which arises from the use of cubic spline interpolation method. However, the proposed method performs better in terms of avoiding mode mixing (energy leakage between modes is lower than the one between the EMD modes).

Future work will involve more extensive experiments with large datasets, comparison with other signal decomposition methods and practical applications such as extraction of features for ship-radiated noise and bearing vibrations.

References

Afzal MAK (2000) Wavelet based multiresolution zero-crossing representations. PhD Dissertation, Iowa State University

Apostolidis G, Hadjileontiadis L (2017) Swarm decomposition: a novel signal analysis using swarm intelligence. Signal Process 132:40–50

Bellizzi S, Sampaio R (2015) The smooth decomposition as a nonlinear modal analysis tool. Mech Syst Signal Proc 64–65:245–256. https://doi.org/10.1016/j.ymssp.2015.04.015

Boashash B (2015) Time-frequency signal analysis and processing: a comprehensive reference, 2nd edn. Academic Press, Cambridge

Cai G, Selesnick IW, Wang S, Dai W, Zhu Z (2018) Sparsity-enhanced signal decomposition via generalized minimax-concave penalty for gearbox fault diagnosis. J Sound Vib 432:213–234. https://doi.org/10.1016/j.jsv.2018.06.037

Chen S, Peng Z, Yang Y, Dong X, Zhang W (2017) Intrinsic chirp component decomposition by using Fourier series representation. Signal Process 137:319–327

Cohen L (1995) Time-frequency analysis. Prentice-Hall, New York

Coifman RR, Steinerberger S, Wu H (2017) Carrier frequencies, holomorphy, and unwinding. SIAM J Math Anal 49(6):4838–4864

Cronbach LJ (1946) A case study of the splithalf reliability coefficient. J Educ Psychol 37(8):473–480. https://doi.org/10.1037/h0054328

Damasevicius R, Vasiljevas M, Martisius I, Jusas V, Birvinskas D, Wozniak M (2015) BoostEMD: an extension of EMD method and its application for denoising of EMG signals. Electron Electr Eng 21(6):57–61

Damaševičius R, Napoli C, Sidekerskienė T, Woźniak M (2017) IMF mode demixing in EMD for jitter analysis. J Comput Sci 22:240–252

Daubechies I, Lu J, Wu H-T (2011) Synchrosqueezed wavelet transforms: an empirical mode decomposition-like tool. Appl Comput Harmon Anal 30(2):243–261

El Centro earthquake data. Available: http://www.vibrationdata.com/elcentro.htm

Fernández Rodríguez A, de Santiago Rodrigo L, López Guillén E, Rodríguez Ascariz JM, Miguel Jiménez JM, Boquete L (2018) Coding Prony's method in MATLAB and applying it to biomedical signal filtering. BMC Bioinform 19(1). https://doi.org/10.1186/s12859-018-2473-y

Hernandez-Ortega MA, Messina AR (2018) Nonlinear power system analysis using Koopman mode decomposition and perturbation theory. IEEE Trans Power Syst 33(5):5124–5134. https://doi.org/10.1109/TPWRS.2018.2815587

Huang NE, Shen Z, Long SR (1998) The empirical mode decomposition and the Hilbert spectrum for non-linear and nonstationary time series analysis. Proc R Soc Lond 454:903–995

Jaber AM, Ismail MT, Altaher AM (2014) Empirical mode decomposition combined with local linear quantile regression for automatic boundary correction. Abstr Appl Anal 8:731827

Kedem B (1994) Time series analysis by higher order crossings. IEEE Press, Piscataway

Kerschen G, Golinval J, Vakakis AF, Bergman LA (2005) The method of proper orthogonal decomposition for dynamical characterization and order reduction of mechanical systems: an overview. Nonlinear Dyn 41(1–3):147–169

Kumaresan R, Wang Y (2001) On the duality between line-spectral frequencies and zero-crossings of signals. IEEE Trans Speech Audio Process 9(4):458–461. https://doi.org/10.1109/89.917690

Lian J, Liu Z, Wang H, Dong X (2018) Adaptive variational mode decomposition method for signal processing based on mode characteristic. Mech Syst Signal Process 107:53–77. https://doi.org/10.1016/j.ymssp.2018.01.019

Liu Z, He Q, Chen S, Dong X, Peng Z, Zhang W (2019) Frequency-domain intrinsic component decomposition for multimodal signals with nonlinear group delays. Signal Process 154:57–63. https://doi.org/10.1016/j.sigpro.2018.07.026

Okamura S (2011) The short time fourier transform and local signals. Dissertations. Paper 58

Raz G, Siegel I (1992) Modeling and simulation of FM detection by zero-crossing in the presence of noise and FH interference. IEEE Trans Broadcast 38(9):192–196

Rios RAĂ, de Mello RF (2012) A systematic literature review on decomposition approaches to estimate time series components. INFOCOMP 11(3–4):31–46

Schmid PJ (2010) Dynamic mode decomposition of numerical and experimental data. J Fluid Mech 656(1):5–28

Sidekerskienė T, Woźniak M, Damaševičius R (2017) Nonnegative matrix factorization based decomposition for time series modeling. In: Computer information systems and industrial management, CISIM 2017, Lecture Notes in Computer Science, pp 604–613

Sidibe S, Niang O, Ngom NF (2019) Auto-adaptive signal segmentation using spectral intrinsic decomposition. In: Third international congress on information and communication technology. Advances in intelligent systems and computing, vol 797. Springer, Singapore, pp 553–561

Singh P, Joshi SD, Patney RK, Saha K (2015) The Hilbert spectrum and the energy preserving empirical mode decomposition. arXiv preprint arXiv:1504.04104

Van Huffel S, Chen H, Decanniere C, Vanhecke P (1994) Algorithm for time-domain NMR data fitting based on total least-squares. J Magn Reson Ser A 110:228–237

Venkatappareddy P, Lall B (2017) Characterizing empirical mode decomposition algorithm using signal processing techniques. Circuits Syst Signal Process 37(7):2969–2996. https://doi.org/10.1007/s00034-017-0701-8

Zheng J, Pan H, Liu T, Liu Q (2018) Extreme-point weighted mode decomposition. Signal Process 142:366–374

Zierhofer CM (2017) A nonrecursive approach for zero crossing based spectral analysis. IEEE Signal Process Lett 24(7):1054–1057. https://doi.org/10.1109/LSP.2017.2706358

DSS—A Class of Evolving Information Systems

Florin Gheorghe Filip

Abstract The paper is intended to describe the evolution of a particular class of information systems called DSS (Decision Support Systems) under the influence of several technologies. It starts with a description of several trends in automation. Decision-making concepts, including consensus building and crowdsourcing-based approaches, are presented afterwards. Then, basic aspects of DSS, which are meant to help the decision-maker to solve complex decision problems that count, are reviewed. Various DSS classifications are described from the perspective of specific criteria, such as: type of support, number of users, decision-maker type, and technological orientation. Several modern I&CT (Information and Communication Technologies) ever more utilized in DSS design are addressed next. Special attention is paid to Artificial Intelligence, including Cognitive Systems, Big Data Analytics, and Cloud and Mobile Computing. Several open problems, concerns and cautious views of scientists are revealed as well.

Keywords Decision support · Cognitive systems · Enabling technologies · Human agent · Service systems

1 Introduction

Several decades ago, Peter Ferdinand Drucker (1909–2005), a respected and influential consultant, educator and author, viewed as "the father of Post War management thinking", stated:

> The computer makes no decisions; it only carries out orders. It's a total moron and therein lies its strength. It forces us to think, to set the criteria. The stupider the tool, the brighter the master has to be—and this is the dumbest tool we have ever had. (Drucker 1967a) [...]
> The computer, being a mechanical moron, can handle only quantifiable data. These it can handle with speed, accuracy, and precision. It will, therefore, grind out hitherto unobtainable quantified information in large volume. One can get, however, by and large quantify only what goes on inside an organization – costs and production figures, patient statistics in the

F. G. Filip (✉)
The Romanian Academy and INCE, Calea Victoriei 125, Bucharest, Romania
e-mail: ffilip@acad.ro

© Springer Nature Switzerland AG 2020 253
G. Dzemyda et al. (eds.), *Data Science: New Issues, Challenges and Applications*, Studies in Computational Intelligence 869,
https://doi.org/10.1007/978-3-030-39250-5_14

hospital, or training reports. The relevant outside events are rarely available in quantifiable form until it is much too late to do anything about them (Drucker 1967b).

Nine years later, in a preface to a book by Claudio Pozzoli, Umberto Eco (1932–2016), a well-known semiotician, philosopher and novelist, articulated a similar opinion:

> Il computer non è una macchina intelligente che aiuta le persone stupide, anzi è una macchina stupida che funziona solo nelle mani delle persone intelligenti ("The computer is not an intelligent machine that helps the stupid people, but is a stupid tool that functions only in the hands of intelligent people") (Eco 1986).

Since then, significant progresses in computer and communication technologies have been made. At present, the computer is characterized by a high degree of intelligence and, together with the modern communication systems which make possible remote and mobile interactions, can be effectively utilised, among other applications, to support the human agent to solve complex and complicated decision problems. At present, speacking about PDA (*Personal Digital Assistants*), Lenat (2016) noticed that

> Almost everyone has a cradle-to-grave general personal assistant application that builds up an integrated model of the person's preferences, abilities, interests, modes of learning, of learning, idiosyncratic use of terms and expressions, experiences (to analogize to), goals, plans, beliefs.

Lenat also predicted phenomena that will show up together with RAI (*Robot and Artificial Intelligence*), such as: (a) *weak telepathy* (when the "AI understands what you have in mind and why, completes that action [...], and accomplishes your actual goal for you"), and (b) *weak immortality* ("PDA's cradle-to-grave model of you is good enough that, even after your death, it can continue to interact with loved ones, friends, business associates, carry on conversations, carry our assigned tasks").

At the same time, business models and the enterprise have evolved and the human agent has acquired more skills in using the computer and knowledge on how to solve the decision problems he/she faces.

This paper deals with several specific modern *Information and Communication Technologies* (I&CT), such as *Computer-Based Cognitive Systems*, *Big Data*, and *Cloud Computing*, and their impact on *Decision Support Systems* (DSS) utilized in management and real-time control applications. It also reveals possible open problems and concerns related to the advent and usage of modern technologies. The rest of the paper is organized as follows. The recent developments in enterprise organization and operation are reviewed together with the evolving meaning of automation concept. Then, several basic notions are addressed on decision-making activities, as well as the specific features of a specific class of information systems, namely the Decision Support Systems. The fourth section contains information about several pacing I&CT, such as Big Data and Analytics, Cloud and Mobile Computing, and Artificial Intelligence based tools that enable the decision-maker to approach and solve complex problems in an ever more effective manner.

2 Context

2.1 Trends in the Evolution of Controlled Objects

Under the impact of several factors of influence, such as new business models, emerging I&CT, and human agent's new values, aspirations, and improved skills and knowledge, the *controlled objects* (production processes, enterprises) are continuously evolving. Some of the trends identified in Filip et al. (2017: pp. 1–4) are: (a) the continuous integration of enterprises among themselves or/and with their suppliers of materials and parts, and product distributors, (b) the diversity of cultures of the people involved and the technologies utilized when the control object consists of several interconnected and interacting subsystems, (c) the new market requirements for increased product quality, associated services, and customization, and the implication of the consumer in product design and production, and (d) the pursuit of a sustainable, environmentally conscious development. Beside the enterprise paradigms of the last two decades of the past century, such as *extended* and *virtual enterprises*, or *double-loop re-manufacturing* chains, a series of new ones have been recently reviewed by Buchholz (2018), Panetto et al. (2019), Vernadat et al. (2018), Zhong et al. (2019):

- *Industry 4.0,* that refers to a set of technologies for the factories of the future, as a logical evolution from the CIM (*Computer Integrated Manufacturing*) movement that started in early '80s;
- *S^3 Enterprise*, that refers to enhancing sensing, smart, and sustainability capabilities of manufacturing or service companies;
- *Cloud Manufacturing*, that takes full advantage of networked organization and Cloud Computing;
- *Symphonic enterprise*, *Cyber Physical Manufacturing Enterprise* and so on.

The above developments have led to several findings:

- There are ever more people involved in the decision-making activities carried-out in management and control. This requires intensive exchange of information and collaboration to solve the decision problems;
- Decision problems have become more and more complex, multi-facet ones, and, in a significant number of cases, should be solved in real time;
- There is a real need and a definite market for advanced and evolving computer-based tools to support collaborative decision-making activities.

2.2 Evolutions in Automation

We speak about *automation* when a computer or another device executes certain functions commonly performed by a human agent. Today, automation is present not only in most safety highly critical systems, such as aviation, power plants, chemical plants, and refineries, but also in autonomous car driving, libraries, medicine, robotized homes, and even intelligent clothes.

In the early days of automation and computer usage, a clear "division of work" was viewed as a prerequisite for designing control and management schemes. The MABA-*MABA* ("Men Are Better At-Machines Are Better At") list of Fitts (1951) has been viewed by many system designers as a new *Decalogue* and it was adopted for the specific needs of allocating the tasks between human and automation equipment of the time. The above approach was named by Inagaki (2003) *comparison allocation* and coexisted with other traditional methods, such as *leftover* and *economic* design. Over time, other allocation methods have been proposed, such as *human centered automation, balanced automation* and so on. As Dekker and Woods (2002), noticed, the early automation attempts were based on the assumption of permanent human and machine strengths and weaknesses, and suggested an often *quantitative* division of work ("I do this much you do this much"). The above authors stated that:

> Quantitative "who does what" allocation does not work because the real effects of automation are qualitative: it transforms human practice and forces people to adapt their skills and routines.

A similar view has been recently expressed by Rouse and Spohrer (2018) who advocate for a "new perspective to the automation-augmentation continuum and create the best cognitive team or cognitive organization to address the problems at hand".

So far, the discussion has addressed several key concepts, such as: (a) division *of work* between human and machine, (b) s*ubstitution* of human by a machine and (c) *adaptation* and *transformation* of the human practice. With the advent of fast and ubiquitous computing and communication means and networks, a new issue deserves consideration, namely the *collaboration* of humans with various agents, such as other humans, robots, computers, and communication systems. In general, two or more entities collaborate, because each one working independently cannot deliver the expected output, such as a product, a service, a decision (Nof et al. 2015; Nof 2017). This type of common work is called by Zhong et al. (2015) *mandatory collaboration*. There are also other reasons for collaboration. Two or more entities might also start collaborating because they aim at improving the quality of their deliverables or/and to achieve higher values for all of them. That later form is called *optional collaboration*. Collaboration has enabled various forms of control, such as *shared control, cooperative control, human-machine cooperation, collaborative automation, adaptive/shared automation* (Flemisch et al. 2016).

There are cases when a fully-automated process neither technically possible, nor desirable and the human in the loop is necessary. Bainbridge (1983) described the *ironies of automation* and, even earlier, Bibby et al. (1975) anticipated that "even highly automated systems need human beings for supervision, adjustment, maintenance and improvement".

Table 1 Original *MABA-MABA* list of Fitts (1951, p. 10) and de Winter and Dodou (2014)

Humans appear to surpass present-day machines in respect to the following	Present-day machines appear to surpass humans in respect to the following
Ability to detect a small amount of visual or acoustic energy Ability to perceive patterns of light or sound Ability to improvise and use flexible procedures Ability to store very large amounts of information for long periods and to recall relevant facts at the appropriate time Ability to reason inductively Ability to exercise judgment	Ability to respond quickly to control signals and to apply great force smoothly and precisely Ability to perform repetitive, routine tasks Ability to store information briefly and then to erase it completely Ability to reason deductively, including computational ability Ability to handle highly complex operations, i.e.to do many different things at once

More recently, Chui et al. (2016), having made a survey of the trends in a large number of enterprises, remarked the limits of current technologies in automating a series of activities:

> The hardest activities to automate with currently available technologies are those that [...] apply expertise to decision making, planning, or creative work (18 percent).

Decision support systems (DSS), addressed in the third section, constitute a specific and evolving class of information systems where the collaboration between humans and I&CT based tools is deployed (Table 1).

2.3   Enabling Technologies

A series of specific I&CT have facilitated the evolutions of the enterprise and control schemes described above. Several are enlisted below following the lines of Vernadat et al. (2018), Weldon (2018), Spohrer (2018), and Zhong et al. (2019),

- Web Technology and Social Media;
- *Internet/Web of Things*, that enables objects and machines, such as sensors, actuators, robots or even users' mobile devices to communicate with each other as well as with human agents, with a view to performing various tasks including those designed to find solutions to decision problems;
- *Internet of Services*, that enables a service to be performed as a genuine transaction where one party gets temporary access to the resources of another party (human workforce or specialized knowledge, IT-based operations) to perform a certain function at a given/agreed cost;
- *Big Data* associated with *Data Sciences*, that enable processing the massive volumes of data, in order to discover new facts or extract relevant and valuable information or knowledge by using technologies, such as data warehousing, data mining, knowledge discovery, machine learning techniques, and data visualization;

- *Cloud Computing*, that enables ubiquitous access to available shared I&CT resources and services which can be quickly provisioned;
- *Artificial Intelligence*, *Deep Learning*, and *NLP&G* (*Natural Language Processing & Generation*);
- *Cyber-Augmented Interaction* and *Collaboration*, that combine I&CT real-time control, brain models to operate with multi-agent systems in a parallel cyber-space, task administration protocols and algorithms, with a view to providing streamlined and optimized workflows for of human-machine interactions.
- *Augmented and Mixed Reality*, that complements real data (live direct and/or indirect video images) with computer-processed data (sound, 2D or 3D images, videos, charts, and so on) to be used via various sensory modalities (visual, auditory, haptic, and somato-sensory sensations);
- *Ambient Intelligence*, defined by ISTAG (*Information Society Technology Program Advisory Group to the European Community*) as "the convergence of ubiquitous computing, ubiquitous communication, and interfaces adapted to the user" (Dukatel et al. 2010);
- Advanced (intelligent/real-time/collaborative) *Decision Support Systems* (DSS).

Several of the above specific technologies, such as DSS, Big Data and Analytics, Cloud and Mobile Computing, and Artificial Intelligence-based tools, will be presented later in the chapter through the perspective of their impact on decision-making activities.

3   Computer-Aided Decision Support

3.1   Decisions and Participants to Decision-Making Processes

According to Filip et al. (2017: p. 19), we can speak about a *decision problem* when a situation that requires action is perceived, there are several possible courses of action, called *alternatives*, composing the set $A = \{a_1, a_2,..., a_{na}\}$ and a *decision unit* is empowered to identify or design the feasible alternatives, and eventually choose an adequate one. The decision unit could be made up of one person—possibly supported by a team of assistants—or of several participants who are accountable for their actions and characterized by several attributes, such as the goals they pursuit, the decision powers, stability of unit composition and so on. We will note the set of participants by $D = \{d_1, d_2, ..., d_{nd}\}$.

3.1.1   Group Collaborative Decisions and Negotiations

There are two classes of specific *multi-participant decision units* made up of several persons: (a) the *collaborative group* and (b) the *decision collectivity*. The attributes of a *collaborative group* of humans (Briggs et al. 2015) are: (a) *congruence* of methods used by members with agreed common goals, (b) group *effectiveness* in attaining the common goals, (c) group *efficiency* in saving member resources consumed, (d) group *cohesion*, viewed as preserving members' willingness to collaborate in the future. While the collaborative group is characterized by a set of relevant common objectives, the group components pursuit, beside their own local ones, and a rather clear stability in time of the group composition, the members of a decision collectivity may or may not necessarily share a common set of goals and, moreover, they take part in a certain activity only occasionally. A particular form of a collectivity quite frequently met is the *negotiation panel*. The mission of such an entity is to carry out activities which should result in problem solutions that are acceptable for all parties involved in the process. Klingour and Eden (2010) state that, in both above mentioned cases, one may identify a set of collaborative and interactive activities meant to reach a collective decision. The cited authors state that there is, however, only a minor differentiation which consists in viewing the negotiations carried out as *soft social* and *psychological* decision activities. Moreover, they remark that, while in collaborative groups the decision problem is shared by more than one participant who must collectively find a solution for which all of them will later bear their specific responsibility, in negotiation panels the concerned parties may reach a collective solution or, it might happen, no decision at all.

3.1.2 Consensus Building and Crowdsourcing

In the third chapter of the monograph on computer-supported collaborative decision making (Filip et al. 2017), a series of methods of social choice for aggregating individual preferences, such as various *voting mechanisms, judgment aggregation, resource allocation, group argumentation* and the basic ideas of *collaborative engineering* are presented.

A frequently used procedure in multi-participant decision making is composed of two main processes: (a) consensus building, and (b) a selection of a recommended solution (or set of acceptable solutions) (Kou et al. 2017). In consensus building, decision-makers might need to modify their opinions by making them more similar or closer to one another in an interactive process so that, eventually, an acceptable level of consensus is reached. Kacprzyk et al. (2008), Herrera-Viedma et al. (2014) survey the usage of fuzzy logic to model participants' preferences and consensus assessment. Dong et al. (2018) describe the consensus reaching process as composed of several activities, such as: (a) individual preference representations of the participants, expressed by using models, such as: cardinal, ordinal, multi-attribute measures, linguistic preference relations and so on, (b) aggregating individual preferences by using various methods, such as: *Weighted Average* (WA), *Ordered WA* (OWA), *Importance-Induced OWA* (I-IOWA) and so on (Zavadskas et al. 2019), (c) measuring the *consensus level* as a distance of individual preferences either to the calculated collective one or as the result of comparing pairs of preferences, (d) implementing a feedback mechanism to increase the consensus level based either on (a) identifying the participants whose contribution to consensus reaching is minor or (b) aiming at minimizing the number preference revisions. The above authors propose several methods based on *social network analyses*, such as: (a) trust relationship-based approaches, and (b) opinion evolutions. Zhong et al. (2019) propose concepts of *soft minimum cost consensus* and explain its economic meaning. Sentiment analysis over comparative expressions when free text is used for expressing individual preference was proposed by Morente-Molinera et al. (2019) in a recent paper. Expressing the individual preferences in non-homogenous representations might cause difficulties in consensus reaching and decision selection. Recent papers (Li et al. 2018; Zhang et al. 2019a, b) provide solutions to such cases by minimizing the information loss between decision- makers' heterogeneous preference information and also by seeking the collective solution with a consensus.

There is an implicit common assumption in multi-participant decision-making approaches based on consensus building systematic procedures: nd, the number of participants, should be limited so that various methods proposed would be technically applicable. Moreover, the alternatives are presumed to form a set of numerable entities, and their attributes take discrete values. This might not be applicable, for example, when deciding upon a price of a product to be released, or upon the distance to travel or the running speed which can take values over a continuous interval. The identification or design of the set of alternatives is a decision problem per se. If such a problem is not solved in a satisfactory way, there is the risk of applying systematic consensus building and selection procedures to an incomplete set and of eventually choosing the best one of several poor candidate alternatives.

There are decision situations when a solution to complex and persistent decision problems faced by the decision unit cannot be found by a limited number of decision-makers even assisted by their permanent support teams or close assistants. In such cases, a working approach could be to ask other people's collaboration, such as external experts, business partners or even anonymous people in case of choosing the location of a production facility, adopting a new technology or deciding on products or services launching and so on.

In recent years, the concept of *crowdsourcing* (Brabham 2013), a new web-based business model meant to enable finding creative and effective solutions from a very large, unlimited group of people in answer to an [usually] open call for proposals, has gained significant traction. The early definition of crowdsourcing was articulated by Howe (2006) as "the act of taking a job traditionally performed by a designated agent (usually an employee) and outsourcing it to an undefined, generally large group of people in the form of an open call". A more integrated definition was later proposed by Estellés-Arolas and Gonzales-Ladron-de Guevara (2012):

> Crowdsourcing is a type of participative online activity in which an individual, an institution, a non-profit organization or a company proposes to a group of individuals of varying knowledge, heterogeneity and number, via flexible open call, the voluntary undertaking a task. The undertaking of the task, of variable complexity and modularity, and in which the crowd should participate bringing their work, money, knowledge and experience, always entails mutual benefit. The user will receive the satisfaction of a given type of need, be it economic, social recognition, self-esteem or the development of individual skills, which the crowdsourcer will obtain and utilize to their advantage that what the user has brought to the venture whose form will depend on the type of activity undertaken.

There are already available several outsourcing platforms used for various purposes, such as: design, development and testing, marketing and sales, and support. An incomplete list would include: Amazon's *Mechanical Turk3*, and *CrowdFlower, Microworker, mCrowd,* and many others (Hirth et al. 2011; Wang et al. 2016a, b). Gadiraju et al. (2015) and Chiu et al. (2014) analysed the concept and field experiences and highlighted the potential of crowdsourcing to support hard problem solving, managerial and political decision-making included. The above cited authors also warn about possible malicious usages of outsourcing platforms and undesirable outcomes, when the ideas and opinions of the crowd are not used in an appropriate way, respectively.

A possible framework of crowdsourcing to support decision-making activities of Herbert Simon's (1960/1977) process model of decision activities was proposed by Chiu et al. (2014). It is composed of the following phases of the model:

- *Identification* of the problem to be solved or the opportunity to be exploited based on opinions and predictions collected from the crowd and definition of the task. This corresponds to *intelligence* phase of Simon's process model;
- *Task broadcasting* to the crowd, performed, in most cases, in the form of an open call. The crowd may be composed of either enterprise employees (as a modern technology-supported variant of the old "suggestion box") or customers and/or other outsiders;
- *Idea generation* by the crowd in the form of the set of various action alternatives, $A = \{a_1, a_2,..., a_{na}\}$ and, possibly, evaluation criteria, $C = \{c_1, c_2,..., c_{nc}\}$. It basically corresponds to the *Design* phase of Simon's process model or to *Generate* collaboration pattern of collaboration engineering (Kolfschoten et al. 2015). The ideas are viewed as cooperative, when they contribute to accumulating knowledge. They can also be of a competitive nature when they serve to find solutions to problems;
- *Evaluation of ideas* by the same members of the crowd that generated the idea or by another crowd or group of experts. It can be viewed as *Reduce* [the size of the set A], *Clarify*, and *Organize* collaboration patterns described in Kolfschoten et al. (2015), Filip et al. (2017: p. 106);
- *Choosing* the solution through a voting mechanism or by using a systematic consensus building and selection procedures carried out as described in the first part of this section.

3.2 Decision Support Systems (DSS)

In Filip (2008: p. 36), Filip and Leiviskä (2009), a DSS is defined as "an anthropocentric and evolving information system which is meant to implement the functions of a human support system or team of assistants that would otherwise be necessary to help the decision-maker to overcome his/her limits and constraints he/she may encounter when trying to solve complex and complicated decision problems that really matters". The literature about DSS is continuously enriching with new titles (Filip et al. 2014; Kaklauskas 2015). An on-line bibliography of the early DSS literature (1947–2007) made by M. Cioca and F. G. Filip and D. J. Power's newsletter and portal containing up-to-date domain information are available at http://intelligent-enterprise.ro/SSD/bibliographySSD.htm and www.dssresources.com, respectively.

3.2.1 Classifications and Generations

It is worth noticing that the set of I&CT tools used to build the system has evolved in time. Powers and Phillips-Wren (2011) identified seven *generations* or *evolution stages* of DSS:

- *DSS 1.0* were built using timesharing systems;
- *DSS 2.0*, were built using mini-computers;
- *DSS 3.0* were built using personal computers and tools like various spreadsheets, such as Visicalc, Lotus, and Excel;
- *DSS 4.0* were built using DB2 and 4th generation languages;
- *DSS 5.0* were built using a client/server technology or LAN (*Local Area Network*);
- 6.0 were built using large scale data warehouses with *OLAP* (*OnLine Analytical Processing) servers;
- DSS 7.0 were built using Web technologies.

At present, one can add *DSS 8.0* of systems which are built by using computer-based cognitive systems and cloud and mobile computing.

The general class of DSS is not a homogenous one. Some systems are designed to work in real time in industrial milieu, others support marketing applications. While some are frequently used by *hands on* decision assistants, some others are to be occasionally used by executives in examining the figures of the enterprise performance and so on (Filip et al. 2017: pp. 42–45). The general DSS class can be systematically decomposed into several specialized subclasses in accordance with several criteria, such as; (a) the user, (b) the usage envisaged, and (c) the main *dominant technology* (Table 2).

3.2.2 Multi-participant DSS

Multi-participant DSS constitute an important specialized subclass of the general class of DSS. They are designed to support the activities of collaborating groups and are characterized by the following attributes (Nunamaker et al. 2015):

- *Parallelism* meant to avoid the waiting time of participants who want to speak in an unsupported meeting by enabling all users to add, in a simultaneous manner, their ideas and points of view;
- *Anonymity* that makes it possible for an idea to be accepted based on its value only, no matter what position or reputation has the person who has proposed it;

Table 2 A taxonomy of DSS

Criterion	DSS subclasses
User envisaged	• Individual • Multi-participant (Group, Organizational) DSS
Type of support provided	• Passive (data analysis) • Traditional ("What if… analyses) • Recommender • Collaborative • Proactive (Smart guidance)
"Dominant" technology	• Data-driven • Document-driven • [Mathematical] Model-driven • Communication-driven (Web-based) • Knowledge-driven (Intelligent) DSS • Service-oriented complex platform

- *Memory of the group* that is based on long term and accurate recording of the ideas expressed by individual participants and conclusions that were reached;
- *Improved precision* of the contributions which were typed-in compared with their oral presentation;
- *Unambiguous display* on computer screen of the ideas;
- *Any time* and/or *any place* operation that enables the participation of all relevant persons, no matter their location.

3.2.3 Evolution Processes in DSS Lifetime

In an inspiring classical paper, Keen (1980) stated that:

> The label "Support system" is meaningful only in situations where the "final" system must evolve through an adaptive process of design and usage.

Peter Keen offered a justification for the above statement and identified the interacting factors: the user (U), the system builder (B) and the system (S) and three influence loops: (a) *cognitive loop*, between the user and the system, (b) *implementation loop*, between the user and the system builder, and (c) evolution *loop*, between the system builder and the system itself. The paper author has come to a similar conclusion (Filip 2012: pp. 51–56). An illustrative example is the evolution of *DISPATCHER*—a family of DSS designed to support the decision-making for production control in the process industries made up of subsystems interconnected via buffer tanks (chemical plants, fertilizer factories, refineries) and related-controlled plants (water systems). It evolved under the influence of several factors, such as the evolving

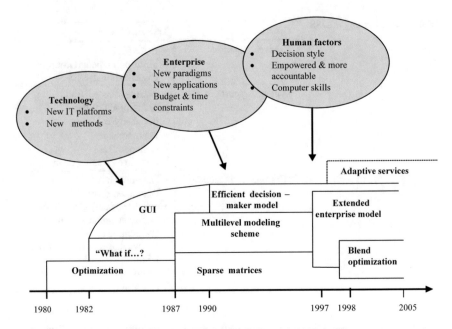

Fig. 1 The evolution of DSS DISPATCHER family (Filip et al. 2017: p. 56)

controlled objects, the technology progress, and the human factor (ever more skilled users and/or more knowledgeable system designers). *DISPATCHER* V1 started, in early '80s, as optimization software running on a minicomputer and evolved in successive generations to a complex solution, which included a rule-based expert system designed to guide the user's operations (Fig. 1).

4 Enabling Information and Communication Technologies (I&CT) and Their Impact

4.1 Artificial Intelligence (AI) and Cognitive Systems

4.1.1 Time for AI

Two decades ago, The Nobel Prize laureate Simon (1987) noticed that Artificial Intelligence-based tools can complement mathematical models in solving decision problems. He stated:

> The MS/OR profession has, in a single generation, grown from birth to a lively adulthood and is playing an important role in the management of our private and public institutions. This success should raise our aspirations. We should aspire to increase the impact of MS/OR

by incorporating the AI kit of tools that can be applied to ill-structured, knowledge rich, non-quantitative decision domains.

Over the past two decades since Simon's above statement, an unprecedented progress has been made in developing AI-based tools and their deployment in practical applications. From the academic community, Stoica et al. (2017) give an explanation of this "perfect storm" emergence which is caused, in their view, by three factors: (a) massive amounts of amassed data collected from various and multiple sources, (b) scalable and powerful computer and specialized software systems, and (c) the broad accessibility of state-of-the-art technologies. In I&CT industry, Google's CEO Sundar Pichai launched the slogan "AI-first". From the business circles, Scott Buchholz of *Deloitte* noticed that the reasons of various companies from all industries to "quickly jumping on the artificial intelligence and machine learning bandwagon is their managers' conviction that there are great gains to be expected in employee productivity and operational cost savings" (Weldon 2019).

4.1.2   Computer-Supported Cognitive Systems

There are several AI application domains, such as: (a) Internet crowd intelligence, (b) Big-Data-based AI (transforming Big Data into knowledge), (c) cross-media intelligence/computing, (d) autonomous-intelligent systems, (e) new forms of hybrid-augmented intelligence, from the pursuit of an intelligent machine to high level human-machine collaboration and fusion (Pan 2016). It is the latest one which is of particular interest in the context of this paper.

Early Definitions

The computer-based (digital) cognitive systems were anticipated by JCR Licklider (1960), the man who was called "Computing's Johnny the Appleseed" for his visionary and inspiring ideas concerning the digital age. He described the collaborative human-machine systems:

> Man-computer symbiosis is an expected development in cooperative interaction between men and electronic computers. It will involve very close coupling between the human and the electronic members of the partnership. The main aims are: to let computers facilitate formulative thinking as they now facilitate the solution of formulated problems, and to enable men and computers to cooperate in making decisions and controlling complex situations without inflexible dependence on predetermined programs […].

Several years later, Hollnagel and Woods (1983) described the characteristic features of a collaborative system the mission of which was to improve the overall operation of the system composed of human and the machine rather than replacing as many as possible functions of the operator. In their view, the envisaged system behaved in a goal-oriented manner, was adaptive and able to view the problem in more than one ways, was able to plan and modify its actions based on knowledge about itself and environment, was data-driven and concept-driven as well.

Current Situation

In the paper, the definition of cognitive computing articulated by Kelly (2015) is adopted

> Cognitive computing refers to systems that a) learn at scale, b) reason with purpose, and c) interact with humans naturally. Rather than being explicitly programmed, they learn and reason from their interactions with us and from their experiences with their environment. They are made possible by advances in a number of scientific fields over the past half-century, and are different in important ways from the information systems that preceded them. Those systems have been deterministic; cognitive systems are probabilistic. They generate not just answers to numerical problems, but hypotheses, reasoned arguments and recommendations about more complex — and meaningful — bodies of data. They can make sense of the 80 percent of the world's data that computer scientists call "unstructured." This enables them to keep pace with the volume, complexity and unpredictability of information and systems in the modern world.

There are already many computer-based cognitive systems. Siddike et al. (2018) list more than 20 currently available systems and analyze the functions provided by some of them such as: *Apple Siri, IBM Watson, Google's Now, Brain, AlphaGo, Home, Assistant, Amazon's Alexa, Microsoft's Adam, Braina & Cortana,* and *Viv*. The cited authors propose a theoretical framework for stimulating people's performance using *cognitive assistants* (CA). In the sequel, only two families of systems will be briefly presented with the view to highlighting their capabilities to be used in decision-making activities:

- *IBM Watson* (High 2012), which is characterized by the following attributes: (a) natural language processing by helping to understand the complexities of unstructured data, (b) hypothesis generation and evaluation by applying advanced analytics to weigh and evaluate a panel of responses based on only relevant evidence, and (c) dynamic learning by helping to improve learning based on outcomes to get smarter with each iteration and interaction;
- *DISCIPLE* methodology and set of cognitive agents have been developed by Tecuci et al. (2016). In DISCIPLE, the agents play the role of intelligent assistants that use the evidence-based reasoning and learn the expertise in problem solving and decision making directly, from human experts, support people and non-experts and the users may not be computer people but possess application and problem solving knowledge.

Toward Cognition as a Service

In Siddike et al. (2018), digital (computer-based) cognitive systems are viewed as the *emerging decision support tools* that enable humans to augment their capabilities and expertise through an improved understanding of the environment. The cited authors forecast that the digital cognitive systems can potentially progress even more from *tools* to *cognitive assistants, collaborators* to *coaches*, and *mediators*. According to Spohrer et al. (2017), the business model employed will be *Cognition as a Service* (CaaS). The above authors also agree with engineers who predict that

> ...by 2055, nearly everyone has 100 cognitive assistants that "work for them [....] and almost all the people including doctors, physicians, patients, bankers, policymakers, tourists,

customers, as well as community people greatly augmented their capabilities by the cognitive mediators or cognition as a service CaaS.

4.1.3 Open Questions, Concerns, and Cautious Opinions

Having read the above lines, one could be tempted to think that all the strong points of the human present in *MABA-MAMA* list must have vanished with the advent of AI and digital cognitive systems. However, Rouse and Spohrer (2018) state that several capabilities that will remain, at least for the foreseeable future, by the human agent, because computers cannot take responsibility for things they were not designed for, do not have consciousness, and are not capable of reflection and cannot have feelings. On the other side, very optimistic opinions come from the business circles. For example, Mark Cuban, a successful investor, predicted, at the SXSW conference and festival in March 2017, that "the world's first trillionaire will be an artificial intelligence entrepreneur" (Clifford 2017). Enterprises and even states have already engaged themselves in a serious competition on developing and using AI-based tools. A certain confusion has been noticed among several IT and business leaders in understanding AI tools, their limitations, and their possible place in the enterprise. Alexander Linden, the *Gartner Research*'s vice-president, identified and explained five myths and misconceptions about AI, so that the new technologies would be more correctly understood and could be used in a proper way. They are the following: (a) AI works as a human brain, (b) AI is unbiased, (c) AI will only replace repetitive jobs, not requiring advanced degrees, (d) intelligent machines will learn on their own, (e), not every business requires an AI strategy, and (f) AI is the next phase of automation (Bhattacharjee 2019).

There are still several other worries and cautious views presented by respected scientists and technology visionaries about the possible future open problems which can be caused by AI. Liu and Shi (2018) set the stage for the AI ethical standards based on the knowledge base of the population. In Helbing (2015: p. 99) and Helbing et al. (2017), the opinions of Elon Musk, Stephen Hawking, Bill Gates, and Steven Wozniak, who are preoccupied by the possible evolution and impact of AI, are cited. For example, Steven Wozniak, the co-founder of Apple, stated:

> Computers are going to take over from humans, no question ... Like people including Stephen Hawking and Elon Musk have predicted, I agree that the future is scary and very bad for people ... If we build these devices to take care of everything for us, eventually they'll think faster than us and they'll get rid of the slow humans to run companies more efficiently ... Will we be the gods? Will we be the family pets? Or will we be ants that get stepped on? I don't know.

A quite balanced view was presented by Sundar Pichai, the CEO of Google:

> AI is one of the most important things humanity is working on. It is more profound than, I dunno, electricity or fire [....] We have learned to harness fire for the benefits of humanity but we had to overcome its downsides too. So my point is, AI is really important, but we have to be concerned about it (Clifford 2018).

4.2 Big Data

4.2.1 Big Data Attributes

Data-driven DSS (Power 2008; Wang et al. 2016a, b) represent an important and ever more developing subclass of the general DSS class. At present, data are accumulated from various internal and external, traditional and nontraditional sources such as: sensors, transactions, web searches, human communication possibly associated with crowdsourcing, Internet/ Web of Things (Blasquez and Domenech 2018). The data contain information useful for the intelligent decision making in finance, trade, manufacturing, education, personal and political life and so on. The amounts of data are already too big for storing and processing capacities of humans alone.

Characteristic Features

There are several generally accepted attributes which qualify a data set as Big Data. They are: (a) v*olume* which measures the amount of data available and accessible to the organization, (b) v*elocity*, that is a measure of the speed of data creation, streaming and aggregation, (c) *variability* of data flows, that might be inconsistent with periodic peaks (an attribute introduced by SAP), (d) v*ariety*, that measures the richness of data representation: numeric, textual, audio, video, structured, unstructured, and so on, (e) v*alue, usefulness,* and *usability* in decision making (an attribute introduced by ORACLE), (f) *complexity*, which measures the degree of interconnectedness, interdependence in data structures and sensitivity of the whole to local changes (an attribute introduced by SAP), (g) v*eracity*, which measures the confidence in the accuracy of data (an attribute introduced by IBM).

There are several challenges concerning Big Data (Shi 2015; Filip and Herrera-Viedma 2014) such as: (a) transforming semi-structured and unstructured data collected into a structured format to be processed by the available data mining tools, (b) exploring the complexity, uncertainty and systematic modeling of big data, (c) exploring the relationship of data heterogeneity, knowledge heterogeneity and decision heterogeneity, (d) engineering decision making by using cross-industry standards, such as the six-step CRISP-DM (*Cross Industry Standard Process for Data Mining*), developed under the EU *Esprit Research Programme* (Wirth and Hipp 2000).

Various Views

A few years ago, the views expressed about Big Data were not congruent. Power (2008) quotes the apparently opposite opinions expressed in the past by two bloggers who have been rather influential within the IT community:

> The excitement around Big Data is huge; the mere fact that the term is capitalized implies a lot of respect (A. Burst in 2012). Big data as a technological category is becoming an increasingly meaningless name (B. Devlin in 2013).

Since those opinions were expressed, Big Data has become a real issue due to the combined influence of various new I&C technologies, which have led to accumulation of data and, at the same time, enable their effective and useful deployment for new technologies and platforms as surveyed by Borlea et al. (2016), Günther et al (2017), Dzemyda (2018), and Oussous et al. (2018). As Shi (2018) pointed out, Big

Data can be viewed as a "new type of strategic resource in the digital era and the key factor to drive innovation, which is changing the way of human being's current production and living". The ever increasing number of successful applications contributed to the well-defined domain of research for which there is an ever developing market.

4.2.2 Big Data Analytics

Big data is based on *Data Science,* that is defined as "the science of data collection, management, transformation, analysis, and application, and its core is to study the acquisition of knowledge from data […] and has begun to gradually replace the known *business intelligence and analysis* (BI&A)" (Shi 2018). A series of particular subclasses compose the general class of *Big Data Analytics* (Gandomi and Haider 2015):

Text Analytics (Mining) subclass, which is the broadest one and includes several tasks: (a) *Information extraction* techniques that extract structured data from unstructured text with two sub-tasks: *Entity recognition* and *Relation extraction,* (b) *Text summarization* techniques meant to automatically produce a succinct summary of a single or multiple documents, (c) *Question answering* techniques that provide answers to questions posed in natural language (in fact, cognitive systems such as *Siri* of Apple and *Watson* of IBM), (d) *Opinion mining* techniques analyze the texts which contain people's opinions toward various entities, such as products, services, organizations, peoples, events and so on;

Audio Analytics subclass, which includes: (a) *Transcript-based* approach (called Large Vocabulary Continuous Speech Recognition-LVCSR), and (b) *Phonetic-based* approach.

Video Content Analytics subclass, which includes several techniques meant to monitor, analyze, and extract meaningful information from video streams.

Social Media Analytics subclass, which includes (a) *Community Detection/Discovery,* meant to identify implicit communities within a network, (b) *Social influence analysis*, and (c) *Prediction* of future links.

Elgendy and Elragal (2016) analyze the support that could be provided by Big Data Analytics to decision making and propose a framework named B-DAD (*Big-Data Analytics and Decisions*) meant to map Big Data tools and analytics to decision-making phases of Simon's process model. Wang et al. (2016a, b) describe a series of intelligent decision-making applications based on Big Data.

4.2.3 Projects in Romania

In Romania, several Big Data oriented projects are undertaken in the public sector (Alexandru et al. 2016):

- *SEAP*: *Electronic Public Procurement System*, a unified I&CT infrastructure which ensures the management and automation of the procurement processes;
- *DEDOC*: *Electronic Document Filing System* used for reporting individual budgets by means of electronic declarations and forms;
- *NTRO*: *Integrated Information System of the National Trade Register Office* intended to provide online services for the business community;
- *RAMP*: *Project for Modernizing Revenue Administration*, implemented by the National Agency for Fiscal Administration (ANAF) and intended to generate prerequisites for achieving a higher level of revenue collection;
- *ORIZON*: *The Integrated System of the National House of Pensions* meant to manage the systems for public pensions and insurances;
- *E-Terra*: *National Cadastral and Estate Registry* system.

4.3 Cloud and Mobile Computing

A decade ago, Richard Stallman of *Free Software Foundation* and Larry Ellison, the founder of *Oracle* expressed their views about Cloud Computing in *The Guardian*, on September 29, 2008 (Johnson 2018):

> It's stupidity. It's worse than stupidity: it's a marketing hype campaign. Somebody is saying this is inevitable — and whenever you hear somebody saying that, it's very likely to be a set of businesses campaigning to make it true (R. Stallman)

> The interesting thing about cloud computing is that we've redefined cloud computing to include everything that we already do […]. The computer industry is the only industry that is more fashion-driven than women's fashion. Maybe I'm an idiot, but I have no idea what anyone is talking about. What is it? It's complete gibberish. It's insane. When is this idiocy going to stop? (L. Ellison).

The academia circles (Ambrust et al. 2010), government authorities (Kundra 2011), and business community perceived the advantages of the new technology and supported it. Two years ago, Baer (2017) admitted that it had been more difficult for Oracle, a major provider of database software and services, to ignore the cloud and, according to company forecasts, about 80% of its client base would consider using cloud in the next decade. Consequently, Oracle launched *Database 12c Release 2 1* as "a poster child" for Oracle's new "cloud-first" product strategy. In October 2018, Oracle announced the release of *Oracle Monetization Cloud* (OMC) 18C (Baer 2018).

4.3.1 Cloud Computing Paradigm

According to NIST (The National Institute of Standards and Technology of US),

> Cloud Computing is a model for enabling ubiquitous, convenient, on-demand network access to a shared pool of configurable computing resources (e.g., networks, servers, storage, applications, and services) that can be rapidly provisioned and released with minimal management effort or service provider interaction (Mell and Grance 2011).

The essential keywords that characterize the general class of cloud computing are: (a) on-demand service, (b) broad network access, (c) resource pooling, (d) rapid elasticity, and (e) measured service. The four Cloud subclasses (or deployment models) defined by NIST (Kundra 2011) are: (a) *Private cloud*, (b) *Community cloud*, (c) *Public cloud, and* (d) *Hybrid cloud.* Hybrid cloud uses an infrastructure which is composed of two or more clouds (private, com-munity, or public) that remain well defined entities, all observing a standardized or proprietary technology, needed to enable data and application portability. It has, apparently, gained traction in the recent years. NIST defined three service models: (a) *Cloud Infrastructure as a Service (IaaS)*, (b) *Cloud Platform as a Service (PaaS), and* (c) *Cloud Software as a Service (SaaS).* Besides the above service models, there are various other service models generically named as XaaS (*Anything as a Service*): *Database as a Service* (DBaaS), *Business as a Service* (BaaS) and so on.

4.3.2   Example

Cloud computing is, among others, a symptom of the ever increasing movement towards *service orientation* in I&CT applications. It has enabled, together with mobile computing, the evolutions of decision support systems. An example is *iDS (intelligent Decision Support),* a family of platforms developed by the Romanian company *Ropardo,* with a view to supporting multi-participant as well as individual decision sessions (Candea et al. 2012; Candea and Filip 2016; Filip et al. 2017: pp. 194–207).

The latest version of iDS is characterized by the following features:

- it takes advantage of web 3.0 technologies and integrates social networks, with a view to supporting collaborative work;
- it allows integrating third party tools via API (*Application Program Interface*) together with its own default set of tools which includes a forum-like *discussion list*, a *voting* module, which is meant for grading or expressing the agreements over a set of issues, an *electronic brain-storming*;
- it facilitates asynchronous decisions through web 2.0 clients or dedicated mobile clients.

The successive versions of iDS developed over a 15-year time period have evolved under the influence of various factors, such as: (a) new I&CT, (b) users' requirements, skills and usage experience, and (c) practical results in various applications such universities, local government units, and digital factory. In Fig. 2, BaaS means the *Business as a Service* model, in which the system provides the requested services to the user and can be managed, so that the business goals could be met. The BaaS service

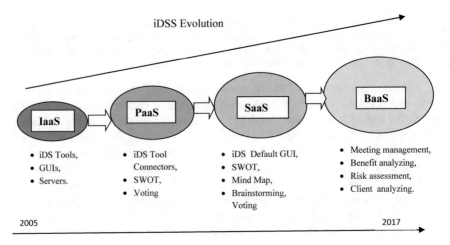

Fig. 2 Steps in *iDS* evolution

model provides companies with a mix of business support, proven methodology, strategy support and hosted technology.

5   Conclusions

Collaborative business models have led to collaborative control and decision making. At the same time, the new I&C technologies are apt to influence the decision-making activities. One can view their impact on the phases of the process model of Simon (1960) as follows:

- Big Data make more effective the *Intelligence* phase of the decision-making process model, in particular decision situation detection and contributes to identifying a valuable set of alternatives to be compared during the *Choice* phase;
- Mobile computing makes possible locating and calling the best experts to perform the evaluation of alternatives during the *Choice* phase;
- Cloud computing enables complex computations during the *Choice* phase;
- Social networks enable consensus building and crowd problem solving;
- Cognitive systems are designed for augmenting the human ability to understand the various complex systems and find the solutions in a faster and more comfortable way.

One should notice, however, that the new I&CT may cause debatable phenomena, such as *Digital Vertigo* or *Big Brother syndrome* and so on (Keen 2012; Power 2016; Susskind 2018).

References

Alexandru A, Alexandru CA, Coardos D et al (2016) Big Data: concepts, technologies and applications in the public sector. Int J Comput Inf Eng 10(10):1670–1676

Ambrust M, Fox A, Griffith R et al (2010) A view of cloud computing. Commun ACM 53(4):50–58

Baer T (2017) The cloud-first strategy of Oracle Database 12c Release 2. Ovum. TMT Intelligence, http://www.oracle.com/us/corporate/analystreports/ovum-cloud-first-strategy-oracle-db-3520721.pdf. Accessed 21 Feb 2019

Baer T (2018) Next-generation cloud capabilities underpin Oracle Monetization Cloud 18C release. Ovum, TMT Intelligence, http://www.oracle.com/us/corporate/analystreports/ovum-next-gen-cloud-capabilities-5212953.pdf. Accessed 21 Feb 2019

Bainbridge L (1983) Ironies of automation. IFAC J Automatica 19(6):775–779

Bibby KS, Margulies F, Rijndorp JE, Whithers RM (1975) Man's role in control systems. In: Proceedings, IFAC 6th triennial world congress, Boston, Cambridge, Mass, pp 24–30

Bhattacharjee S (2019). Five artificial intelligence misconceptions you must know in 2019. Viansider, https://www.viainsider.com/artificialintelligence-misconceptions/. Accessed 1 Mar 2019

Blazquez D, Domenech J (2018) Big Data sources and methods for social and economic analyses. Technol Forecast Soc Chang 130:99–113

Borlea I-D, Precup R-E, Dragan F (2016) On the architecture of a clustering platform for the analysis of big volumes of data. In: IEEE 11th international symposium on applied computational intelligence and informatics (SACI), pp 145–150. https://doi.org/10.1109/saci.2016.7507335

Brabham DC (2013) Crowdsourcing. MIT Press, Cambridge, Massachusetts

Briggs RO, Kolfschoten GL, de Vrede G-J et al (2015) A six-layer model of collaboration. In: Nunamaker JF, Romero NC Jr, Briggs RO (eds) Collaborative systems: concept, value, and use. Routledge, Taylor & Francis Group, London, pp 211–227

Buchholz S (2018) Tech trends 2018: the symphonic enterprise. https://www.din.de/blob/271286/9dcd4b604a3fbf8c3c3ecf67eb75fce0/01-keynote-speech-scott-buchholz-data.pdf. Accessed 22 Feb 2019

Candea C, Filip FG (2016) Towards intelligent collaborative decision support platforms. Stud Inf Control 25(2):143–152

Candea C, Candea G, Filip FG (2012) iDecisionSupport – web-based framework for decision support systems In: Borangiu T et al (eds) Proceedings of 14th IFAC INCOM symposium, pp 1117–1122 http://doi.org/10.3182/20120523-3-RO-2023.00332. Accessed 12 Mar 2019

Chiu CM, Liang TP, Turban E (2014) What can crowdsourcing do for decision support? Decis Support Syst 65:40–49

Chui JM, Manyika J, Miremadi J (2016) Where machines could replace humans—and where they can't (yet). McKinsey Q 30(2):1–9

Clifford C (2017) Mark Cuban: the world's first trillionaire will be an artificial intelligence entrepreneur. MAKE IT, https://www.cnbc.com/2017/03/13/mark-cuban-the-worlds-first-trillionaire-will-be-an-ai-entrepreneur.html. Accessed 21 Feb 2019

Clifford C (2018) Google CEO: A.I. is more important than fire or electricity. CNBC. https://www.cnbc.com/2018/02/01/google-ceo-sundar-pichai-ai-is-more-important-than-fire-electricity.html. Accessed 20 Sept 2018

de Winter JCF, Dodou D (2014) Why the Fitts list has persisted throughout the history of function allocation. Cogn Tech Work 16:1–11. https://doi.org/10.1007/s10111-011-0188-110

Dekker SW, Woods DD (2002) MABA-MABA or abracadabra? Progress in human–automation co-ordination. Cogn Technol Work 4(4):240–244

Dong Y, Zha Q, Zhang H, Kou G, Fujita H, Chiclana F, Herrera-Viedma E (2018) Consensus reaching in social network group decision making: research paradigms and challenges. Knowl-Based Syst 162:3–13

Drucker PF (1967a) The manager and the moron. In: Drucker P (ed) Technology, management and society: essays by Peter F. Drucker. Harper & Row, New York, pp 166–177

Drucker PF (1967b/2011) The effective executive. Butterworth-Heinemann, republished by Rutledge (2011), New York, p 15

Dukatel K, Bogdanowicz M, Scapolo F et al (2010) Scenario for ambient intelligence in 2010. Final Report. IPTS Seville. http://www.ist.hu/doctar/fp5/istagscenarios2010.pdf. Accessed 20 Feb 2019

Dzemyda G (2018) Data science and advanced digital technologies. In: Lupeikiene A., Vasilecas O, Dzemyda G (eds) Databases and information systems. DB&IS 2018. Communications in Computer and Information Science, vol 838. Springer, Cham, pp 3–7

Eco U (1986) Prefazione. Pozzoli. Come scrivere una tesi di laurea di laurea con il personal computer. RCS Rizzoli Libri, Milano, pp 5–7

Elgendy N, Elragal A (2016) Big Data analytics in support of the decision-making process. Proceedia Comput Sci 100(2016):1071–1084

Estellés-Arolas E, Gonzales-Ladron-de-Guevara F (2012) Towards an integrated crowdsourcing definition. J Inf Sci 38(2):189–200

Filip FG (2008) Decision support and control for large-scale complex systems. Annu Rev Control 32(1):62–70

Filip FG (2012) A decision-making perspective for designing and building information systems. Int J Comput Commun Control 7(2):264–272

Filip FG, Herrera-Viedma E (2014) Big Data in Europe. The Bridge, Winter, pp 33–37

Filip FG, Leiviskä K (2009) Large-scale complex systems. In: Nof SY (ed) Springer handbook of automation. Springer Handbooks. Springer, Berlin, Heidelberg, pp 619–638. https://link.springer.com/chapter/10.1007/978-3-540-78831-7_36

Filip FG, Suduc AM, Bizoi M (2014) DSS in numbers. Technol Econ Dev Econ 20(1):154–164

Filip FG, Zamfirescu CB, Ciurea C (2017) Computer supported collaborative decision-making. Springer, Cham

Flemish F, Abbink D, Itoh M, Pacaux-Lemoigne MP, Weßel G (2016) Shared control is the sharp end of cooperation: towards a common framework of joint action, shared control and human machine cooperation. IFAC-Papers OnLine 49(19):072–077

Fitts PM (1951) Human engineering for an effective air navigation and traffic control system. Nat. Res, Council, Washington, DC

Gadiraju U, Kawase R, Dietze S et al (2015) Understanding malicious behavior in crowdsourcing platforms: the case of online surveys. In: Begole B, Kim J et al (eds) CHI '15 Proceedings of the 33rd annual ACM conference on human factors in computing systems, 18th–23rd Apr 2015, Seoul, Korea. ACM, pp 1631–1640

Gandomi A, Haider M (2015) Beyond the hype: big data concepts, methods, and analytics. Int J Inf Manage 35:137–144

Günther WA, Mehrizi MHR et al (2017) Debating big data: a literature review on realizing value from big data. J Strateg Inf Syst 26:191–209

Herrera-Viedma E, Caprerizo FJ, Kacprzyk J et al (2014) A review of soft consensus models in a fuzzy environment. Inf Fusion 17:4–13

High R (2012) The era of cognitive systems: an inside look at IBM Watson and how it works. http://johncreid.com/wp-content/uploads/2014/12/The-Era-of-Cognitive-Systems-An-Inside-Look-at-IBM-Watson-and-How-it-Works_.pdf. Accessed 23 Feb 2019

Helbing, D (2015) The automation of society is next: how to survive the digital revolution. Available at SSRN: http://dx.doi.org/10.2139/ssrn.269431. Accessed 10 Mar 2019

Helbing D, Frey BS, Gigerenzer G et al (2017) Will democracy survive big data and artificial intelligence? Scientific American. https://www.scientificamerican.com/article/will-democracy-survive-big-data-and-artificial-intelligence/. Accessed 28 Feb 2019

Hirth M, Hoßfeld T, Phuoc Tran-Gia P (2011) Anatomy of a crowdsourcing platform—using the example of Microworkers.com. In: 2011 Fifth international conference on innovative mobile and internet services in ubiquitous computing, 30 June–2 July 2011, Seoul, Korea. https://doi.org/10.1109/imis.2011.89

Hollnagel E, Woods DD (1983/1999) Cognitive systems engineering: new wine in new bottles. Int J Man-Mach Stud 18(6):583–600 (Intern J Human-Comp Stud 51:339–356)

Howe J (2006) The rise of crowdsourcing. Wired 14(6):176–183

Inagaki T (2003) Adaptive automation: sharing and trading of control. In: Hollnagel E (ed) Handbook of cognitive task design, LEA, pp 147–169

Johnson B (2018) Cloud computing is a trap, warns GNU founder Richard Stallman. The Guardian, 29. https://www.theguardian.com/technology/2008/sep/29/cloud.computing.richard. stallman. Accessed 3 Mar 2019

Kacprzyk J, Zadrożny S, Fedrizzi M et al (2008) On group decision making, consensus reaching, voting and voting paradoxes under fuzzy preferences and a fuzzy majority: a survey and some perspectives. In: Bustince H, Herrera F, Montero J (eds) Fuzzy sets and their extensions: representation, aggregation and models. Studies in Fuzziness and Soft Computing, vol 220. Springer, Berlin, Heidelberg, pp 263–295

Kaklauskas A (2015) Biometric and intelligent decision making support. Springer, Cham, Heidelberg

Keen A (2012) Digital Vertigo: how today's online social revolution is dividing, diminishing, and disorienting us. Mc Millan, New York

Kelly III JE (2015) Computing, cognition and the future of knowing. How humans and machines are forging a new age of understanding. IBM Global Services

Kou G, Chao X, Peng Y et al (2017) Intelligent collaborative support system for AHP-group decision making. Stud Inf Control 26(2):131–142

Kundra V (2011) Federal cloud computing strategy. https://obamawhitehouse.archives.gov/sites/ default/files/omb/assets/egov_docs/federal-cloud-computing-strategy.pdf. Accessed 21 Feb 2019

Keen PGW (1980) Adaptive design for decision support systems. In: ACM SIGOA Newsletter— Selected papers on decision support systems from the 13th Hawaii international conference on system sciences, vol 1(4–5), pp 15–25

Klingour M, Eden C (2010) Introduction to the handbook of group decision and negotiation. In: Klingour M, Eden C (eds) Handbook of group decision and negotiation. Springer Science + Business Models, Dordrecht, pp 1–7

Kolfschoten GL, Nunamaker JF Jr (2015) Organizing the theoretical foundation of collaboration engineering. In: Nunamaker JF Jr, Romero NC Jr, Briggs RO (eds) Collaboration systems: concept, value, and use. Routledge, Taylor and Francis Group, London, pp 27–41

Kolfschoten GL, Lowry P B, Dean DL, de Vreede G-J, Briggs RO (2015) Patterns in collaboration. In: Nunamaker Jr JF, Romero Jr NC, Briggs RO (eds) Collaboration systems: concept, value, and use. Routledge, Taylor & Francis Group, London, pp 83–105

Lenat DB (2016) WWTS (what would Turing say?). AI Magazine, Spring 37(1):97–101

Li G, Kou G, Yi P (2018) A group decision making model for integrating heterogeneous information. IEEE Trans Syst Man Cybern Syst 48(6):982–992. https://doi.org/10.1016/j.ejor.2019.03.009

Licklider JCR (1960) Man-computer symbiosis. IRE Trans Hum Factors Electron HFE-1(1):4–11

Liu F, Shi Y (2018) Research on artificial intelligence ethics based on the evolution of population knowledge base. In: Shi Z, Pennartz C, Huang T (eds) intelligence science II. ICIS 2018. IFIP Advances in information and communication technology, vol 539. Springer, Cham. https://arxiv. org/ftp/arxiv/papers/1806/1806.10095.pdf. Accessed 2 Mar 2019

Mell P, Grance T (2011) The NIST definition of cloud computing. Special publication 800-145. http://csrc.nist.gov/publications/nistpubs/800-145/SP800-145.pdf. Accessed 15 Sept 2018

Morente-Molinera JA, Kou G, Samuylov K, Ureña R, Herrera-Viedma E (2019) Carrying out consensual group decision making processes under social networks using sentiment analysis over comparative expressions. Knowl-Based Syst 165:335–345

Nof SY (2017) Collaborative control theory and decision support systems. Comput Sci J Moldova 25(2):15–144

Nof SY, Ceroni J, Jeong W, Moghaddam M (2015) Revolutionizing collaboration through e-work, e-business, and e-service. Springer

Nunamaker JF Jr, Romero NC Jr, Briggs RO (2015) Collaboration systems. Part II: foundations. In: Nunamaker JF Jr, Romero NC Jr, Briggs RO (eds) Collaboration systems: concept, value and use. Routledge, London, pp 9–23

Oussous A, Benjelloun F-Z, Lahcen AA et al (2018) Big data technologies: a survey. J King Saud Univ Comput Inf Sci 30:431–448

Panetto H, Iung B, Ivanov D, Weichhart G, Wang X (2019) Challenges for the cyber-physical manufacturing enterprises of the future. Annu Rev Control. https://doi.org/10.1016/j.arcontrol. 2019.02.002

Pan Y (2016) Heading toward artificial intelligence 2.0. Engineering 2:400–413

Power DJ (2008) Understanding data-driven decision support systems. Inf Syst Manage 25:149–157

Power DJ (2016) "Big Brother" can watch us. J Decis Syst 25:578–588

Power DJ, Phillips-Wren G (2011) Impact of social media and Web 2.0 on decision-making. J Decis Syst 20(3):249–261

Rouse WB, Spohrer JC (2018) Automating versus augmenting intelligence. J Enterp Transform. https://doi.org/10.1080/19488289.2018.1424059. Accessed 22 Feb 2019

Shi Y (2015) Challenges to engineering management in the big data era. Front Eng Manage 2(3):293–303

Shi Y (2018) Big data analysis and the belt and road initiative. The 2018 Corporation Forum on "One-Belt and One-Road Digital Economy", Chengdu, China, 21 Sept 2018

Siddike MAK, Spohrer J, Demirkan H, Kohda J (2018) People's interactions with cognitive assistants for enhanced performances. In: Proceedings of the 51st Hawaii international conference on system sciences 2018, pp 1640–1648

Simon H (1960/1977) The new science of management decisions. Harper & Row, New York (revised edition in Prentice Hall, Englewood Cliffs, N.J., 1977)

Simon H (1987) Two heads are better than one; the collaboration between AI and OR. Interfaces 17(4):8–15

Spohrer JC (2018) Open technology, innovation, and service system evolution. ITQM 2018 Keynote, Omaha NE USA. 20 Oct 2018. URL: https://www.slideshare.net/spohrer/itqm-20181020-v2. Accessed 22 Feb 2019

Spohrer J, Siddike MAK, Khda Y (2017) Rebuilding evolution: a service science perspective. In: Proceedings of the 50th Hawaii international conference on system sciences, pp 1663–1667

Stoica I, Song D, Popa RA et al (2017) A Berkeley view of systems challenges for AI. https://arxiv. org/pdf/1712.05855.pdf. Accessed 22 Feb 2019

Susskind J (2018) Future politics: living together in a world transformed by tech. Oxford University Press, Oxford

Tecuci G, Marcu D, Boicu M, Schum DA (2016) Knowledge engineering: building cognitive assistants for evidence-based reasoning. Cambridge University Press, New York

Vernadat FB, Chan FTS, Molina A, Nof SY, Panetto H (2018) Information systems and knowledge management in industrial engineering: recent advances and new perspectives. Int J Prod Res 56(8):2707–2713

Wang, Jia X, Jin Q, Ma J (2016) Mobile crowdsourcing: framework, challenges, and solutions. https://doi.org/10.1002/cpe.3789

Wang H, Xu Z, Fujita H, Liu S (2016b) Towards felicitous decision making: an overview on challenges and trends of Big Data. Inf Sci 367–368:747–765

Weldon D (2018) 12 top emerging technologies. In: Information management, 20 July. https://www.information-management.com/slideshow/12-top-emerging-technologies-that-will-impact-organizations. Accessed 20 Feb 2019

Weldon D (2019) 2019 is the year AI investments will distinguish leaders from laggards. In: Information management https://www.dig-in.com/news/2019-is-the-year-ai-investments-will-distinguish-leaders-from-laggards. Accessed 23 Feb 2019

Wirth R, Hipp D (2000) CRISP-DM: towards a standard process model for data mining. https://citeseerx.ist.psu.edu/viewdoc/download?doi=10.1.1.198.5133&rep=rep1&type=pdf. Accessed 28 Mar 2019

Zavadskas EK, Antucheviciene J, Chatterjee P (2019) Multiple-criteria decision-making (MCDM) techniques for business processes information management. Information 10(4). https://doi.org/10.3390/info10010004

Zhang B, Dong Y, Herrera-Viedma E (2019a) Group decision making with heterogeneous preference structures: an automatic mechanism to support consensus reaching. Group Decis Negot. https://doi.org/10.1007/s10726-018-09609-yAccessed21Febr2019

Zhang H, Kou G, Yi P (2019b) Soft consensus cost models for group decision making and economic interpretation. Eur J Oper Res 277:264–280. https://doi.org/10.1016/j.ejor.2019.03.009

Zhong H, Reyes Levalle R, Moghaddam M, Nof SY (2015) Collaborative intelligence - definition and measured impacts on internetworked e-work. Manage Prod Eng Rev 6(1):67–78

Zhong R, Xu X, Klotz E, Newman S (2019) Intelligent manufacturing in the context of Industry 4.0: a review. Frontiers Mech Eng. https://doi.org/10.1007/s11465-000-0000-0

A Deep Knowledge-Based Evaluation of Enterprise Applications Interoperability

Andrius Valatavičius and Saulius Gudas

Abstract Enterprise is a dynamic and self-managed system, and the applications are an integral part of this complex system. The integration and interoperability of enterprise software are two essential aspects that are at the core of system efficiency. This research focuses on the interoperability evaluation methods for the sole purpose of evaluating multiple enterprise applications interoperability capabilities in the model-driven software development environment. The peculiarity of the method is that it links the causality modeling of the real world (domain) with the traditional MDA. The discovered domain causal knowledge transferring to CIM layer of MDA form the basis for designing application software that is integrated and interoperable. The causal (deep) knowledge of the subject domain is used to evaluate the capability of interoperability between software components. The management transaction concept reveals causal dependencies and the goal-driven in-formation transformations of the enterprise management activities (an in-depth knowledge). An assumption is that autonomic interoperability is achievable by gathering knowledge from different sources in an organization, particularly enterprise architecture, and software architecture analysis through web services can help gather required knowledge for automated solutions. In this interoperability capability evaluation research, 13 different enterprise applications were surveyed. Initially, the interoperability capability evaluation was performed using four know edit distance calculations: Levenshtein, Jaro-Winkler, Longest common subsequence, and Jaccard. These research results are a good indicator of software interoperability capability. Combining these results with a bag of words library gathered from "Schema.org" and included as an addition to the evaluation system, we improve our method by moving more closely to semantic similarity analysis. The prototype version for testing of enterprise applications integration solution is under development, but it already allows us to collect data and help research this domain. This research paper summarizes the conclusions of our research towards the autonomic evaluation of interoperability capability between different

A. Valatavičius · S. Gudas (✉)
Institute of Data Science and Digital Technologies, Vilnius University, Vilnius, Lithuania
e-mail: saulius.gudas@mif.vu.lt

© Springer Nature Switzerland AG 2020
G. Dzemyda et al. (eds.), *Data Science: New Issues, Challenges and Applications*, Studies in Computational Intelligence 869,
https://doi.org/10.1007/978-3-030-39250-5_15

enterprise applications. It reveals basic concepts on which we proved our assumption that enterprise application could be evaluated in a more objective, calculable manner.

Keywords Enterprise application interoperability · Measurement of interoperability capability · Edit distance calculation · Autonomic interoperability component

1 Introduction

The changing nature of the business processes causes many problems with the already developed enterprise architecture and business process models, as well as with implemented (legacy) applications. A most common scenario occurs when changes in business forces to replace outdated legacy enterprise applications by one or multiple new applications designed for some specific business process (i.e., bookkeeping software, enterprise resource planning system or e-commerce software). Changes in legacy software cause the problem of an Enterprise Application Software (EAS) integrity and interoperability within the business domain. EAS interoperability evaluation methods are highly needed. The optimization of the business process, reduction of data inconsistencies, reduction of redundant tasks can be achieved when applications are integrated and interoperable. There are some theoretical works concerning enterprise application interoperability measurement, but no deterministic or probabilistic methods have been recognized as suitable. Most approaches are based on empirical observations rather than a detailed computational analysis of EAS properties. Although in Levels of Information Systems Interoperability (LISI) approach (Chen et al. 2008; Kasunic and Anderson 2004) interoperability measurement scope is overviewed from a broader perspective and not explained by exact data structure and data exchange examples. This research relies on the premise that it should be possible to evaluate EAS interoperability using a deterministic measurement of applications and business processes. The measurement of applications interoperability capability should give the primary indicators for establishing or improving interoperability between the applications. In this approach, the interoperability evaluation of the interoperability solution carried out by comparing the names of the web service operation using existing edit distance methods. The interoperability capability of software systems (SuiteCRM, ExactOnline, NMBRS, PrestaShop) has been measured experimentally by comparing web service operations using edit distance calculations (Levenshtein, Jaro-Winkler, Jaccard, and Longest Common Subsequence). Experiments confirmed that interoperability capability evaluation could be carried out using analysis of web service architecture design and comparison of the architecture between multiple applications. Web service API operations of the application considered similar if the operation identifier of application 1 is the same as the operation identifier of application 2, and then the capability for interoperability on the operations level estimate is 100%.

This research is limited to enterprise applications developed using service-oriented architecture and focuses on EAS that uses web services, SOAP, and REST-ful protocol for data transfer when meta-data is described using standardized documents. REST web service meta-data description is not standardized; therefore, it is more complicated to extract meta-data for interoperability evaluation. The primary assumption in this research paper, that interoperability should be evaluated by comparing web service meta-data (i.e., operation names, objects, object field names, object types, and finally object values) using edit distance calculations. The EAS interoperability measurement serves as a basis for improving interoperability methods.

The goal of this research was a preliminary evaluation of the interoperability of different enterprise applications. The usage of the obtained data and metadata can be applied for further research in the automation of interoperability. This goal of evaluating capability of interoperability was achieved successfully and can be applied in the control loop or as knowledge for the autonomic interoperability component.

This paper is structured as follows. In the second section, we provide the basic concepts of interoperability evaluation. In the third section, recent findings in interoperability solutions are described. In the fourth section, the main ideas are described as evaluating interoperability using deep knowledge (external and internal data sources) to enhance the efficiency of evaluation. Interoperability measurement experiment is explained in Sect. 5. In the sixth section, we provide an example and results of the experiment. Finally, conclusions, cover the brief overview of results and summarize the experiment.

2 Enterprise Applications Interoperability

According to ISO/IEC 2382 (International Organisation for Standardization 2015) standards interoperability is a distributed data processing or, in other words, the capability of two or more functional units to process data cooperatively. Also, the other meaning in this standard is the capability to communicate, execute programs, or transfer data among applications, and the user has little to no knowledge of the characteristics of those units. By this definition of interoperability, all systems considered for interoperability must be designed using SOA (Service-oriented architecture). The central principle of SOA is to design system in such a way that it would be a black box for a user, but also provide a description about its inputs and outputs so that user of such system would be able to interact with it (Krafzig et al. 2005, 330 p). Also, the end-user might not even need to know inputs and outputs, because interoperability problems are solved by interoperability architect who designs a middleware application to support this interoperability between applications in an organization.

In the New European Interoperability Framework (European Commission 2019), the interoperability layer recommendations are re-defined and can be applied when gathering knowledge of why EAS are not interoperable. European integration framework (EIF) identifies interoperability layers: technical, semantical, organizational, legal, integrated public service governance, and governance (European Commission 2019). This research focuses on semantic and technical layers of interoperability, but also, takes into consideration legal layer recommendations for digital check:

- collection of web service policies allows checking suitability for physical and digital worlds
- web service policies allow identifying barriers for digital exchange, such as the amount of data exchanged, exchange frequency, the limit of several daily exchanges.

From a legal perspective in this research state, it is not possible to identify and assess ICT impact on the stakeholders. On the organizational interoperability layer from EIF for this research only partially applied because the focus of the research is inter-organizational application interoperability for multiple applications. However, documenting the business process using EA and MDA as a purpose to gain knowledge for solving interoperability problems of lower levels of interoperability (mainly semantical and technical). According to new EIF semantic interoperability covers both semantic and syntactic aspects, where semantic aspect focuses on the meaning and description of data elements and relations, and the syntactic aspect refers to the formatting of data elements (European Commission 2019). This study focuses on the semantic aspect of data by collecting and comparing data by its counterparts. After calculation of the specific comparison score for the data element, and if it is high enough to a certain threshold, it is considered semantically like its counterpart in another application. From a technical level of EIF, the assumption is apparent that most of the applications in the organization are legacy and fragmented. Although sharing the same data and, therefore, could interoperate. For technical interoperability layer implementation, we use API reference and Web Service description (WSDL) documents provided by legacy applications. Further, we overview interoperability problems and areas.

In a dynamic organization, there could be multiple obstacles that do not allow legacy and new applications to interoperate automatically. Mainly these obstacles are:

- Business processes change when new applications are introduced
- Applications are dynamic; their schema might be changed over time
- Multiple applications are used in a single domain
- There are no common methods to describe collaboration among multiple different applications
- Application changes usually impact business process. Therefore, previous business process models become invalid and cannot be used for knowledge extraction.
- To ensure interoperability, the integration expert needs to perform the following tasks:

  - Perform schema alignment (Hohpe and Woolf 2002; Mccann 2005; Peukert et al. 2012; Rahm and Bernstein 2001; Silverston et al. 1997);
  - Ensure record linkage and data fusion (Dong and Srivastava 2013; Kutsche and Milanovic 2008);

- Ensure orchestration—the timing of each data migration;
- The choreography of application services and data objects—sequence and order in which applications would share data.

- Lack of skills and knowledge.

Lack of necessary skills is a barrier to implementing interoperability solutions. Lack of necessary knowledge on used applications is also a barrier to implementing interoperability solutions. The full tree of interoperability obstacles is represented in Fig. 1. Previously interoperability layers were called barriers in earlier documents of EIF (IDABC 2004). Data from one system cannot be interoperable with similar data in another system without passing these barriers, according to Chen et al. (2008). The five layers of interoperability:

- Governance layer—decisions on interoperability structures, roles, responsibilities policies, and agreements.
- Organizational layer—these barriers relate to the structure of the organization and how an organization is dealing with constant and rapid changes. Usually, a structure of organizations and especially its processes must be discovered and evaluated. Some integration solutions can help improve business processes and therefore, get over the organizational barriers (Valatavičius and Gudas 2015).
- Legal layer—ensure that the data will not be abused or leaked to the public during the interoperability operations. This layer also might include, for example, new general data protection regulation (GDPR) that allows people to get all related data from business applications.

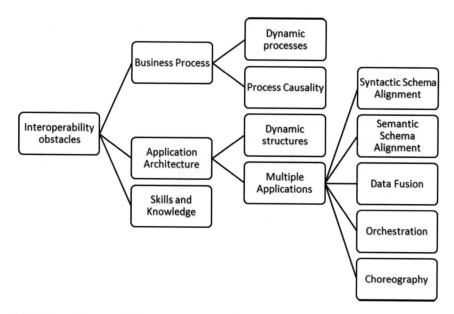

Fig. 1 Tree of interoperability obstacles

- Semantic layer—Semantic or conceptual layers cover semantic differences of information, for example, the use of different software systems leads to the semantic differences.
- Technical layer—is a layer in which interface specifications, communication medium, interconnection services, data integration services, and other aspects are analyzed.

The Interoperability area describes the object of the interoperability solution. As there could be multiple layers of interoperability, a different aggregation and granularity of data are taken into perspective. Interoperability areas investigated by other researchers are as follows (Chen et al. 2008): data, services, processes, and business. The interoperability of data covers different issues of the complex data integration from diverse sources with different schemas. The interoperability of services covers different issues of the heterogeneous data enveloped to the shell of web-services of applications that designed and implemented independently. In this level of interoperability, it might be easier to deal with different schemas and solve semantic issues. The interoperability of processes solves the problem of process sharing or optimizing a value chain for a company. Processes are optimized by developing good interoperability of services/data that are used in these processes. Recent research showed that it might be possible to get internal models from the business process and apply it as knowledge in integration solutions (Valatavičius and Gudas 2015). The interoperability of business cover B2B integration problems and focuses on issues of data sharing between businesses, but all previous interoperability options must be assured to have a successful business.

This research is limited to the interoperability of data and web services of the applications in a domain. In this research knowledge gathered from the models of a business process that is common in the interoperability of processes solutions.

3   Recent Findings in Interoperability Solutions

Various application interoperability methods are applied to create and maintain the interoperability of enterprise applications. The research varies among layers (e.g., organizational, legal, semantic, and technical) and levels (system specific, documented data, aligned static data, aligned dynamic data, harmonized data) of conceptual interoperability model (Tolk and Muguira 2003). Most researchers of integration subject use advanced methods such as agent technologies (Cintuglu et al. 2018; Overeinder et al. 2008), which usually cover self-describing services which cannot be applied in RESTful protocol in applications. Moreover, as RESTful protocol becomes increasingly popular API protocol in business applications, this provides a difficulty to create automated bindings between different systems. Although even with good protocol description usually, lack of semantics could also be a blocking point for successful interoperability. Ontology-based technologies (Li et al. 2005;

Table 1 Selected systems interoperability capability measure by LISI method (Kasunic and Anderson 2004)

(a) Technical view, technical interoperability scorecard	(b) Systems view, systems interoperability scorecard				
Source	Compliance to standards	S1	S2	S3	S4
S1 ExactOnline	Y		Y	Y	G
S2 PrestaShop	Y	Y		G	Y
S3 SuiteCRM	Y	Y	G		Y
S4 NMBRS	G	G	Y	Y	

Shvaiko and Euzenat 2013). However, sophisticated methods of the process integration already exist (El-Halwagi 2016), just not being applied in the application area. In a dynamic environment, business processes often need optimizing, similar to (El-Halwagi 2016) examples of business process integration (El-Halwagi 2016; Pavlin et al. 2010).

Some researchers underline the guidelines of measurements and give propositions of what methods should be used, but they are not presented in such a way that could be easily replicated. One of the favorite inspirers for this research Kasunic and Anderson (2004) proposed to evaluate systems interoperability using three views: Technical, Operational, and Systems. A similar approach to the business and information systems alignment measurement introduced in Morkevičius (2014).

Technical view table indicates that it needs more effort than anticipated to extract meta-data (Kasunic and Anderson 2004). Colors represent the usage of standards in Table 1 in-adequate (R), marginal (Y), or adequate (G). Conclusions: Such evaluation method could be biased by ones understanding on whether the system is standardized, and on thought how easily it could integrate providing interoperability.

The enterprise application (EA) interoperability measurement (between services) is the basis for improving interoperability methods. Some interoperability evaluation methods are known: Scorecard—DoD in Kasunic and Anderson (2004), I—Score in Ford et al. (2008), and Comparison by functionality in Dzemydienė and Naujikienė (2009).

These EA interoperability evaluation methods are not enough because of the assessments obtained through questionnaires and expert judgment. We strive to develop a method that evaluates the characteristics of the systems being integrated—without using personal opinions tests, questionnaires, and experiences. We aim to use only characteristics of software: metadata and systems network service architectures. It is reasonable to use structured (internal) models of systems than to fill out questionnaires. We are looking for a deterministic method that can evaluate or measure the capability of interoperability.

The principles of the second order cybernetics provide the methodological basis for the internal viewpoint (Dzemydienė and Naujikienė 2009) and aim to disclose internal causal relationships of the domain. In our case, we need to explore the causal relationships between application software and no access to use the questionnaires as stated by Kasunic and Anderson (2004). The principles of the second order cybernetics provide the basis for the latter viewpoint, and discloses internal causal relationhips of the domain (Heylighen and Joslyn 2001).

4 Interoperability Evaluation Using MDA, EA Approach

Our study is based on a few assumptions. First, internal modeling with the MDA approach help determines the influence of domain causality to the interoperability of applications. Second, it is possible to create an architecture of interoperable enterprise applications using only the enterprise architecture model and data for each service for enterprise software. Another assumption is as follows: interoperability should be evaluated by comparing web service operation names using edit distance calculations. The measurement of EAS interoperability capability serves as a basis for improving interoperability methods. In case that interoperability is required between these applications, how should one know whether these systems can have interoperability at all? The capability of interoperability of applications can be evaluated using their architectural design by comparing web service operation names using edit distance calculations (Fig. 2).

Levenshtein calculates edit distance by a minimum number of single-character edits required to change one word into the other. Levenshtein algorithm was the first known method developed to compare string distances in 1965 (ЛЕВЕНШТЕН, Владимир Иосифович 1965). For a given two strings b and a with a total character count of m and n. For each character pair from two strings, if they not equal, take the minimum amount of changes required to make them similar. Jaro-Winkler algorithm uses a formula out of 4 values that calculate similarity. Longest common subsequence edit distance, as the name suggests, calculates edit distance removing characters and counting how many characters removed to leave the longest common subsequence. Jaccard edit distance calculates how many similar attributes are in both compared sets for an n-gram. For a given character sequence of each string, a character matrix is formed where characters for each set represent the total number of characters that have the same value (matched).

Although string distance algorithms only provide syntactic similarity evaluation capabilities. For semantic evaluation capabilities, we have developed an ontology

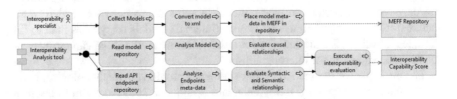

Fig. 2 Analysis of models from MDA cycle to produce interoperability capability score

library describing data structure with semantic meaning. The steps to calculate inter-operability capability (potentiality): (1) locate web-service reference documentation; (2) extract and parse meta-data of web service reference files; (3) categorize parsed metadata into operations, methods, objects, field names, and field types; (4) select operations and create meta-data for each operation: (a) get source same; (b) get service name; (c) extract method GET, POST, PUT, DELETE, PATCH, HEAD); (d) extract operation to the related method; (e) Strip redundant information from operation (repeating meaningless keywords; (5) Save operation meta-data to Microsoft SQL Server database; (6) Using master data services and prepared SQL procedure scan through operations in the database table and compare it with other operations from different source; (7) Save each comparison for different method in a new table; (8) Visualize and explore results.

For the following systems (OpenCart, PrestaShop, Lemon-Stand, NMBRS_ReportService, NMBRS_DebtorService, Zen Cart, NMBRS_CompanyService, NMBRS_Employees, SuiteCRM, Kona-Kart_StoreFront, KonaKart_Administration, MIVA, ExactOnline) used in the experiment, we describe its web-service interface protocol and complexity to extract data automatically (Fig. 3). According to the documentation SOAP and REST, development should follow design recommendations, but there are already many systems developed without the SOA approach. Once a system implements web

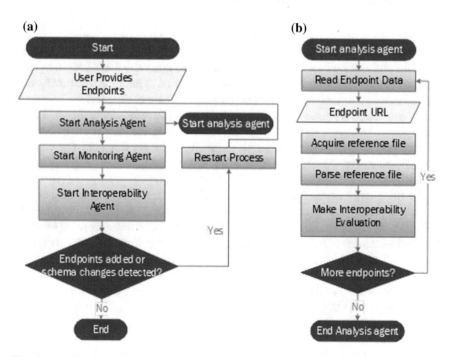

(a) **(b)**

Fig. 3 Activity diagram of a proposed solution of interoperability capability analysis and interoperability tool

services, it is required to have an API that is not always created using common recommendations. Therefore, it is harder to automate data extraction. Additional steps are needed to get to the objects of web services—it is not enough to get the initial structure described in web-service for meta-data analysis. During the experiment, additional steps were carried invoking web service—for returning a list of objects related to the operations described in SOAP WSDL files. The REST web service meta-data description is not standardized, and it is more challenging to extract meta-data. A lack of common pattern following the description of objects exists, therefore it needs additional procedures to extract and parse meta-data from API. The web service meta-data for each system data extracted to the database using a custom C# algorithm and manual data entry from web service reference documentation. Data storage was set up using the Microsoft SQL Server database. From the database, data was analyzed, cleaned, and formed in such a way that it is usable with edit distance measurement algorithms. Edit distance algorithms were executed using Microsoft SQL Server Master Data Services to produce enterprise software system compatibility for interoperability results. Further results and data described in Sect. 6.

5 Experiment Environment Setups

This research is limited to enterprise applications developed using service-oriented architecture and mostly focuses on software that uses web services and SOAP and RESTful protocol for data transfer, which meta-data is usually described using standardized documents. Web service operations compared to multiple software system applications for the enterprise shows the difference in similarity scoring. Randomly picked applications presented in Table 2. Each application has some different roles and aspects of an enterprise. Although this research is limited to a few applications, the intention is to expand the research to involve more applications. The core set of applications are On-site e-commerce applications and some on-site accounting applications.

For these applications and their services (Table 2), API reference data is collected and parsed to evaluate interoperability. Microsoft SQL Server, PostgreSQL, R, Microsoft Visual Studio, and Tableau was used to acquire data from web services. We used Microsoft SQL Server for collecting initial data from C# script written to extract and parse API reference descriptions. C# reference parser was good for a limited amount of applications, but more time needed to expand to enable it to work with a more extensive data set. C# script loaded meta-data from API, parsed, and stored in Microsoft SQL Server. Later for edit distance analysis, R script was used to determine similarities between operations, objects, and fields of sets between multiple applications. Data stored in the PostgreSQL server. Data was finally analyzed and represented using Tableau software. The activity diagram below depicts a proposed solution of the interoperability capability analysis tool (Fig. 3).

Table 2 Randomly picked software applications for analysis

Software application	API protocol	Objects	Description
OpenCart	REST	24	On-site e-commerce application
PrestaShop	REST	49	On-site e-commerce application
LemonStand	REST	76	On-site e-commerce application
NMBRS_ReportService	SOAP	80	On-site accounting application
NMBRS_DebtorService	SOAP	106	On-site accounting application
Zen Cart	REST	208	On-site e-commerce application
NMBRS_CompanyService	SOAP	444	On-site accounting application
NMBRS_Employees	SOAP	1107	On-site accounting application
SuiteCRM	SOAP	1426	On-site CRM application
KonaKart_StoreFront	SOAP	1644	On-site e-commerce application
KonaKart_Administration	SOAP	2425	On-site e-commerce application
MIVA	REST	4322	Cloud e-commerce application
ExactOnline	REST	6043	Cloud accounting application

From the figure above (Fig. 3b)—a simple process of analysis agent depicted. This agent takes part in the job done manually by a data integration specialist. It reads endpoint data from the endpoint URL, acquires reference file, then parse it and runs evaluation scripts, then repeats all the process for more endpoints. In the holistic view for software interoperability, there should be three steps: Analysis, Monitoring, and Action (interoperability); Hence, the three blocks in activity diagram (Fig. 3). The interrelation between activity diagrams in (a) and (b) in figure file is that sub-activities of analysis agent might be running independently from any other agent activity, such as monitoring or interoperability.

6 Experiment Results

For each enterprise application, it is possible to gather meta-data of web service and API descriptions. Some meta-data are automatically extracted from these services (therefore can be automated), other EA require more efforts to do the extraction, but with careful re-thinking, the meta-data extraction can be automated as well. Section 5 describes the interoperability capability (potentiality) evaluation experiment of 9 different enterprise software applications (see Sect. 5). Some of the applications are repeated in the list (Table 2) because web services have several descriptions for different packages with different endpoints. Using the meta-data of web services, we counted for each system how many operations can be carried out using its web services (Fig. 4).

The largest analyzed enterprise application is MIVA—a cloud computing based e-commerce application. Automated parsing determined 3908 data related operations

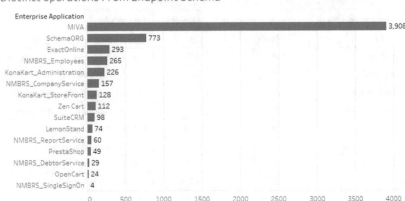

Fig. 4 The number of distinct operations in EA packages

for this specific application. For ExactOnline and NMBRS (employees related web service), the script counted operations 293 and 265, respectively. KonaKart, ZenCart, SuiteCRM contained a smaller number of web service operations below 150.

Additionally, in Fig. 4, meta-data from Schema.org added background knowledge and semantics for other applications.

Considering only the number of operations can be carried out by EA packages, some conclusions can be drawn:

- MIVA—the most extensive software package from a test set;
- MIVA—contains more modules and data management points than other systems;
- Other systems are smaller, or their web services are limited or split (e.g., NMBRS) than the others (e.g., MIVA).

There are 5323 distinct operations overall EA used in the experiment. On average EA has 116 operations per system provided by their web service (excluding SchemaOrg and MIVA). The experiment results are the analysis of similarity for each operation name in each enterprise application. If the edit distance for each operation name is high enough, this indicates that most operations are similar in that pair of EAS packages. Results in Fig. 5 summarize the outcome of the edit distance calculations for e-commerce packages. The heatmap of possible interoperability (Fig. 5) shows the edit distance score of operations. Consider the "Prestashop" to "KonaKart_StoreFront" interoperability comparison. Red spots indicate <50% operation similarity as opposed to other operations (green). The white area indicates around 50% similarity. Red spots also indicate a higher probability of operations being similar. For example, "PrestaShop" operation "categories" match "KonaKart_StoreFront" operation "category" by 75% using an ensemble of edit distance calculation.

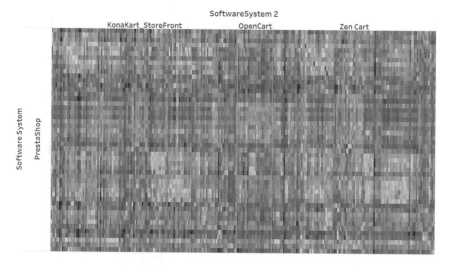

Fig. 5 Operation interoperability scoring—a heat map using the average ensembled score of edit distance algorithms, green spots indicate above 50% similarities

In Fig. 5 apparent similarity of operations of e-commerce products presented. In this example, syntactic overlap can compare and evaluate syntactic overlap of operations between software applications. Results from multiple edit distance methods (Levenshtein, Jaccard, Jaro-Winkler, Longest Common Subsequence) presented further in text. An average score of all selected methods taken as it was not in the scope of this research to evaluate edit distance methods, but rather provide an overview of the capability of evaluation.

Interoperability evaluation using ensemble method

Evaluation of the next results is presented using the ensemble method. The ensemble method is the average of all similarity scores from the edit distance algorithms. After looking at the results from the operation level, we see that operations of web services are similar to each application: accounts, absences, addresses (Fig. 6). The results of the operations interoperability scoring leads to conclusions as follows: In ExactOnline (E) and NMBRS (N) there exist operations that are similar: E Addresses—N Address (85%); E BankAccounts—N BankAccount (91%); E Cost centers—N Cost-Center (90%); E Cost units—N CostUnit (88%); E Departments—N Department (90%); E Employees—N Employee (88%); E Schedules—N Schedule (88%).

In Exact Online (E) and NMBRS (N) there exist operations that are confused: E Contacts—N Contract (76%); E Contacts—N ContractPerson (72%)—these share some similar data, but need to evaluate from data structure perspective for this operation; E Contacts—N ContractV2 (70%);

Exact Online with NMBRS has 20 operations with a result higher than 65%. We can analyze and determine thresholds by semantic meaning while trying to avoid mismatching. As can be seen, Exact Online 285 NMBRS 130 operations have only 20 operations possible interoperability with score >65%. Further, compared Exact

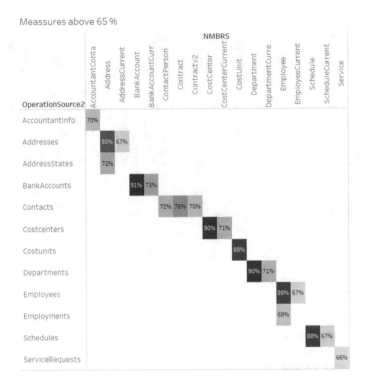

Fig. 6 Similarity results greater than or equal to 65% (Exact Online, NMBRS)

Online (E) and PrestaShop (P), where similarity results are above or equal to 70%. In research results exist cases with full similarity (100%) between a few objects: Addresses; Contacts; Currencies; Employees; Warehouses. However, the algorithms are not precise, so some confusion can be found, for example, at (74%): E Projects—P products (74%).

Exact online with PrestaShop has 18 operations with a result higher than 70%. As can be seen, Exact Online 285 PrestaShop 72 operations have only 18 operations possible interoperability with score >70%. Other results are overviewed as follows and presented in Table 3. The experiment confirms that it is possible to evaluate the interoperability capability, i.e., identify the pairs of specific operations that potentially can be interoperable.

In Fig. 7 The similarity of sources using edit distance calculations (a) Levenshtein, (b) Jaro-Winkler, (c) Jaccard, (d) Longest common subsequence, e) ensemble the similarity of applications using different edit distance calculations is depicted. All edit distance algorithms determine the same similarity between the EAS (Fig. 7).

The scoring amplitudes are somewhat shifted (a—[13; 21], b—[46; 53], c—[2; 10], d—[23; 33], e—[21; 29]) because of the difference of the edit distance calculation methods. The method can compare the different amount of procedures. The lower the percentage—the more procedures tried to compare, but the score was lower because

Table 3 Count of operations with a given score for each software interoperability combination

	Similarity ≥ 100%						
	60%	70%	Ensemble	Levenshtein	Jaro-Winkler	Jaccard	Longest Common Subsequence
ExactOnline X NMBRS	40	20	–	–	–	–	–
ExactOnline X Prestashop	54	18	5	5	5	5	5
ExactOnline X SuiteCRM	48	12	–	–	–	8	–
NMBRS X Prestashop	11	6	1	1	1	1	1
MMBRS X SuiteCRM	7	–	–	–	–	–	–
SuiteCRM X Prestashop	13	6	1	1	1	5	1

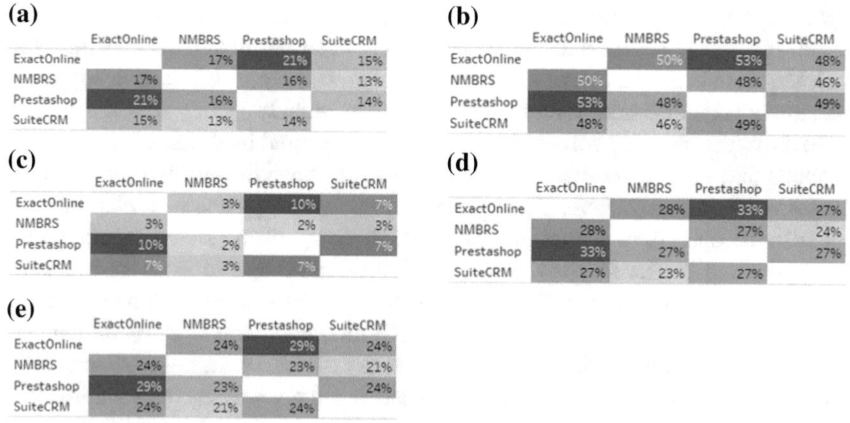

(a)

	ExactOnline	NMBRS	Prestashop	SuiteCRM
ExactOnline		17%	21%	15%
NMBRS	17%		16%	13%
Prestashop	21%	16%		14%
SuiteCRM	15%	13%	14%	

(b)

	ExactOnline	NMBRS	Prestashop	SuiteCRM
ExactOnline		50%	53%	48%
NMBRS	50%		48%	46%
Prestashop	53%	48%		49%
SuiteCRM	48%	46%	49%	

(c)

	ExactOnline	NMBRS	Prestashop	SuiteCRM
ExactOnline		3%	10%	7%
NMBRS	3%		2%	3%
Prestashop	10%	2%		7%
SuiteCRM	7%	3%	7%	

(d)

	ExactOnline	NMBRS	Prestashop	SuiteCRM
ExactOnline		28%	33%	27%
NMBRS	28%		27%	24%
Prestashop	33%	27%		27%
SuiteCRM	27%	23%	27%	

(e)

	ExactOnline	NMBRS	Prestashop	SuiteCRM
ExactOnline		24%	29%	24%
NMBRS	24%		23%	21%
Prestashop	29%	23%		24%
SuiteCRM	24%	21%	24%	

Fig. 7 The similarity of applications using edit distance calculations **a** Levenshtein, **b** Jaro-Winkler, **c** Jaccard, **d** Longest common subsequence, **e** ensemble

of the different amounts. It is still more important to check per each comparison method rather than looking for a difference in each of them.

7 Further Work

This research is an experimental part of an investigation on autonomic solutions for application integration in the dynamic business environment using in-depth domain knowledge. Comprehensive research is still in progress, and this experimental part reveals essential knowledge on how autonomic component can evaluate whether its managed application systems are interoperable. What is more, this research provides the basis for supporting Business Process alignment to Application Processes and may impact the quality of application interoperability when using business process models. The idea is that after measuring whether software systems are interoperable, it is possible to measure the alignment to business processes and see which operation falls outside of the business process model.

8 Conclusions

In this paper, more enterprise applications where analyzed and evaluated the level of capability to be interoperable. The novelty and difficulty of this work are that the goal to asses interoperability through the knowledge available by automated algorithms has not yet been covered in the available solutions. This research opens a possibility for a machine to machine interaction evaluation, helping people that work on integration projects. For now, the results might be helpful as decision support to gain knowledge of compatibility between systems quickly.

In the experiment, 13 software systems were compared by different edit-distance methods and give the output of evaluation of the capability of interoperability in the form of similarity score. Similarity score in percentage show at what percentage all API operations names are like each in comparison sets by Application 1 × Application 2. The results showed that ExactOnline operations were more like Prestashop than any other software within the comparison. SuiteCRM operations where much closer to Prestashop and ExactOnline, both with slight variation. The negative side of such scoring is that the summary of API operation similarity score does not provide a full picture of similar objects, and operations count difference in all applications might affect this scoring method. For the latter reason in this research, filtered comparison results were also covered (see Figs. 5 and 6).

The measurements of the capability of interoperability conducted using the edit distance calculation methods: Jaccard, Jaro-Winkler, Levenshtein, and Longest Common Subsequence. Methods show the same separation of interoperability measures. Methods have a different level of precision estimating, not such similar strings (below 60%).

This method suggests drilling down to characteristics of software systems and discovers web service operations would improve similarity scoring results. However, this approach does not include analysis for data structures, which could provide even better results and help evaluate the possible schema—matching issues.

References

ЛЕВЕНШТЕН, Владимир Иосифович. Двоичные коды с исправлением выпадений, вставок и замещений символов. In: Доклады Академии наук. Российская академия наук, 1965, pp 845–848

Chen D, Doumeingts G, Vernadat F (2008) Architectures for enterprise integration and interoperability: past, present and future. Comput Ind 59(7):647–659

Cintuglu MH, Youssef T, Mohammed OA (2018) Development and application of a real-time testbed for multiagent system interoperability: a case study on hierarchical microgrid control. IEEE Trans Smart Grid 9(3):1759–1768

Dong XL, Srivastava D (2013) Big data integration. In: IEEE 29th International conference on Data engineering (ICDE), pp. 1245–1248

Dzemydienė D, Naujikienė R (2009) Elektroninių viešųjų paslaugų naudojimo ir informacinių sistemų sąveikumo vertinimas. Informacijos mokslai, 50

El-Halwagi MM (2016) Process integration, vol 7. Academic Press. ISBN 0-12-370532-0

European Commission. New European Interoperability Framework. Interoperability solutions for public administrations, businesses and citizens (ISA2). https://ec.europa.eu/isa2/sites/isa/files/eif_brochure_final.pdf. Accessed 9 Mar 2019

Ford T, Colombi J, Graham S, Jacques D (2008) Measuring system interoperability. In: Proceedings CSER

Heylighen F, Joslyn C (2001) Cybernetics, and second-order cybernetics. In: Encyclopedia of physical science and technology, vol 4, pp 155–170

Hohpe G, Woolf B (2002) Enterprise integration patterns. In: 9th Conference on pattern language of programs, pp 1–9

IDABC E, Industry DG (2004) European interoperability framework for pan-European e-government services. European Communities. http://ec.europa.eu/idabc/servlets/Docd552.pdf. Accessed June 3 2017. ISBN 92-894-8389-X

International Organisation for Standardization, ISO/IEC 2382:2015 Information technology—Vocabulary, 2015. https://www.iso.org/obp/ui/#iso:std:iso-iec:2382:ed-1:v1:en. Accessed 9 Mar 2019

Kasunic M, Anderson W (2004) Measuring systems interoperability: challenges and opportunities. Carnegie-Mellon Univ Pittsburgh Pa Software Engineering Inst

Krafzig D, Banke K, Slama D (2005) Enterprise SOA: service-oriented architecture best practices. Prentice Hall Professional

Kutsche R-D, Milanovic N (eds) (2008) Model-based software and data integration: first international workshop. In: Proceedings, vol 8. Springer Science & Business Media. MBSDI

Li L, Wu B, Yang Y (2005) Agent-based ontology integration for ontology-based applications. In: Proceedings of the 2005 Australasian ontology workshop-volume 58. Australian Computer Society, Inc., pp 53–59

Mccann R et al (2005) Mapping maintenance for data integration systems. In: Proceedings of the 31st international conference on very large data bases. VLDB Endowment, pp 1018–1029

Morkevičius A (2014) Business and information systems alignment method based on enterprise architecture models. Doctoral dissertation, Kaunas

Overeinder BJ, Verkaik PD, Brazier FMT(2008) Web service access management for integration with agent systems. In: Proceedings of the 2008 ACM symposium on applied computing. ACM, pp 1854–1860

Pavlin G, Kamermans M, Scafes M (2010) Dynamic process integration framework: toward efficient information processing in complex distributed systems. Informatica 34(4):477–490

Peukert E, Eberius J, Rahm E (2012) A self-configuring schema matching system. In: 2012 IEEE 28th international conference on data engineering. IEEE, pp 306–317

Rahm E, Bernstein PA (2001) A survey of approaches to automatic schema matching. VLDB J 10(4):334–350

Shvaiko P, Euzenat J (2013) Ontology matching: state of the art and future challenges. IEEE Trans Knowl Data Eng 25(1):158–176

Silverston L, Inmon WH, Graziano K (1997) The data model resource book: a library of logical data models and data warehouse designs. Wiley & Sons, Inc, ISBN: 0471153672

Tolk A, Muguira JA (2003) The levels of conceptual interoperability model. In: Proceedings of the 2003 fall simulation interoperability workshop, vol 7, pp 1–11. Citeseer

Valatavičius A, Gudas S (2015) Enterprise software system integration using autonomic computing. CEUR-WS. org, 1420, pp 156–163

Sentiment-Based Decision Making Model for Financial Markets

Marius Liutvinavicius, Virgilijus Sakalauskas and Dalia Kriksciuniene

Abstract The effect of sentiment information for evoking unexpected decisions of investors and incurring anomalies of financial market behaviour is an intensively explored object of research. The recent scientific research works include big variety of approaches for processing sentiment information and embedding it into investment models. The proposed model implies that the sentiment information is not only influential to investment decisions, but it has a varying impact for different financial securities and time frames. The algorithm and simulation tool are developed for including the composite indicator and designing adapted investment strategies. The results of simulations by applying different ratios of financial versus sentiment indicators and investment parameters enabled selecting efficient investment strategies, outperforming the S&P financial index approach.

Keywords Behavioural finance · Sentiment indicators · Composite index

1 Introduction

The fluctuations of financial markets and recent financial crises ignite intensive search for new insights and methods in the asset pricing and risk management spheres. The research works admit the effect of irrationality in financial markets versus the idea of its efficiency (Barberis 2018; Fang et al. 2018; Petita et al. 2019).

Behaviour of financial markets is a dynamic process, where the comparatively stable periods are interchanged by the volatile ones. During the periods of high volatility, the efficiency of markets falls sharply and the possibility of crisis increases. Therefore, the design of advanced methods enabling evaluation of the processes of financial

M. Liutvinavicius · V. Sakalauskas · D. Kriksciuniene (✉)
Vilnius University, Vilnius, Lithuania
e-mail: dalia.kriksciuniene@knf.vu.lt

M. Liutvinavicius
e-mail: marius.liutvinavicius@knf.vu.lt

V. Sakalauskas
e-mail: virgilijus.sakalauskas@knf.vu.lt

© Springer Nature Switzerland AG 2020
G. Dzemyda et al. (eds.), *Data Science: New Issues, Challenges and Applications*, Studies in Computational Intelligence 869,
https://doi.org/10.1007/978-3-030-39250-5_16

markets is a very important task, which can be solved by new intellectual methods and tools that integrate dynamic process simulation and evaluation of rational and irrational factors affecting the risk and profit level of financial markets.

The research works analysing modern financial markets (Johnman 2018; Chiong et al. 2018; Jiang et al. 2019) highlight various aspects of irrationality, which could be explained by the influence of sentiment information coming from various sources. They are not only affected by economic, geopolitical, social factors, but also by human fears, greed, intuition and even manipulation of content, intensity or channels of information flows (Fang et al. 2018; Fear and Greed Index 2019). Each factor can affect financial markets in different ways, which have to be taken into account for making financial decisions and designing investment strategies.

The existing methods of financial market analysis investigate the price fluctuations according to historical price changes or in relationship to economic and financial factors, however, the asset prices are difficult to predict due to irrational behaviour of investors, as the gap between the fundamentally correct and actual price is always present and its level changes over time (Liutvinavicius et al. 2017). The proposed method addresses the problem of inclusion sentiment information to investors' decision process. The essential characteristics of novelty and originality include exploring dynamics of sentiment impact for particular securities and time frames by computing composite indicator, and possibility to maximize the outcomes of investment strategies by adjusting ratios and parameters, determined by financial/economic and sentiment information.

The article is organized as follows. The second section presents the research of applying sentiment information for market forecasting and investment decision making. In the third section we present the methodology for development of proposed composite indicator and its inclusion to investment decision making. In the fourth section, the algorithm is proposed for investment strategy and the simulation tool is designed for its investigation. The investment decisions for portfolio management are made according to the upper and lowered bounds determined by the behaviour of the composite index. The performance of the model is compared to the financial indicator based approach. The experimental testing of the proposed algorithm is based on both manual parameter adjustment and application of standard optimization procedure for maximizing outcomes of different investment strategies (evaluated by profit/loss and Sharpe ratios). The conclusion section presents analysis of the sentiment–based investment model, results of the experimental research and discussion of the limitations and reliability of the proposed approach.

2   Sentiment Based Market Forecasting

The task of applying sentiment information for market forecasting and investment decision making is addressed by numerous researchers from various perspectives. The emotional and psychological effect of sentiment information takes important role

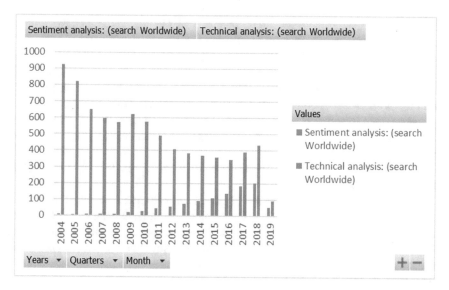

Fig. 1 The interest of topics "Sentiment analysis" and "Technical analysis". *Source* Google trends

in the research, as behavioural economics tells us that emotions can affect individual behaviour and decision making.

The influence of psychological factors, which can be based on beliefs, preferences, and cognitive limits are explored in behavioural finance (Barberis 2018). Here it was concluded that a few simple assumptions about investor psychology capture a wide range of facts about prices and risk and lead to better price changes predictions.

The quality of sentiment information is one of the major concerns. In Bollen et al. (2010) it is stated that news may be unpredictable, yet useful information can be extracted from online social media to predict changes in various economic and commercial indicators.

The three major classes of customer behaviour data are highlighted by Ranco et al. (2015): web news, search engine queries and social media. They study the stock price reaction to news, the role of macroeconomic news in stock returns, the role of news in trading actions, the relation between the sentiment of news and return predictability.

When trying to capture the patterns of investors' behaviour social media and news channels can be very useful (Houlihan and Creame 2014). The research works demonstrate the advantages of the methods that use online search to predict stock markets, as the search intensity in the previous period enabled to forecast abnormal returns and increased trading volumes in the current period and demonstrated strong correlation with the market risk factor (Kissan et al. 2011).

The approach of investigating collective behaviour, especially in online channels, aims to identify anomalous behaviour of investors and can help to prevent losses. According to Google trends, the interest of topic "Sentiment analysis" has been constantly increasing in recent years. In Fig. 1 of Google trends analysis reveals increasing popularity of search terms "Sentiment analysis" and "Technical analysis".

Houlihan and Creame (2014) investigated sentiment extraction from social media and options market and found its correlation with future asset prices. They used both social media sentiment and investors' sentiment captured through the call-put ratio with several predictive models to forecast market price direction. Authors concluded that injection of news into the marketplace in conjunction with various trader behaviour of the options market help explain both the volatility and evolution of assets price.

Trends of Web search queries was also used in researches made by Bordino (2012) and Kristoufek (2013).

The relations between Twitter and financial markets was researched by Ranco et al. (2015), Gu and Kurov (2018). They found a significant dependence between the Twitter sentiment and abnormal returns during the peaks of Twitter volume and showed that sentiment polarity of Twitter peaks implies the direction of cumulative abnormal returns. Gruhl et al. (2005) showed that blogs can predict "real-world" behaviour. Bollen et al. (2010) extracted the mood state of a large number of users on a stock blogging site and used it to predict moves of the Dow Jones Industrial Average index.

Li et al. (2019) analysed Twitter signals as a medium for user sentiment to predict the price fluctuations of a small-cap alternative cryptocurrency called \emph{ZClassic}. They extracted tweets on an hourly basis for a period of 3.5 weeks, classifying each tweet as positive, neutral, or negative. They compiled these tweets into an hourly sentiment index, creating an unweighted and weighted index, with the latter giving larger weight to retweets. These two indices, alongside the raw summations of positive, negative, and neutral sentiment were juxtaposed to ~400 data points of hourly pricing data to train an Extreme Gradient Boosting Regression Tree Model. Price predictions produced from this model were compared to historical price data, with the resulting predictions having a 0.81 correlation with the testing data. Their model's predictive data yielded statistical significance at the $p < 0.0001$ level. Authors claimed that their model proofs that social media platforms such as Twitter can serve as powerful social signals for predicting price movements in the highly speculative alternative cryptocurrency, or "alt-coin", market.

The factors explored in the scientific literature for market analysis and prediction were summarized for positioning the role of sentiment information (Liutvinavicius et al. 2017) and presented in Fig. 2. Here we can conclude that the irrational factors of sentiment and media are very important in market movement prediction.

According to Chiong et al. (2018), many investors rely on news disclosures to make their decisions in buying or selling stocks. However, accurate modelling of stock market trends via news disclosures is a challenging task, considering the complexity and ambiguity of natural languages used. Unlike previous work along this line of research, which typically applies bag-of-words to extract tens of thousands of features to build a prediction model, authors proposed a sentiment analysis-based approach for financial market prediction using news disclosures. Specifically, sentiment analysis is carried out in the pre-processing phase to extract sentiment-related

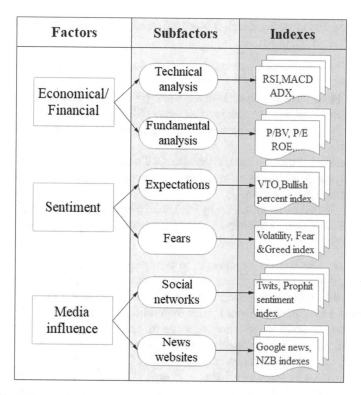

Factors	Subfactors	Indexes
Economical/ Financial	Technical analysis	RSI,MACD ADX, ...
	Fundamental analysis	P/BV, P/E ROE,...
Sentiment	Expectations	VTO,Bullish percent index
	Fears	Volatility, Fear &Greed index
Media influence	Social networks	Twits, Prophit sentiment index
	News websites	Google news, NZB indexes

Fig. 2 Factors for market analysis and forecasting

features from financial news. Historical stock market data from the perspective of time series analysis is also included as an input feature. With the extracted features, they used a support vector machine (SVM) to build the prediction model, with its parameters optimised through particle swarm optimisation (PSO). Experimental results showed that proposed SVM and PSO-based model is able to obtain better results than a deep learning model in terms of time and accuracy.

New technologies empower analysis of big data generated by social networks or media streams, either by applying textual analysis of positive and negative keywords, or the user activities related to the intensity of news flow and providing feedback. Application of advanced analytics extends the scope of explored factors and enables creating new ones.

In parallel to text and search related approaches, the quantification methods for presenting sentiment information in the form of indices are explored. One of sentiment index examples is CNN Money Fear and Greed Index (Fear AND Greed index) which combine seven indicators to show primary emotions that drive investors: fear and greed. Each of these seven indicators is measured on a scale from 0 to 100, with 50 denoting a neutral reading, and a higher reading signalling more greed. The index is then computed by taking an equal-weighted average of the seven market indicators (Fear and Greed Index 2019):

1. Stock Price Momentum—as measured by the S&P 500 versus its 125-day moving average;
2. Stock Price Strength—based on the number of stocks hitting 52-week highs versus those hitting 52-week lows on the NYSE;
3. Stock Price Breadth—as measured by trading volumes in rising stocks against declining stocks;
4. Put and Call Options—based on the Put/Call ratio;
5. Junk Bond Demand—as measured by the spread between yields on investment grade bonds and junk bonds;
6. Market Volatilit—as measured by the CBOE Volatility Index or VIX;
7. Safe Haven Demand—based on the difference in returns for stocks versus Treasuries.

There are also successful attempts to create integrated quantitative indicators of investors' expectations by their social networks activities, e.g. StockTwits sentiment index (http://www.downsidehedge.com/twitter-indicators/), HedgeChatter social media sentiment dashboard (https://www.hedgechatter.com/).

The research works propose investigations for creating new indices, such as composite index of investor sentiment on the time varying long-term correlation of the U.S. stock and bond markets based on the DCC-MIDAS model (Fang et al. 2018). A web-based investor sentiment index is shown to be influential for investment decisions in (Petita 2019). The search sentiment index by utilizing the return difference between a portfolio of high search intensity stocks and one of low search intensity stocks, has differnet impact to low volatility and high volatility stocks (Kissan 2011). The role of manager in the forecasting of stock market returns is explored by proposing manager sentiment index based on the aggregated textual tone of corporate financial disclosures (Jiang et al. 2019). Most of the research works focus on short term (e.g. next day) prediction of specific securities, however there is lack of research to adapt these methods for risk level identification and long term investment decision making.

3 Development of Composite Indicator and Its Inclusion to Investment Decision Making Process

In order to capture long term information for building investment strategies we present the methodology for development composite indicator and its inclusion to investment decision making. The algorithm and simulation for designing the investment strategies and evaluating their performance is further described.

As the influence of sentiment information can vary for the securities of different volatility we explore the behaviour of the price of the security in different time frames

Fig. 3 Values of selected indicator (Fear & Greed Index)

in relationship to financial/economic and sentiment based information in order to find the best combination of the indicators for designing composite indicator.

The composite indicator is calculated as a weighted sum of financial/economic and sentiment indicators I_i ($i = 1, 2, ..., n$), where the weights w_i are adjusted for best describing volatility of each particular security within selected time window

$$C = I_1 * w_1 + I_2 * w_2 + \cdots + I_n * w_n$$

The important aspect of designing composite indicator is taking into consideration different sentiment indicators and combining them. In Fig. 3 the behaviour of Fear and Greed indicator (Fear & Greed Index 2019) is plotted for each time moment.

In Fig. 4 the behaviour of composite indicator in respect to its compound indicators are visualised.

Figure 4 shows the dynamics of selected and composite index for two different sets of weights: in the top figure the composite indicators is calculated as 50% Fear and greed index and 50% Volatility index; the bottom figure shows composite indicator consisting of 30% Fear and greed index and 70% Volatility index. The second case better fits the behaviour of volatility index and can be applied for taking investment decisions based on composite index behaviour. Profit/Loss and Sharpe ratio values are used for evaluating the investment strategy.

It is assumed that by applying the proposed method we can find indicators and weights which secure the highest profit and Sharpe ratio values for selected financial asset. In Fig. 5 the algorithm for designing the investment strategies is presented. The algorithm includes the stage of exploring and defining the weights and compounds for building composite indicators adjusted for particular securities and time frames. The investment decisions for portfolio management are made according to the upper and lowered bounds determined by the behaviour of the composite index.

The proposed algorithm can be applied for long term and consistent market analysis which could ensure high performance of investment decisions, ad this approach

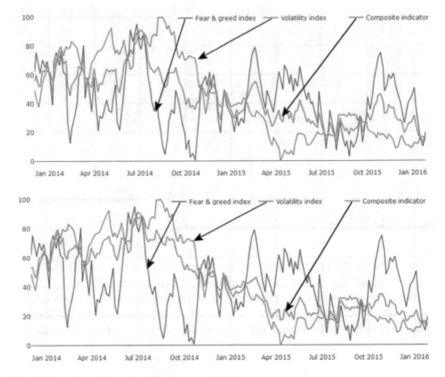

Fig. 4 Dynamics of the selected and composite indicators

enables multi-factor analysis of financial markets, patterns detection and market movement predictions. The long term analysis included the following steps:

1. Load historical price values of particular financial asset;
2. Select sentiment, technical indicators and set initial weights for each of them;
3. Select the initial parameters investment strategy and composite indicator;
4. Make simulation and provide initial results;
5. Explore the influence of parameter change for the investment outcomes and apply the particular optimization method for maximizing them;
6. Explore possibilities to determine the optimal weights of indicators and optimal values of investment strategy parameters;
7. Save results to repository and make investment decisions for portfolio structure and buy-sell processes for selected securities.

Fig. 5 Algorithm of
composite indicator based
method

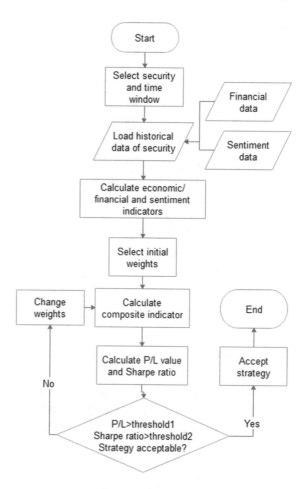

4 The Tool and Simulation of the Investment Process by Applying Composite Indicator

The simulation tool SBT (Sentiment-Based Tool) was created by authors for exploring influence of the parameters of the proposed algorithm and designing investment strategies. The experimental testing of the proposed algorithm is based on both manual parameter adjustment and application of standard optimization procedure for maximizing outcomes of different investment strategies (evaluated by profit/loss and Sharpe ratios). The performance of the model is compared to the S&P financial index investment strategy (without inclusion of sentiment).

The SBT tool software prototype was developed by applying R language, the R-Shinny package was used to create interactive user interface. By applying the software, we can explore real data sets of historical prices of financial asset (security).

Fig. 6 Selection of financial
asset

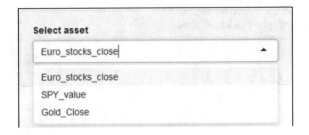

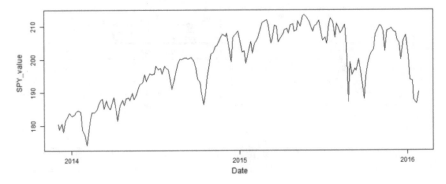

Fig. 7 Historical price changes of SPY

In Fig. 6 the historical prices of SPDR S&P 500 ETF Trust (SPY), SPDR
EURO STOXX 50 ETF (FEZ) and SPDR Gold Trust (GLD) were downloaded from
finance.yahoo.com (Yahoo Finance. SPDR Gold shares) and uploaded to the SBT
tool.

The historical price changes of selected asset for time frame of 2014–2016 are
plotted in SBT (Fig. 7).

The following step of building the composite indicators requires selecting the
indicator(s) that will be used for decision making. In SBT the selection of several sentiment indices is presented (Fear and Greed index, Volatility index) and
economic/financial indicators (Retail sales volume, Consumer confidence index).
Figure 8 illustrates the SBT interface for selecting sentiment, economic/financial
indicators and setting the relevant weights.

The further advancement of the SBT will include possibility for automated
(instead of manual) download of the data files containing values of indicators.
It will be implemented by the integration with Trading Economics Application
Programming Interface, to get the direct access to various indicators.

For different cases (different financial securities, different time frame) the set
of indicators and their weights will differ. The parameters of investment strategy
will differ as well. There is no one general rule which could work in all the cases.
Therefore, the main goal of proposed method and tool SBT is the possibility to find

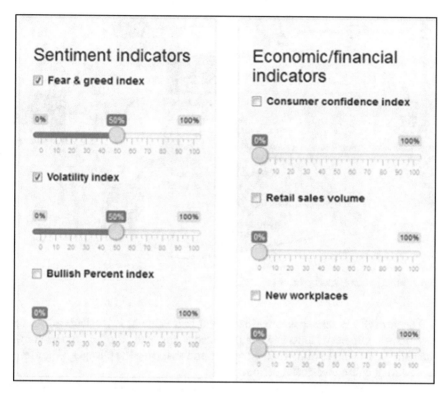

Fig. 8 Selection of sentiment indicator and weights

how different factors affect different securities and use this information for market pattern prediction and decision making.

The software SBT simulates the changes of portfolio value depending on securities' price. The simulation model is designed in Powersim environment (Fig. 9).

Security prices correspond to real historical data of selected financial asset. All decisions to buy or sell assets are made accordingly to composite indicator values accessed from corresponding historical data sources.

The steps of investment simulation process (Fig. 9) are further performed. Firstly, the amount of premium is calculated in accordance to indicator. When indicator value is extremely low—below Lower bound—we can increase the amount of premium: initial premium is multiplied by leverage. Accordingly, the amount of premium is lowered when indicator value is extremely high—above Upper bound. Then initial premium is multiplied by leverage.

This amount is used to purchase the units of security for the current price of the asset. The newly purchased units are added to the accumulated ones. The total amount of accumulated units increases after each buy operation, but it decreases if the units are sold.

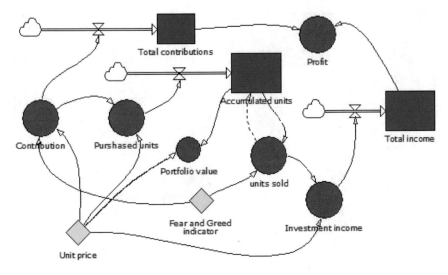

Fig. 9 Model of investment process

The portfolio value depends on the current unit value. It is recalculated for the accumulated number of units and demonstrated at any time. The SBT model allows simulating the changes of investment returns and analysing the changes of portfolio value during entire investment period.

It can be noticed that due to inclusion of the sentiment and other indicators the portfolio assets are sold and the profit is taken when strong greed level is observed, when it exceeds the Sell threshold value. This is done to prevent sharp drops of portfolio value. At these moments we get Investment income that is accumulated during entire investment period.

Finally, the return of all investment process is determined. It is estimated by deducting all premiums from the total accumulated sum.

In Fig. 10 we can see the interface of the SBT, where the initial parameters panel is shown on the left side, and the simulation results are in the right panel.

The initial parameter panel enables us to set and change the parameters of investment strategy by manually using the slider or numerical input. The plot on the right immediately shows the simulation results. This feature enables to make an informed decision to buy or sell selected amount of assets depending on initial parameters and indicators weights selection. The line plot compares the performance of the designed strategy with the inclusion of composite indicator versus benchmark performance (SPDR S&P 500) and performance of periodical investment to ETF without using composite indicator.

The performance of selected indicators can be compared graphically or in numerical way by calculating coefficients: total invested sum, total accumulated sum, profit (net value), profit (percentage), Sharpe ratio. Sharpe ratio allows to compare the strategies having similar profit ratio (Fig. 11).

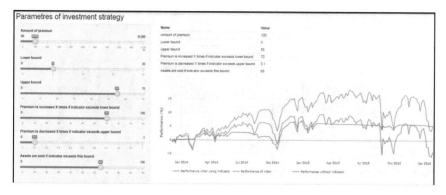

Fig. 10 Parameter selection and simulation results

Strategy 1

	Using_indicators	Not_using_indicators
Total sum paid	40840.00	18100.00
Total accumulated sum	43019.79	17367.40
Profit/loss	2179.79	-732.60
Profit/loss (%)	5.34	-4.05
Sharpe ratio	0.78	

Strategy 2

	Using_indicators	Not_using_indicators
Total sum paid	33500.00	18100.00
Total accumulated sum	35627.89	17367.40
Profit/loss	2127.89	-732.60
Profit/loss (%)	6.35	-4.05
Sharpe ratio	0.97	

Strategy 3

	Using_indicators	Not_using_indicators
Total sum paid	40000.00	18100.00
Total accumulated sum	43217.54	17367.40
Profit/loss	3217.54	-732.60
Profit/loss (%)	8.04	-4.05
Sharpe ratio	1.27	

Fig. 11 The statistics of particular strategy

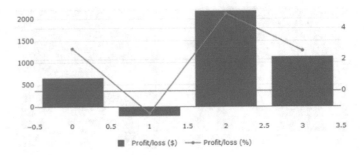

Fig. 12 Comparison of results of different strategies

In Fig. 11 the outcomes of simulated three trading strategies are presented. The strategies were selected by manual experimenting for observing behaviour of the model and selection of the parameters. According to the composite indicator changes and trading strategy selected we can calculate the predictable profit, and compare the profitability and risk of the investment strategies.

Each time of changing the parameter by the user, the results are recalculated and can be saved to the result repository, including net profit, percentage profit, the initial parameters and other characteristics of corresponding strategy.

The results of selected strategies are presented on the profit/loss plot for comparative evaluation (Fig. 12).

Manual selection of composite indicator structure (indicators and their weights) and values of investment strategy allows investigating the impact of different factors (fundamental indicators, sentiment based indicators) while simulating the investment to various assets. However, the further step of simulation process by SBT software aims for automatic solution finding.

In the present research stage the basic optimization procedure (Simplex LP) is applied to find out the weights of indicators and values of investment strategy parameters, where profit and Sharpe ratio values were the highest.

The results of various simulation cases served as an input for making decisions which indicator is preferable for different group of assets and financial markets. The parameters found while analysing historical data can be used for making real-time decisions.

The experimental analysis enabled to explore the performance of the algorithm, get the insights on the impact of sentiment information and the role of parameters of the investment strategy. The main limitations included manual operations, such as data upload, the parameter selection and basic use of optimization model for maximizing P/L and Sharpe ratio.

Our next step is to enable real-time interface to quotes for various currencies, commodities, stocks and bonds. The composite indicator created while analysing historical data then could trigger alert of the approaching anomalous situation in present time and help to make investment decision. The simulation of the investment process revealed the potential of the proposed approach for automated experimenting and optimization of the model performance.

5 Conclusion

The theories emphasizing the rationality and efficiency of the financial markets lack instruments to explain variety of real financial market phenomena, such as high volatility, unexpected decisions of investors, bubbles, crises or market anomalies.

The increasing number of attempts to include sentiment information for investment strategies confirm, that the financial indicators are not sufficiently reliable for predicting market behaviour, as well as reducing investment risks, and maximizing profits.

The attempts to quantify sentiment information for embedding it to the investor decision making process include broad variety of approaches, such as text analysis in news, blogs, twitter (for detection of positive and negative keywords), exploring intensity of web search for particular investment-related terms and companies, measuring characteristics of news flows on particular topics, discussion patterns in social networks. The absence of reliable characteristics enabling to quantify sentiment information lead to proposing several indices, such as CNN Money Fear and Greed index, StockTwits sentiment index or HedgeChatter social media sentiment dashboard. The indices aim to join variety of approaches and create measures of sentiment impact.

Different kinds of indicators recommended for financial markets analysis are summarized in the article, covering entire scale from rational measures to sentiment-based factors. The analysis of research literature revealed that the influence of sentiment information for market prediction complements financial indicators, however the rate of its influence varies for different types of securities, and depends on the time window as well.

The composite indicator is elaborated as a weighted sum combining selected indicators of different types, such as sentiment, financial, economic, political and others. The adapted composite indicators can be elaborated for behaviour of each selected security within different time frames and further used for making investment decisions and portfolio strategies.

The algorithm is proposed, based on the composite indicator for designing the investment strategies. The algorithm selects the best structure of composite indicator, applies it for defining the lower and upper thresholds for investment (buy/sell) decisions and estimates the investment strategy for selected portfolio. The P/L (profit-loss) and Sharpe ratio were taken for evaluating the profitability of the proposed approach. The performance of the designed investment strategies was compared to S&P financial indicator-based approach.

The presented approach and algorithm were made operational by developing the investment simulation model Sentiment-Based Tool (SBT). The simulation prototype and software was designed by applying R language and R Shinny package. The factors characterizing financial and behavioural patterns are explored and integrated to the dynamic dashboard simulation, the experimental analysis is done by uploading real data sets. The tool enables to experiment and design investment strategies for different securities and indices (sentiment, financial, economic).

The composite indicator and parameters of investment strategies can be explored either manually by applying slider interface for observing instant change of parameters and investment strategy outcomes, or by maximizing P/L and Sharpe ratios with the help optimization techniques. The presented research included applying standard optimization (simplex LP) for maximizing investment outcomes. However, the computational part of the tool will be enhanced for more optimization methods to be applied. It will enable automated finding of optimal investment decisions maximizing the P/L and Sharpe ratio criteria for strategy evaluation.

In the article the performance of the proposed approach is experimentally tested for exploring real financial data of SPY, FEZ and GLD securities and illustrated by several cases of composite indicator structure (50/50 and 30/70 ratio of sentiment and financial indicators). The three examples of investment strategies in all cases outperformed the strategy without using composite indicator.

The experimental research summarized in the article demonstrated ability of the proposed method and tool to increase performance of investments decision making based on multi-factor analysis of financial markets, pattern prediction and sentiment-based market forecasting.

References

Barberis NC (2018) Psychology-based models of asset prices and trading volume. Available at https://www.nber.org/papers/w24723

Bollen J, Mao H, Zeng X (2010) Twitter mood predicts the stock market. http://arxiv.org/abs/1010.3003

Bordino I, Battiston S, Caldarelli G, Cristelli M, Ukkonen A, Weber I (2012) Web search queries can predict stock market volumes. PLoS ONE 7(7):e40014

Chiong R, Fan Z et al (2018) A sentiment analysis-based machine learning approach for financial market prediction via news disclosures. Available at https://dl.acm.org/citation.cfm?id=3205682

Fang L, Yu H, Huang L (2018) The role of investor sentiment in the long-term correlation between U.S. stock and bond markets. Finance J. http://dx.doi.org/10.1016/j.iref.2018.03.005

Fear & Greed Index (2019) Available at SSRN: http://money.cnn.com/data/fear-and-greed/

Fear and Greed Index (2019) Investopedia. https://www.investopedia.com/terms/f/fear-and-greed-index.asp

Gruhl, D, Guha R, Kumar R, Novak J, Tomkins A (2005) The predictive power of online chatter. In: Proceedings of the eleventh ACM SIGKDD international conference on knowledge discovery in data mining (KDD '05), pp 78–87. New York, NY, USA

Gu C, Kurov A (2018) Informational role of social media: evidence from twitter sentiment. Available at https://ssrn.com/abstract=3206093

HedgeChatter. Available at https://www.hedgechatter.com/

Houlihan P, Creame GG (2014) Can social media and the options market predict the stock market behavior? Stevens Institute of Technology. Available https://editorialexpress.com/cgi-bin/conference/download.cgi?db_name=CEF2015&paper_id=521

Jiang F, Lee J, Martin X, Zhou G (2019) Manager sentiment and stock returns. J Finance Econ 132:126–149. https://doi.org/10.1016/j.jfineco.2018.10.001

Johnman M, James B (2018) Predicting FTSE 100 returns and volatility using sentiment analysis. Available at https://doi.org/10.1111/acfi.12373

Kissan J, Wintoki MB, Zhang Z (2011) Forecasting abnormal stock returns and trading volume using investor sentiment: evidence from online search. Int J Forecast 27:1116–1127. https://doi.org/10.1016/j.ijforecast.2010.11.001

Kristoufek L (2013) Can Google Trends search queries contribute to risk diversification? Sci Rep 3:2713. https://doi.org/10.1038/srep02713

Li TR, Chamrajnagar AS, Fong XR, Rizik NR, Fu F (2019) Sentiment-based prediction of alternative cryptocurrency price fluctuations using gradient boosting tree model. Front Phys 7:98. https://doi.org/10.3389/fphy.2019.00098

Liutvinavicius M, Zubova J, Sakalauskas V (2017) Behavioural economics approach: using investors sentiment indicator for financial markets forecasting. Baltic J Mod Comput 5(3):275–294. https://doi.org/10.22364/bjmc.2017.5.3.03

Petita JG, Lafuenteb EV, Vieites AR (2019) How information technologies shape investor sentiment: A web-based investor sentiment index. Borsa Istanbul Rev 19–2:95–105. https://doi.org/10.1016/j.bir.2019.01.001

Ranco G, Aleksovski D, Caldarelli G, Grčar M, Mozetič I (2015) The effects of twitter sentiment on stock price returns. PLoS ONE 10(9):e0138441. https://doi.org/10.1371/journal.pone.0138441

SPDR Gold Shared ETF. Available https://www.spdrs.com/product/fund.seam?ticker=GLD

Stocktwits (Social Network for Investors and Traders). Available at http://stocktwits.com/

Trading Economics. Available at https://tradingeconomics.com/

Yahoo finance. SPDR Gold Shares. Historical prices